Quantum Entanglements

Rob Clifton, 1964–2002

Quantum Entanglements

Selected Papers

ROB CLIFTON

Edited by

JEREMY BUTTERFIELD

and

HANS HALVORSON

OXFORD

UNIVERSITY PRESS

OXFORD
UNIVERSITY PRESS

Great Clarendon Street, Oxford OX2 6DP

Oxford University Press is a department of the University of Oxford.
It furthers the University's objective of excellence in research, scholarship,
and education by publishing worldwide in

Oxford New York

Auckland Bangkok Buenos Aires Cape Town Chennai
Dar es Salaam Delhi Hong Kong Istanbul Karachi Kolkata
Kuala Lumpur Madrid Melbourne Mexico City Mumbai Nairobi
São Paulo Shanghai Taipei Tokyo Toronto

Oxford is a registered trade mark of Oxford University Press
in the UK and certain other countries

Published in the United States
by Oxford University Press Inc., New York

First published 2004

British Library Cataloguing in Publication Data
Data available

Library of Congress Cataloging in Publication Data
Data available

ISBN 0–19–927015–5

1 3 5 7 9 10 8 6 4 2

Typeset by J. Butterfield and H. Halvorson
Printed in Great Britain
on acid-free paper by
Biddles Ltd.,
King's Lynn, Norfolk

Contents

List of Figures

Preface

Rob Clifton's career as a philosopher of physics was tragically cut short in the summer of 2002 when, after a year-long battle, he succumbed to cancer. In the fifteen years before he died, Clifton published more than fifty research articles, an edited volume, and other review articles and book reviews: a body of work which pushed the field to a new level of sophistication and depth. Personally, he was a wonderful man, as all who knew him would attest: generous, modest and brave.

So the philosophy of physics community has suffered a grievous loss. To commemorate Rob, a large memorial conference was held at the American Institute of Physics in College Park, Maryland, in 2003; the papers will be published in a special issue of the journal Rob edited, *Studies in the History and Philosophy of Modern Physics*.

In the same spirit, we are proud to honour Rob's memory with this selection of his papers. We have chosen fourteen papers, several co-authored, from the second half of his career; we describe the papers more fully in the Introduction. We also include a full bibliography of Rob's publications. We hope this volume will serve both as a memorial to Rob's achievement, and as a convenient source for scholars of some of his papers.

It is a pleasure to thank several people. First and foremost: Marilyn Clifton for all of the help and support she has given us, not only in preparing this volume, but over the last several years. We also thank: Peter Momtchiloff at OUP, both for his initial enthusiasm for this project, and for unfailing help from then on; Jacqueline Baker and Rebecca Bryant from OUP for help with the layout of the text;

Humaira Erfan-Ahmed for helping type Chapters 1 and 2; Jeff Bub for help obtaining the source for Chapter 3; Dennis Dieks and Pieter Vermaas for help obtaining the source for Chapter 4; and of course Rob's co-authors, Jeff Bub, Michael Dickson, Shelly Goldstein, and Adrian Kent for agreeing to be reprinted here.

Jeremy Butterfield
Oxford

Hans Halvorson
Princeton

Acknowledgments

We thank the original publishers of the papers reprinted here, for their kind permission; as follows.

The American Institute of Physics, for permission to reprint Chapter 6, which was originally published in *Journal of Mathematical Physics*, volume 41 (2000), pp. 1711–17.

The American Physical Society, for permission to reprint Chapter 10, which was originally published in *Physical Review A*, volume 61 (2000), p. 042101.

Elsevier, for permission to reprint:

(i): Chapters 2, 3 and 7, which were originally published in *Studies in the History and Philosophy of Modern Physics* as, respectively: volume 27 (1996), pp. 181–219; volume 31 (2000), pp. 95–8; volume 32 (2001), pp. 1–31;

(ii): Chapters 5 and 11, which were originally published in *Physics Letters A* as, respectively: volume 271 (2000), pp. 167–77, pp. 1–7.

Kluwer Academic Publishers, for permission to reprint Chapters 4 and 12, which were originally published in, respectively: *The Modal Interpretation of Quantum Mechanics*, D. Dieks and P. Vermaas (eds), Kluwer 1998, pp. 9–47; *Non-locality and Modality*, T. Placek and J. Butterfield (eds), Kluwer 2002, pp. 3–18.

The Royal Society of London, for permission to reprint Chapter 13, which was originally published in *The Proceedings of the Royal Society of London A*, volume 456 (2000), pp. 2101–24.

Oxford University Press: for permission to reprint Chapters 1 and 9, which were originally published in *The British Journal for Phi-*

losophy of Science as, respectively: volume 46 (1995) pp. 33–57; volume 52 (2001), pp. 417–70.

The Philosophy of Science Association: for permission to reprint Chapters 8 and 14, which were originally published in *Philosophy of Science* as, respectively: volume 69 (2002), pp. 1–28; volume 69 (2002), S150–S167.

Editors' Introduction

The aim of this volume is to provide a representative sample of Clifton's research articles during the second half of his career, from 1995 to 2002. (A full bibliography of Clifton's publications is included at the end of the volume.) We have chosen not to follow a strict chronological order, but have divided the papers under four major headings: I. Modal interpretations; II. Foundations of algebraic quantum field theory; III. The concept of a particle; IV. New light on complementarity, hidden variables and entanglement.

Chronologically, Section I corresponds to a phase of Clifton's career (approximately 1994 to 1997) during which he focused on modal interpretations of elementary quantum mechanics. Sections II and III contain papers drawn from a short, but highly productive period (approximately 1999 to 2001) when Clifton was working primarily on the foundations of quantum field theory: Section II contains papers on nonlocality, and on the modal interpretation in quantum field theory; and Section III contains papers on the particle aspect of quantum field theory. Section IV collects some of Clifton's papers from 1999 to 2002 that are not on the topic of quantum field theory.

Of the fourteen papers in this volume, ten have one or more co-authors. This high proportion is evidence of Clifton's penchant and gift for collaborative research. It was not only that his talent and enthusiasm — and his temperamental inclination to dive in and struggle with any problem that came up — attracted many people, especially aspiring doctoral students and younger colleagues (or e-correspondents!), to work with him. Also, he had a knack for articulating joint projects appropriate to his and the other person's interests and strengths. Furthermore, the enterprise was always underpinned by his sense of fun, and his great intellectual and personal generosity.

We will now proceed to give a slightly more detailed description of the papers in this volume, and of their place in the development of the philosophical foundations of contemporary physics.

1 The modal interpretation of quantum mechanics

The so-called 'orthodox' interpretation of quantum mechanics faces a serious difficulty in its description of measurement: If we assume that the standard law of dynamical evolution (i.e., the Schrödinger equation) holds universally, and if we talk in the normal way about when quantities possess values, then it follows that measurements do not usually have results — contrary to our experience. Attempts to solve this measurement problem have been the driving force in the development of contemporary interpretations of quantum mechanics.

There are essentially two routes toward a solution of the measurement problem: One can modify the standard dynamics by introducing a collapse of the wavefunction, or one can keep the standard dynamics but supplement the standard rule for assigning values to quantities. Each route comes with its own perils. On the one hand, those who want to modify the dynamics must not only devise a new dynamical law that is approximated by the Schrödinger equation, but they must also show that this new dynamical law is consistent with the relativity of simultaneity. On the other hand, modifying the orthodox interpretation's way of assigning values to quantities is a nontrivial task, since the Kochen-Specker theorem entails that not all quantities can possess values simultaneously.

The modal interpretation of quantum mechanics follows the second route — i.e., it is a 'no collapse' interpretation. It goes back to the work of van Fraassen in the 1970s, but came to fruition in papers by Kochen (1985) and Dieks (1989), and in the monograph by Healey (1989). However, the modal interpretation soon faced several philosophical challenges. For example, Arntzenius (1990) pointed out some of the more bizarre metaphysical consequences of Kochen's modal interpretation, and Albert & Loewer (1990) argued that the modal interpretation cannot explain why 'error-prone' measurements have outcomes. Similarly, Elby (1993) argued against the modal interpretation's ability to solve the measurement problem.

These challenges were met by a 'second generation' of workers on the modal interpretation, who developed it to a higher level of sophistication. First, Bub (1993), Dieks (1993), Healey (1993), and Dickson (1994) all defended the modal interpretation's solution of the measurement problem. These responses were then followed by papers in which it was argued that decoherence considerations are sufficient to ensure that the modal interpretation gives the right predictions for the outcomes of measurements (Bacciagaluppi & Hemmo 1996), and in which it was shown that the modal interpretation's assignment of definite values can be extended to deal with troublesome degenerate cases (Bacciagaluppi *et al.* 1995).

Clifton was obviously adept at working out intricate technical details. But what really set his work apart was his ability to translate 'big' philosophical questions into tractable technical problems. For example, while some researchers were continuing to fine-tune the modal interpretation's account of measurement, Clifton raised a more fundamental motivational question: Is there an independent reason, besides its potential for solving the measurement problem, to adopt the modal interpretation? The result of Clifton's investigations into this question is the first chapter of this volume: 'Independently motivating the Kochen-Dieks modal interpretation of quantum mechanics' (1995). The style of argument in this paper is characteristic of Clifton's approach: He first transforms intuitive desiderata (in this case, desiderata for an interpretation of quantum mechanics) into precise mathematical conditions; and he then proves — via an existence and uniqueness theorem — that there is precisely one 'object' (in this case, the Kochen-Dieks (KD) modal interpretation) that satisfies these desiderata.

Of Clifton's desiderata for interpretations of quantum mechanics, the foremost is the definability condition. An interpretation satisfies the definability condition just in case its choice of definite-valued observables can be defined in terms of the quantum state alone. But what is the physical or metaphysical motivation for this condition? It is highly doubtful that Clifton thought that there is some *a priori* metaphysical warrant for the definability condition. It is more likely that Clifton chose the definability condition as a means for avoiding metaphysical disputes about which quantities are the most 'fundamental'. For example, the best known hidden variable theory — the de Broglie-Bohm theory — assigns a privileged role to *position*: it asserts that, no matter what the quantum state is, there are point particles with definite positions, and the distribution of these particles agrees with the probabilities predicted by the quantum state. Granted that the de Broglie-Bohm theory does in fact solve the measurement problem, what is *its* motivation? A quick perusal of the literature indicates that the motivations are usually of either an ontological or of an epistemological sort. For example, it is sometimes claimed that an ontology of particles is more 'intelligible' than, say, the ontology we would get if we required that momentum is always definite. Or, it might be argued that measurements have results if and only if there are particles with definite positions. But since these arguments are controversial and philosophically nontrivial, it might be seen as desirable to find a purely mathematical criterion for selection of the definite-valued observables.

Clifton's 1995 paper on the KD modal interpretation was soon followed by a paper, 'A uniqueness theorem for 'no collapse' interpretations of quantum mechanics' (Chap. 2; 1996), written jointly with Jeffrey Bub. Bub, one of the pioneering advocates of modal interpretations, had recently argued that any interpretation which solves the measurement problem (without abandoning Schrödinger dynamics) must be a 'Bohm-like' hidden variable interpretation (Bub 1994). Bub and Clifton provide support for this claim by means of a characterization theorem which parameterizes 'no collapse' interpretations by the observable that they choose to privilege.

The Kochen-Specker theorem shows that we cannot consistently assign determinate values to all quantum mechanical observables; if we try, we will eventually run into an 'obstruction'. Bub and Clifton turn the Kochen-Specker theorem on its head by asking: What are the *largest* sets of observables that *can* be assigned determinate values without generating a Kochen-Specker contradiction? They then supply a recipe which permits them, for any chosen privileged observable, to construct one of these maximal sets which contains that observable; and, conversely, they show that each such maximal set is generated by a privileged observable.

One of the potential motivations for the KD modal interpretation was the hope that it would provide a 'realistic' interpretation of quantum mechanics that upholds the completeness of the theory, in spirit if not in letter. (The contrast here is with the de Broglie-Bohm theory, which is the paradigm example of an interpretation that concedes the incompleteness of quantum mechanics.) However, the Bub-Clifton theorem — in combination with Clifton's 1995 theorem — shows that this motivation is not completely well-founded. For both the KD modal interpretation and the de Broglie-Bohm theory concede the incompleteness of quantum mechanics in the sense that the quantum state itself is not taken to determine what is actual (i.e., the so-called value state).

In Bub and Clifton's paper we also encounter a thread that runs throughout Clifton's later work: Clifton claims — contrary to the traditional view — that Bohr's complementarity interpretation should be thought of as a modal interpretation (like the KD interpretation, or the de Broglie-Bohm theory). This description of Bohr's interpretation might seem surprising since Bohr was the primary defender of the completeness of quantum mechanics, and since Bohr is often thought to have endorsed (if only tacitly) the collapse of the wavefunction upon measurement. (For a more detailed explanation of why Clifton rejected the traditional view of Bohr's interpretation, see Chap. 12.)

After Bub and Clifton's paper had appeared, they joined forces with Sheldon Goldstein to give a simplified proof of the classification theorem. Since the resulting proof clarifies the conceptual situation, we chose to include it as Chapter 3 of this volume.

In his 1995 paper, Clifton motivates the KD modal interpretation by showing that it satisfies several intuitively plausible criteria. However, there was one motivation for the KD modal interpretation that Clifton does not mention explicitly — viz., the hope that it would prove (unlike the de Broglie-Bohm theory) to be strictly consistent with the relativity of simultaneity. This question — or more specifically, the question whether the KD modal interpretation is consistent with Lorentz invariance — was explored in a collaboration between Clifton and Michael Dickson, which resulted in the 1998 paper 'Lorentz-invariance in modal interpretations' (Chap. 4). Clifton and Dickson argue that the KD modal interpretation must assume that there is a preferred frame of reference, and therefore cannot maintain 'fundamental' Lorentz invariance. They go on to argue, however, that the KD modal interpretation is 'empirically' Lorentz invariant in the sense that no observer can determine which frame is privileged; and they conclude that the KD modal interpretation is on exactly the same footing — in terms of its relationship to relativity theory — as the de Broglie-Bohm theory.

Clifton went on to write a few more papers on the modal interpretation, focusing primarily on extending it to the context of algebraic quantum theory (see papers 20 and 21 of Clifton's publication list on p. 445). However, it appears that after realizing that the KD interpretation violates Lorentz invariance, Clifton ceased to think of it as providing a promising approach to solving the measurement problem.

2 Foundations of algebraic quantum field theory

After 1997, Clifton directed much of his effort at quantum field theory (QFT), especially algebraic QFT: an approach to the theory that is mathematically rigorous and so well suited to foundational investigations. In this endeavor, Clifton was striking out into a territory largely uncharted by philosophers of physics. For though philosophers of physics have written reams about the foundations of elementary non-relativistic quantum mechanics, especially about the measurement problem and nonlocality, the philosophical literature on QFT is small. There is just a handful of books: for example, Brown & Harré (1988) and Teller (1995). This situation is understandable, since QFT is a much more technically demanding theory than elementary quantum mechanics. But it is regrettable, not least because QFT is, after all, the most predictively accurate theory in history, and so seems the obvious place to seek the raw materials for a 'metaphysic of nature'.

Our selection of Clifton's work on QFT comprises Parts II and III, and even the first chapter of Part IV. To introduce this material, it will be clearest not to discuss the papers *seriatim*, but to discuss in order the following topics: nonlocality and the vacuum; the Reeh-Schlieder theorem; the modal interpretation in QFT; and finally, the concept of a particle in QFT.

2.1 Nonlocality and the vacuum

The vacuum state of a relativistic quantum field is very different from our intuitive notion of a state in which 'nothing is happening'. To describe this, let us begin with the article 'More ado about nothing', by Redhead, who had been Clifton's graduate adviser (Redhead 1995). In this article, Redhead discusses how in the relativistic vacuum state, any local event has a nonzero probability of occurrence, and two events in mutually distant regions can display strong correlations.

However, Redhead did not establish that the vacuum state is *nonlocally* correlated in the sense of Bell (1964). Recall that Bell provides a rigorous method for displaying the nonlocality of states: there is a family of inequalities (Bell's inequalities) about the expectation values of local measurements, such that, if a state's predictions violate one of these inequalities, then these predictions cannot be reproduced by a local hidden variable model. Such a state is now called *Bell correlated*.

The question whether the vacuum state is Bell correlated had been floating around for some years before Redhead's paper. For example, in the mid 1980s and early 1990s, the mathematical physicists Summers and Werner obtained a series of deep mathematical results on Bell's inequality in rigorous QFT (Summers & Werner 1985, 1987*a*,*b*,*c*, 1988, 1995). For example, Summers and Werner showed that the vacuum state does violate Bell's inequality (in fact, violates it maximally) relative to measurements that can be performed in tangent wedges of Minkowski spacetime.

Despite the power of Summers and Werner's results, they leave open some interesting questions about nonlocality in QFT. According to oral history, Malament pointed out to Clifton in 1998 that they do not settle the question whether the vacuum state violates Bell's inequality relative to measurements performed in *any* pair of spacelike separated regions (no matter how small these regions are, and no matter how far apart they are). Clifton immediately realized the interest of this question; and he was in a good position to see the difficulties in proving 'Malament's conjecture'. For Mermin (1996) had recently emphasized (at a conference at the University of Western Ontario, organized by Clifton) the foundational relevance of 'Werner states', which had recently been discovered by Werner (1989). To see the importance of Werner states, recall that a state of a composite system in quantum mechanics is called *separable* if it can be decomposed as a mixture of product states (i.e., states for which the two subsystems are completely uncorrelated); if a state is not separable, it is called *nonseparable*. Now, all states that are nonseparable and *pure* (i.e., vector states) violate a Bell's inequality (Popescu & Rohrlich 1992). But Werner discovered that some nonseparable mixed states do *not* violate a Bell's inequality: such states are now called 'Werner states' in honour of their discoverer.

Clifton and Malament knew that the vacuum state is nonseparable. However, when restricted to bounded regions of spacetime, the vacuum state is a mixed state — so that it might be a Werner state, i.e., a state that does not violate Bell's inequality.

History shows that many of the most interesting scientific discoveries are the by-products of failure. For example, it has been said that most advances in number theory over the past 300 years were the results of failed attempts to prove Fermat's last theorem. A similar thing might be said about Clifton's work on Malament's conjecture: Although Clifton was frustrated in his attempts to solve the conjecture, his attempts led to a number of unexpected, fruitful, and perhaps even more interesting results.

For example, in the paper 'Nonlocal correlations are generic in infinite dimensional bipartite systems' (2000; Chap. 10), Clifton, along with collaborators Hans Halvorson and Adrian Kent, shows that a pair of infinite dimensional elementary quantum systems has a dense set of Bell correlated states. (Recall that a subset S of X is dense just in case any element of X can be approximated arbitrarily closely — in the relevant topology — by elements of S.) As Clifton *et al.* point out, this density result is not only foundationally interesting (since it shows that the 'quantumness' of a system 'varies in proportion' to the number of degrees of freedom): It is also of practical interest for quantum information theory, where one wishes to use entanglement as a physical resource to perform certain information processing tasks.

Nonetheless, this Clifton-Halvorson-Kent result does not provide any information directly relevant to Malament's conjecture — which is specifically about the vacuum state in relativistic QFT. Most particularly, the algebras of local observables in relativistic QFT are not isomorphic to the algebra of all bounded operators on a Hilbert space, and so the Clifton-Halvorson-Kent result is simply inapplicable to relativistic QFT. (In fact, the algebras of local observables in relativistic QFT are typically type III von Neumann algebras, whereas the algebras of local observables in nonrelativistic QM are type I von Neumann algebras.)

Very shortly after the appearance of the Clifton-Halvorson-Kent paper, a second paper (by Clifton and Halvorson) appeared: 'Generic Bell correlation between arbitrary local algebras in quantum field theory' (2000; Chap. 6). In this paper, Clifton and Halvorson generalize the Clifton-Halvorson-Kent argument to the case of algebras of local observables that are von Neumann algebras 'of infinite type'. (For the proof of the result, the essential property of these algebras is that they have infinitely many orthogonal projections that are pairwise 'equivalent'.) The infinite von Neumann algebras include the algebra of all bounded operators on an infinite-dimensional Hilbert space (which is used to represent elementary quantum systems whose observables can take infinitely many different values) as well as the type III von Neumann algebras. Thus, Clifton and Halvorson show that whenever local algebras of observables are of infinite type, then the Bell correlated states are dense. Although this generalized result still does not prove Malament's conjecture, it does show that the vacuum is approximated arbitrarily closely by nonlocal, i.e. Bell correlated, states.

So Clifton's work on nonlocality shows that there is a similarity between relativistic QFT and nonrelativistic quantum mechanics on infinite-dimensional Hilbert spaces — viz., in both cases there is a dense set of states that violate a Bell's inequality. But in his next major publication ('Entanglement and open systems in algebraic quantum field theory'; 2001, Chap. 7: again with Halvorson), Clifton shows that there is a sense in which nonlocality is *worse* in relativistic QFT than it is in nonrelativistic QM! In particular, in nonrelativistic QM, one can always disentangle (or 'isolate') a system by performing a measurement of a maximal observable — the resulting state will always be a product state, and so will have no correlations. But Clifton and Halvorson show that when the algebras of local observables are type III, then local observers cannot perform such disentangling operations. So, not only is the generic state of relativistic QFT nonlocally correlated: also, it is impossible to disentangle a local system from its environment.

However, this paper's central topic is not disentanglement or its impossibility, but rather the interpretation of one of the most fundamental results of rigorous quantum field theory: the Reeh-Schlieder theorem, proved in 1961.

2.2 The Reeh-Schlieder theorem

This theorem says, roughly speaking, that any state of the quantum field can be approximated arbitrarily closely by applying to the vacuum state operators from any fixed local algebra of observables (no matter how small the local region is). This is a very striking result. Indeed, many of the interesting features of the relativistic vacuum, and even of relativistic QFT in general, are closely related to it — or even derivable from it.

But there is also an interpretative danger. The theorem seems to suggest that actions in a spacetime region O can have instantaneous effects in a distant, i.e. spacelike separated, region O': viz., changing the state from the vacuum to another state, with expectation values for chosen observables associated with O' that are arbitrarily close to prescribed values. If that were correct, then the Reeh-Schlieder theorem would suggest a serious conflict between relativistic QFT and the fundamental principles of special relativity. This suggestion has of course been addressed before; for example by Segal (1964) and Fleming (2000). But Clifton and Halvorson argue in Chapter 7 (especially Sec. 3) against the suggestion (cf. also Halvorson 2001). The main idea is to apply the distinction between *selective* and *nonselective* operations: a distinction which is also applied in discussion of elementary quantum mechanics, to reconcile nonlocal correlations with the no-signaling theorem. Thus in a selective operation (represented, e.g., by applying a projection operator to a state vector), an observer performs a measurement and then ignores, or destroys, the elements of the original ensemble that do not correspond to a certain set of results. And if we are concerned with an ensemble that is spread over two spacelike separated regions, then one can maintain that such a selection should not be thought of as a physical *action* occurring in just one region that has *effects* in the other (spacelike separated) region.

2.3 The modal interpretation of AQFT

One of Clifton's main goals in the period between 1999 and 2001 was to translate interpretive questions from the context of elementary quantum mechanics to the (more technically demanding) context of relativistic QFT. Of course, Clifton was not alone in this goal. For example, in a 2000 paper, Dennis Dieks put forward a proposal for extending the modal interpretation to relativistic QFT. As Dieks points out, making such an extension is not completely straightforward. In particular, the modal interpretation's rule for picking determinate observables makes use of the fact that there is a one-to-one correspondence between quantum states and a certain kind of operator on a Hilbert space (viz., density operators); but in

QFT this correspondence between states and observables no longer holds (since type III algebras cannot contain density operators). To bypass this problem, Dieks proposes making use of the split property, according to which any local algebra of observables in QFT can be approximated by a type I algebra (i.e., the algebra of all bounded operators on a Hilbert space). Since the approximating type I algebra does contain density operators, Dieks proposes that we apply the standard KD rule to the approximating algebra in order to find an approximation of the set of definite-valued observables for the relevant spacetime region.

Clifton fired off a quick response to Dieks in his paper, 'The modal interpretation of algebraic quantum field theory' (2000; Chap. 5). In this paper, Clifton argues that Dieks' method of using the approximating type I algebra does not provide a non-arbitrary method for determining the definite-valued observables associated with a spacetime region. Clifton then goes on to supply a rule for picking the definite-valued observables within a general von Neumann algebra, and he proves that this rule is a generalization of the Kochen-Dieks rule for type I factors. However, Clifton also shows that this generalized rule entails that very frequently (viz., in all ergodic states, which form a dense subset of the state space), there will be no non-trivial definite-valued quantities associated with any given spacetime region. Thus, it seems very unlikely that the KD modal interpretation can be successfully extended to provide a solution of the measurement problem in the context of relativistic QFT.

3 The concept of a particle

One of the oldest debates in natural philosophy concerns the composition of matter: Is there a limit to the divisibility of matter ('particle ontology'), or is matter a continuum that can be subdivided *ad infinitum* ('field ontology'). Different attitudes toward this question have driven competing research programs in physics; and philosophers similarly have tried either to resolve the issue, or (like Kant in the Second Antinomy) to show that the issue cannot be resolved.

The debate over particle vs. field ontology becomes more interesting in the context of quantum mechanics, and even more so in relativistic quantum theory. First, it seems impossible to build a consistent relativistic quantum theory of particles; and, so it looks like relativity forces us to a quantum *field* theory. However, as Teller (1995, p. 93) points out, 'the subject matter of so-called 'quantum field theory' does not need to be presented as a field theory'. In fact, Fock showed in 1932 that the states of a quantum field have (in many important cases) a completely natural interpretation as particle states, and so it seems that QFT does not immediately settle the particle vs. field dispute.

However, there are a number of difficulties with the particle interpretation of QFT. Part III of this volume consists of two papers that each take up one of these difficulties. The first concerns the localization of particles, the second the appearance of particles in the vacuum.

In the first paper, 'No place for particles in relativistic quantum theories' (2002; Chap. 8), Clifton and Halvorson investigate arguments which purport to show that there can be no localized particles in any relativistic quantum theory.[1] The tensions between relativity and the notion of a localized quantum state have been known since at least the 1940s. They were then made precise in a series of results by Newton and Wigner, Hegerfeldt and others: results whose broad thrust is that localized quantum states would violate relativity's prohibition of superluminal velocities (e.g. Hegerfeldt 1974, Hegerfeldt & Ruijsenaars 1980).[2]

Clifton and Halvorson's point of departure is a theorem of Malament (1996), to the effect that if in a relativistic quantum theory there are no 'act-outcome' correlations at spacelike separation, then there are no localized particles. Although Malament bases his ar-

[1]The adjective 'localized' may seem redundant. However, we will suppose that 'particle' simply means an individual indivisible entity, whether or not that entity is spatially extended.

[2]Since superluminal velocities are consistent with Lorentz covariance, it is a separate issue whether there is a Lorentz covariant notion of localization. Extensive work has been done on this latter topic, most particularly by Fleming; see, e.g., Fleming (1989), Fleming & Butterfield (1999).

gument on a clearly valid mathematical proof, the soundness of his argument has been questioned by a number of researchers — including Barrett, Dickson, and Fleming. In 'No place for particles', Clifton and Halvorson defend Malament's arguments against these criticisms, and they supply variations of his mathematical proof to thwart potential further criticisms. In particular, Malament's proof is a *reductio ad absurdum* which derives a contradiction from the assumption that there is a 'standard' (i.e., self-adjoint) position operator that satisfies certain relativistic conditions. Clifton and Halvorson strengthen Malament's result by showing that the assumption of an 'unsharp' position operator leads to a contradiction, and similarly that the assumption of a system of local number operators (which would quantify the number of particles in a local spacetime region) leads to a contradiction.

Aside from the project of clarifying the ontological commitments of relativistic quantum theories, there is an independent reason for being concerned about the existence of localized particles: namely, the simple fact that we seem to see localized objects. So, if our best theory entails that there are no such objects, then either our best theory is false, or our experience is radically misleading. But even if we can get past this first worry, there is a second worry about the empirical coherence of a physical theory which entails that there are no localized objects. Such a theory seems to preclude determinate perceptions, and hence its own confirmation. In Section 7 of 'No place for particles', Clifton and Halvorson deal with these issues by arguing that the *appearance* of localized objects, which is consistent with relativistic QFT (despite its entailing the nonexistence of strictly localized particles), is sufficient to provide an empirical basis against which we can test theories.

In 'Are Rindler quanta real?' (2001; Chap. 9) Clifton and Halvorson discuss a second difficulty about particles in relativistic QFT: namely, the effect discovered and analysed by Fulling, Unruh and others — and now usually called the 'Unruh effect'. An observer traveling at a constant (nonzero) acceleration in the Minkowski vacuum state will find that her particle detectors click wildly, as if she were immersed in a thermal bath of particles. (An observer traveling at a constant velocity will not detect any particles.) The question, then, is whether the clicks in the accelerating observer's detector should be taken at face value as indicating the presence of particles in the vacuum — which are called 'Rindler quanta' — or whether the response of the detector should be interpreted in some other way.

A number of different interpretations of the Unruh effect have been proposed. For example, Arageorgis (1995) argues that Rindler quanta do not qualify as genuine particles. On the other hand, Arageorgis *et al.* (2002) argue that the inertial and accelerating observers' different descriptions of the quantum field should be thought of as incommensurable theories, in the manner of Kuhn. Finally, Davies (1984) argues that the two observers' descriptions are complementary in exactly the same way that measurements of position and momentum are. From this, Davies goes on to draw the radical conclusion that there is no objective (i.e., observer-independent) fact about whether Rindler quanta exist in the vacuum.

Clifton and Halvorson argue that these three interpretations, and others, become much clearer — and their merits and demerits more visible — if one adopts the framework of algebraic QFT. They go on to adjudicate between the interpretations, proving several theorems *en route*. Their analysis begins by discussing how the two observers' descriptions correspond to distinct, and inequivalent, representations of the canonical commutation relations. Accordingly, the claim that Rindler quanta are not 'real' corresponds in some way to a claim that the Rindler representation is the 'wrong' representation (and the Minkowski representation is the 'right' representation). The claim of incommensurability amounts to claiming that the states in the two representations make incomparable predictions. And the claim that the Minkowski and Rindler pictures are complementary amounts to a claim that the corresponding representations are complementary.

Broadly speaking, Clifton and Halvorson argue that the complementarity interpretation is superior to its rivals, some of which are in any case incompatible with previously established results or with Clifton and Halvorson's theorems. But the complementarity involved is formally more sophisticated than usually envisaged. For usually, complementarity is formally expressed in terms of non-commuting operators on a single Hilbert space; but here the two inequivalent representations are defined on two different Hilbert spaces.

4 Complementarity, hidden variables, and entanglement

Niels Bohr struggled throughout his life to create a philosophical framework that would solve the conceptual puzzles of quantum theory. As is well known, at the root of Bohr's 'solution' is the idea of complementarity: quantum mechanical systems do not admit a single unified description, but instead require the use of mutually exclusive and jointly exhaustive descriptions. But Bohr also claimed that quantum mechanics is 'complete' in the sense that no future theory could resolve these complementary descriptions into a single unified description.

Bohr's idea of complementarity had a huge impact on the first generation of physicists working on quantum theory. However, professional philosophers — at least since the wane of logical positivism — have usually taken a dim view of Bohr's contribution to foundational issues. Early examples include the critiques by Bunge (1955*a*,*b*) and Popper (1967). More recent criticisms include Cushing (1994), Fine & Beller (1994) and Beller (1999). A common theme of these criticisms is that Bohr's views can only be defended, or even made sense of, by invoking some or other positivist doctrine.

As we noted earlier, sometime in the mid-1990s Clifton came to the conclusion that Bohr's interpretation is best thought of as a version of the modal interpretation. Since the modal interpretation is obviously consistent with the negation of positivism — indeed, it was developed by philosophers (such as van Fraassen and Dieks) who explicitly reject positivistic semantic principles — Clifton of course became bothered by claims that complementarity presupposes positivism. Some philosophers and historians had indeed taken a more favorable view of the complementarity interpretation. In particular, a series of papers by Howard (for example Howard (1994) and Howard (2003)) argue for the ongoing relevance of Bohr's complementarity interpretation for the project of interpreting quantum mechanics; and for this interpretation bearing little resemblance to the 'Copenhagen interpretation' — which was only constructed in the mid-1950s onwards, by people other than Bohr.[3] Clifton's work in this area, i.e., reviving the complementarity interpretation, yielded two papers (in addition to Chap. 9's interpreting Rindler quanta in terms of complementarity); both are reprinted in Part IV.

In 'Reconsidering Bohr's reply to EPR' (2002; Chap. 12), Clifton and Halvorson explicate and defend Bohr's reply to EPR's argument for the incompleteness of quantum mechanics. As is well known, EPR argue that Bohr is committed to saying that a measurement on one system can make a difference to 'what is real' at a distant location. The challenge for Bohr — and the challenge for contemporary interpreters of quantum theory — is to explain how this fact is consistent with the claim that physical causes operate locally.

In his reply to EPR, Bohr famously agrees that 'there is ... no question of a mechanical disturbance of the [distant] system'. But notoriously, he goes on:

> But ... there is essentially the question of an influence on the very conditions that define the possible types of predictions regarding future behavior of the system.

It is of this passage that Bell (1987, p. 155) says, 'I have very little idea what it means', and which Fine (1981, pp. 34–5) interprets as an example of 'virtually textbook neopositivism', and as a sort of 'semantic disturbance' without a 'plausible or intuitive physical basis'.

However, Clifton and Halvorson claim that Bohr's statement makes sense as a claim about which quantities of the distant sys-

[3]The 1980s also saw the appearance of three sympathetic expositions of Bohr's philosophy (Murdoch 1987, Folse 1985, Honner 1987). However, these books are not aimed at developing the complementarity interpretation within the technical foundations of quantum theory.

tem can (without falling into a Kochen-Specker contradiction) possess values simultaneously with the quantity that is measured on the local system.

The general idea that Bohr's statement can be explicated via the modal interpretation had been floated earlier by Howard (1979), Bub (1989*a*), and by Clifton himself (in the 1996 paper with Bub). However, Clifton and Halvorson provide a number of novel results (such as Theorem 1, on page 384), with special emphasis on using the tools of algebraic quantum theory to reconstruct Bohr's reply to the *original* EPR argument (which employs continuous spectrum observables).

As mentioned earlier, critics of the complementarity interpretation have claimed that it is based on invalid positivistic reasoning. Clifton, of course, agreed that deriving ontological conclusions from epistemic premises is generally invalid, and so cannot establish complementarity. In 'Complementarity between position and momentum as a consequence of Kochen-Specker arguments' (2000: Chap. 11), Clifton attempts to provide a more solid foundation for complementarity (specifically, between position and momentum) by means of a Kochen-Specker-type no-hidden-variables theorem. One should be clear, however, that Clifton's argument is not just another simplification of the original Kochen-Specker theorem. Rather, the original Kochen-Specker theorem (and the many simplified versions of it that have been proven over the years) does not establish position-momentum complementarity, since the assumption of the Kochen-Specker *reductio* is that *all* observables (and not just position and momentum) possess values. Clifton, however, shows that a contradiction can be derived from the weaker assumption that position and momentum possess values, at least for the case of two and three degrees of freedom. (It is still an open question whether the result holds for the case of a single degree of freedom.)

Though in Chapter 11, Clifton argues for a strengthened version of the Kochen-Specker theorem, in 'Simulating quantum mechanics by non-contextual hidden variables' (2000; Chap. 13), Clifton and Kent argue that Kochen-Specker-type theorems cannot decisively rule out an explanation of quantum statistics by means of hidden variables. More precisely, Clifton and Kent show that hidden variables can explain the results of a set of measurements that is dense in the 'space of measurements' (where each measurement is represented by a projection-valued resolution of the identity, or more generally by a positive operator-valued resolution of the identity); and, thus, that any quantum mechanical measurement can be 'simulated' by a measurement for which there is a hidden variable explanation. More precisely, the original Kochen-Specker theorem proceeds by choosing a (finite) set of measurements for which there is no explanation by a hidden variable theory. However, Clifton and Kent's result shows that the Kochen-Specker contradiction must always choose at least one measurement M that falls outside of Clifton and Kent's set of classically explainable set of measurements. So, since our measurements are not infinitely precise, one can always maintain that instead of M, some classically explainable measurement M' was performed.

In the final years of Clifton's career, he grew increasingly interested in the rapidly growing field of quantum information theory. Here, we have chosen 'The subtleties of entanglement and its role in quantum information theory' (2002; Chap. 14), which is especially helpful in setting out the new questions and research topics that this field offers philosophers of science. As the title suggests, the overarching theme of this paper is that entanglement is subtle. More specifically, it has theoretical features, and suggests experimental possibilities, which have not been addressed by the philosophical analysis of quantum nonlocality. In particular, these features and possibilities suggest various classifications of kinds of entanglement or nonlocality which are different from, and typically more fine-grained than, the traditional philosophical splitting of stochastic hidden variable models' main assumption of factorizability into parameter independence and outcome independence (i.e., into the prohibition of act-outcome correlations and of outcome-outcome correlations, respectively).

Clifton discusses two such experimental possibilities — dense coding and teleportation — and two theoretical features: hidden nonlocality and entanglement thermodynamics. We shall leave him to speak for himself about the experimental possibilities; but we will introduce the theoretical features.

'Hidden nonlocality' refers to the Werner states mentioned in Section 2.1 above: mixed nonseparable states that obey Bell's inequality and indeed admit a local hidden variables model. It emerged in the mid-1990s through the work of Popescu, Peres and others that in general these states, despite admitting a local hidden variables model, had a hidden nonlocality, in the sense that a suitable sequence of local operations (including measurements) on the component systems A and B, together with ordinary classical communication between the parties (always called 'Alice' and 'Bob'), could yield a Bell correlated state. In short: investigating Werner states engendered a new classification of different kinds of nonlocality.

Finally, Clifton discusses entanglement thermodynamics, in which one quantifies the amount of pure entangled states needed to 'form' a given (in general, mixed) state, and the amount of pure entangled states that can be 'distilled' from a given (in general, mixed) state. As Clifton points out, the measures of entanglement of formation and entanglement of distillation provides a finer-grained classification of nonlocality than the measure provided by Bell's inequality. Clifton concludes by exploring the analogy between entanglement thermodynamics and classical thermodynamics, with a view to shedding new light on the question of whether quantum theory needs to be underpinned by a more fundamental (i.e., hidden variable) theory.

5 Conclusion

If the reader was hoping that this Introduction might provide a picture of a unified 'philosophy of Rob Clifton', then we are afraid we must disappoint her. Clifton had a supreme talent for condensing a philosophical claim into a precise technical proposition — and then proving or refuting it. This talent, combined with the fact that he worked across the whole range of the philosophy and foundations of quantum theory, makes for a picture of great achievement — but also of diversity. For example, in some chapters Clifton develops the modal interpretation, while in others he defends complementarity. And in Chapter 13, he shows that hidden variables cannot be completely ruled out by Kochen-Specker arguments. So what was really Clifton's favored interpretation of quantum mechanics? In our opinion, it is not really helpful to try to pin Clifton down like this. It is almost as if Clifton was governed by the 'uncertainty principle for interpreters of quantum mechanics' stated some years ago by Jeff Bub (1989*b*, p. 191):

> ... *as soon as you've found a position, you lose your momentum.*

Rob Clifton may never have found a definite position, but he never lost momentum.

Bibliography

Albert, D. Z. & Loewer, B. (1990), Wanted dead or alive: Two attempts to solve Schrödinger's paradox, in A. Fine, M. Forbes & L. Wessels, eds, 'Proceedings of the 1990 biennial meeting of the Philosophy of Science Association', Philosophy of Science Association, East Lansing, MI, pp. 277–85.

Arageorgis, A. (1995), Fields, particles, and curvature: Foundations and philosophical aspects of quantum field theory on curved spacetime, PhD thesis, University of Pittsburgh.

Arageorgis, A., Earman, J. & Ruetsche, L. (2002), 'Weyling the time away: the non-unitary implementability of quantum field dynamics on curved spacetime', *Studies in History and Philosophy of Science. B. Studies in History and Philosophy of Modern Physics* **33**, 151–84.

Arntzenius, F. (1990), Kochen's interpretation of quantum mechanics, in A. Fine, M. Forbes & L. Wessels, eds, 'Proceedings of the 1990 biennial meeting of the Philosophy of Science Association', Philosophy of Science Association, East Lansing, MI, pp. 241–9.

Bacciagaluppi, G. & Hemmo, M. (1996), 'Modal interpretations, decoherence and measurements', *Stud. Hist. Philos. Sci. B Stud. Hist. Philos. Modern Phys.* **27**, 239–77.

Bacciagaluppi, G., Donald, M. J. & Vermaas, P. E. (1995), 'Continuity and discontinuity of definite properties in the modal interpretation', *Helv. Phys. Acta* **68**, 679–704.

Bell, J. (1964), 'On the Einstein-Podolsky-Rosen paradox', *Physics* **1**, 195–200.

Bell, J. S. (1987), *Speakable and unspeakable in quantum mechanics*, Cambridge University Press, Cambridge. Collected papers on quantum philosophy.

Beller, M. (1999), *Quantum Dialogue*, University of Chicago Press, Chicago.

Brown, H. R. & Harré, R. (1988), *Philosophical foundations of quantum field theory*, Clarendon Press, Oxford.

Bub, J. (1989*a*), 'On Bohr's response to EPR: A quantum logical analysis', *Foundations of physics* **19**, 793–805.

Bub, J. (1989*b*), 'The philosophy of quantum mechanics', *The British Journal for the Philosophy of Science* **40**, 191–211.

Bub, J. (1993), 'Measurement: it ain't over till it's over', *Foundations of Physics Letters* **6**, 21–35.

Bub, J. (1994), 'How to interpret quantum mechanics', *Erkenntnis* **41**, 253–73.

Bunge, M. (1955*a*), 'Strife about complementarity, part I', *The British Journal for the Philosophy of Science* **6**, 1–12.

Bunge, M. (1955*b*), 'Strife about complementarity, part II', *The British Journal for the Philosophy of Science* **6**, 141–54.

Cushing, J. T. (1994), *Quantum mechanics: Historical contingency and the Copenhagen hegemony*, University of Chicago Press, Chicago, IL.

Davies, P. C. W. (1984), Particles do not exist, in S. M. Christensen, ed., 'Quantum theory of gravity', Adam-Hilger, Bristol, pp. 66–77.

Dickson, W. M. (1994), Wavefunction tails in the modal interpretation, in 'PSA 1994', Philosophy of Science Assocation, East Lansing, MI, pp. 366–76.

Dieks, D. (1989), 'Quantum mechanics without the projection postulate and its realistic interpretation', *Foundations of Physics* **19**, 1397–423.

Dieks, D. (1993), 'The modal interpretation of quantum mechanics, measurements and macroscopic behavior', *Physical Review A* **49**, 439–46.

Dieks, D. (2000), 'Consistent histories and relativistic invariance in the modal interpretation of quantum mechanics', *Physics Letters A* **265**, 317–25.

Elby, A. (1993), 'Why "modal" interpretations of quantum mechanics don't solve the measurement problem', *Foundations of Physics Letters* **6**, 5–19.

Fine, A. (1981), Einstein's critique of quantum theory: The roots and significance of EPR, in P. Barker & C. G. Shugart, eds, 'After Einstein', Memphis State University Press, Memphis.

Fine, A. & Beller, M. (1994), Bohr's response to EPR, in J. Faye & H. Folse, eds, 'Niels Bohr and Contemporary Philosophy', Kluwer, NY, pp. 1–31.

Fleming, G. (2000), 'Reeh-Schlieder meets Newton-Wigner', *Philosophy of Science* **67**, S495–S515.

Fleming, G. N. (1989), Lorentz invariant state reduction, and localization, in A. Fine & J. Leplin, eds, 'Proceedings of the 1988 biennial meeting of the Philosophy of Science Assocation', Philosophy of Science Association, East Lansing, MI, pp. 112–26.

Fleming, G. N. & Butterfield, J. N. (1999), Strange positions, in J. Butterfield & C. Pagonis, eds, 'From physics to philosophy', Cambridge University Press, New York.

Folse, H. (1985), *The philosophy of Niels Bohr*, Elsevier Science, New York.

Halvorson, H. (2001), 'Reeh-Schlieder defeats Newton-Wigner: On alternative localization schemes in relativistic quantum field theory', *Philosophy of Science* **68**, 111–33.

Healey, R. (1989), *The philosophy of quantum mechanics: An interactive interpretation*, Cambridge University Press, Cambridge.

Healey, R. (1993), 'Why error-prone quantum measurements have outcomes', *Foundations of Physics Letters* **6**, 37–54.

Hegerfeldt, G. (1974), 'Remark on causality and particle localization', *Physical Review D* **10**, 3320.

Hegerfeldt, G. C. & Ruijsenaars, S. N. M. (1980), 'Remarks on causality, localization, and spreading of wave packets', *Physical Review D* **22**, 377–84.

Honner, J. (1987), *The description of nature: Niels Bohr and the philosophy of quantum physics*, Oxford University Press, New York.

Howard, D. (1979), Complementarity and ontology: Niels Bohr and the problem of scientific realism in quantum physics, PhD thesis, Boston University.

Howard, D. (1994), What makes a classical concept classical?, in J. Faye & H. Folse, eds, 'Niels Bohr and contemporary philosophy', Kluwer, NY, pp. 201–29.

Howard, D. (2003), 'Who invented the "Copenhagen interpretation"? a study in mythology', *Philosophy of Science*. Forthcoming.

Kochen, S. (1985), A new interpretation of quantum mechanics, in 'Symposium on the foundations of modern physics', World Sci. Publishing, Singapore, pp. 151–169.

Malament, D. B. (1996), In defense of dogma: Why there cannot be a relativistic quantum mechanics of (localizable) particles, in R. Clifton, ed., 'Perspectives on Quantum Reality', Kluwer, Dordrecht, pp. 1–10.

Mermin, N. D. (1996), Hidden quantum non-locality, in R. Clifton, ed., 'Perspectives on quantum reality', Vol. 57, Kluwer, Dordrecht, pp. 57–71.

Murdoch, D. (1987), *Niels Bohr's philosophy of physics*, Cambridge University Press, Cambridge.

Popescu, S. & Rohrlich, D. (1992), 'Generic quantum nonlocality', *Physics Letters A* **166**, 293–7.

Popper, K. R. (1967), Quantum mechanics without "the observer", in M. Bunge, ed., 'Quantum Theory and Reality', Springer, Berlin.

Redhead, M. L. G. (1982), Quantum field theory for philosophers, in 'Proceedings of the 1982 biennial meeting of the Philosophy of Science Assocation', Philosophy of Science Association, East Lansing, MI, pp. 57–99.

Redhead, M. L. G. (1995), 'More ado about nothing', *Foundations of Physics* **25**, 123–37.

Segal, I. E. (1964), Quantum fields and analysis in the solution manifolds of differential equations, in W. T. Martin & I. E. Segal, eds, 'Proceedings of a Conference on the Theory and Applications of Analysis in Function Space', MIT Press, Cambridge, MA, pp. 129–53.

Summers, S. J. & Werner, R. (1985), 'The vacuum violates Bell's inequalities', *Physics Letters. A* **110**, 257–9.

Summers, S. J. & Werner, R. (1987*a*), 'Bell's inequalities and quantum field theory. I. General setting', *Journal of Mathematical Physics* **28**, 2440–7.

Summers, S. J. & Werner, R. (1987*b*), 'Bell's inequalities and quantum field theory. II. Bell's inequalities are maximally violated in the vacuum', *Journal of Mathematical Physics* **28**, 2448–56.

Summers, S. J. & Werner, R. (1987*c*), 'Maximal violation of Bell's inequalities is generic in quantum field theory', *Communications in Mathematical Physics* **110**, 247–59.

Summers, S. J. & Werner, R. (1988), 'Maximal violation of Bell's inequalities for algebras of observables in tangent spacetime regions', *Annales de l'Institut Henri Poincaré. Physique Théorique* **49**, 215–43.

Summers, S. J. & Werner, R. F. (1995), 'On Bell's inequalities and algebraic invariants', *Letters in Mathematical Physics* **33**, 321–34.

Teller, P. (1995), *An interpretive introduction to quantum field theory*, Princeton University Press, Princeton, NJ.

Werner, R. F. (1989), 'Quantum states with Einstein-Podolsky-Rosen correlations admitting a hidden-variable model', *Physical Review A* **40**, 4277–81.

Part I

Modal Interpretations

Chapter 1

Independently motivating the Kochen-Dieks modal interpretation of quantum mechanics

1 Kochen-Dieks in context

All interpretations of quantum mechanics still face the issue vigorously debated by Einstein and Bohr in the 1930s: do the theory's mere probabilistic predictions for measurement outcomes indicate that observables lack definite values prior to measurement? Or is it just that the theory is lacking, making it currently not possible to know with certainty what the true values are until they are measured?

Kochen (1985) and Dieks (1989), or 'KD' for short, embrace the

For countless useful comments and often penetrating criticism, I would like to thank Guido Bacciagaluppi, John L. Bell, Harvey Brown, Jeffrey Bub, Bill Demopoulos, Michael Dickson, Dennis Dieks, Arthur Fine, Martin Jones, Simon Kochen, Richard Healey, Meir Hemmo, David Malament, Itamar Pitowsky, Laura Ruetsche, Abner Shimony, and Howard Stein. (This is *not* a list of 'endorsements'.) And I wish to express gratitude to the Social Sciences and Humanities Research Council of Canada for their continued generous support.

first horn of this dilemma, but do not go as far as orthodoxy. For KD, lots of observables can have definite values, at least more than orthodoxy allows; but when they do, those values will usually be acquired in an irreducibly stochastic way, making predictions about them probabilistic at best.

So which observables have definite values? Here a second dilemma is confronted. If too many observables are ascribed values on any given quantum system — as in a naïve realist view, which would ascribe all observables values, regardless of the system's quantum state — then we run the risk of having to answer to the Bell-Kochen-Specker 'no-hidden-variables' theorem and introduce contextualism. But if too few observables are ascribed values — as with orthodoxy, where having a value is linked to the system being in the appropriate eigenstate[1] — then Schrödinger's cat comes under threat.

KD dissolve this second dilemma by carefully charting a middle way between these two extremes. Values are attributed to enough observables to avoid the measurement problem, but not to so many as to force contextualism. Consistent with their desire to uphold the in principle predictive completeness of quantum theory, KD walk this tightrope without introducing any new elements into its formalism — not even the projection postulate.[2]

[1] I do not want to pretend that this simple eigenstate-eigenvalue rule for the possession of a definite value completely encapsulates the Bohrian strain in orthodoxy. But it is at least implicit in von Neumann's treatment of measurement, supplying him with the motivation for introducing his projection postulate. And the eigenstate-eigenvalue rule is certainly explicit in Dirac's book (1947), pp. 46–7):

> The expression that an observable 'has a particular value' for a particular state is permissible in quantum mechanics in the special case when a measurement of the observable is certain to lead to the particular value, so that the state is an eigenstate of the observable. In the general case we cannot speak of an observable having a value for a particular state.

[2] It is helpful to contrast KD's approach to interpreting quantum theory to the approaches of Bohm and Ghirardi, Rimini, and Weber, or 'GRW' for short, probably the two most popular antidotes to naïve realism and to orthodoxy, respectively. These other interpretations are not trying to walk a tightrope at all, but remain at opposite ends of the spectrum of value definiteness. Using the guiding equation Bohm postulates for the evolution of particle positions, all observables in principle have measurement results predictable with certainty, so in a sense they all have

How do KD accomplish such a feat? They jettison orthodoxy's eigenstate-eigenvalue rule and offer us a new, more liberal (but not too liberal!) interpretative rule, which I shall spell out later, for picking out which observables have definite values in which quantum states. The particular values those observables can possess are then taken to be statistically distributed in accordance with the usual quantum algorithm. This is the element of stochasticity in the interpretation, but it is 'stochasticity with a twist'. For, since the quantum state picks out the set of definite-valued observables at any given time, through time that set will have to change deterministically to a new set due to the deterministic Schrödinger evolution of the state.[3]

2 The ad hoc charge

Most critics of KD focus on the results you get when their new interpretative rule is applied, rather than asking where the rule itself comes from. That is, critics argue that the set of definite-valued observables picked out using KD's new rule has certain counterintuitive properties — the worst being that it allegedly fails to yield the definiteness of the observables that need to be definite if the measurement problem is really to be solved in general.[4]

However no critic, as far as I am aware, has asked whether the KD rule in itself has any independent justification, independent, that is, from a mere ad hoc desire to 'come out with the right answers' to contextualism and the problem of measurement. Assum-

---

definite values. But the trap of naïve realism is overcome by the fact that the results of measurement are contextual in a physically natural way. On the other hand, for GRW, only those observables of which the system is an eigenstate get definite values. But the trap of orthodoxy is overcome by modifying the Schrödinger equation to deliver collapse into eigenstates when appropriate, say, for saving Schrödinger's cat.

[3]This dual statistical/dynamical role the quantum state plays for KD is analogous to how the quantum state functions in both the Bohm and GRW interpretations (see the previous footnote). It would certainly be worth seeing how far one can pursue that analogy, though I shall not do so here.

[4]For example, see Albert and Loewer (1990) and Arntzenius (1990).

ing the KD rule does deliver the goods, how do we know that other rules, singling out other sets of observables as having definite values, won't just as satisfactorily resolve those conceptual problems? We could neutralize this charge if we knew the KD rule were not only sufficient, but in some sense necessary for resolving the problems in the intended way. And that is my project: to prove, under four independently natural requirements on the set of observables picked out by the state as having definite values, that this is indeed the case.

I do not claim that this proof will completely exonerate KD's approach, for one is still at liberty to reject one of the four requirements I shall impose.[5] Rather, my claim is merely that there is at least some plausible independent motivation for adopting their approach. I leave open whether this motivation can be strengthened still further by appeal to different, fewer, or even weaker independent requirements on rules for picking out definite-valued observables.[6]

Furthermore, by motivating KD's approach I will, of course, be leading them straight back into the hands of their critics. But to the extent that the points the critics make against KD are still debatable, and as far as I can see they are,[7] the KD interpretation should at least

[5]Or, indeed, deny (with GRW) that the measurement problem can be satisfactorily resolved without modifying the formalism to incorporate collapse, or deny (with Bohm) that contextualism is something worth taking pains to avoid (see footnote 2).

[6]For a slightly different, more 'streamlined' way of motivating the KD rule than the one I will present here, see my 1995*a*. That alternative motivation avoids any reference to the need to circumvent the measurement problem. It also replaces the requirement that contextualism be avoided with a locality requirement that bars definite-valued observables from sustaining Bell-inequality violating correlations, and so from having their values depend on space like separated measurement events. Both these alterations substantially strengthen the argument of the present paper, since it frees the motivation up from having to take a controversial stance on how best to respond to the threat of contextualism and the problem of measurement (see footnote 5). But these advantages come at a price: For my argument to go through, a stronger — and perhaps objectionable — version of one of my conditions below is needed, viz., condition (4). (Condition (1) also needs to be strengthened, but in an uncontroversial way.).

[7]See the numerous replies by Dieks (1993), Healey (1993), Hemmo (1993), Bac-

be shored up on ground more secure than it started on.

At this point, I should acknowledge that there are other modal interpretations due to van Fraassen (1991), Healey (1989), and Bub (1991) which bear many formal similarities to the KD approach. In fact the term 'modal' is due originally to van Fraassen, who introduced it for his own specific reasons, which I shall not elaborate on. For in line with other recent literature, I am simply using 'modal' as an umbrella term to cover a class of interpretations which, in various ways, modify orthodoxy's eigenstate-eigenvalue rule in order to avoid the conceptual problems of quantum mechanics. These other modal interpretations may well be subject to similar independent motivation to that which I shall supply for KD's.[8] But their differences with KD are great enough as to render my particular motivation inapplicable to them, so I shall not discuss them further here.

I now pause to provide a road map for what follows that is more than simply the list of section titles provided at the beginning.

Section 3 will detail the general framework KD employ for picking out which observables have definite values, and will set out exactly which observables KD want to pick out with their new rule. Section 4 spends some time motivating the natural requirements I will impose on any such approach to avoiding the measurement problem and contextualism. Section 5 makes precise what it is actually going to take to avoid those problems within a KD-type approach. So, together, Sections 4 and 5 lay down the requirements to be imposed on rules for picking out which observables have definite values. Section 6 then presents and discusses the main motivation theorem: that KD's rule is the only rule satisfying all these requirements. Because it is somewhat technical, the proof of this motivation theorem has been banished to an appendix.

---

ciagaluppi and Hemmo (1994), and Dickson (1994).

[8]Indeed, Bub has recently been able to build on some of the ideas below to shore up his own modal interpretation.

3 Kochen-Dieks in detail

Candidate statistical states for a quantum system describable by a Hilbert space $\mathcal{H}$ with $\dim\mathcal{H} > 2$ — whether that system be isolated, or part of a larger system with associated Hilbert space $\mathcal{H} \otimes \mathcal{H}' \otimes \mathcal{H}'' \otimes \cdots$ — are exhaustively represented by the set of all Hermitian, trace 1 'statistical' operators W on $\mathcal{H}$. For that reason, and because the contextualism problem arises only for such systems, I shall restrict myself to the case $\dim\mathcal{H} > 2$; and I shall assume for simplicity that $\dim\mathcal{H} < \infty$ (since I see no obvious reason why the considerations below would not generalize). When the system *is* part of a larger one with a given statistical operator, its statistical operator will get fixed (uniquely) by the larger one's via partial tracing.

To start things off, any KD-type modal interpretation must adopt some rule specifying, for any system and its associated $\mathcal{H}$, a collection of sets $\{\mathrm{Def}(W)\}_{\mathcal{H}}$ ($\equiv \{\mathrm{Def}(W) | W : \mathcal{H} \to \mathcal{H}\}$), where

$$\mathrm{Def}(W) = \{P | P \text{ has definite value } [P]_W, \text{ in state } W \text{ equal to 0 or 1}\},$$

and Ps denote projection operators. Thus, such interpretations need to take a stand on what idempotent observables have definite values for any given statistical state of any system.

Of course, we also need to know the full set of observables with definite values as well. That is, such interpretations will also need to specify the collection $\{\underline{\mathrm{Def}}(W)\}_{\mathcal{H}}$ where $\underline{\mathrm{Def}}(W) = \{O | O$ has a definite value, $[O]_W$, in state W equal to one of O's eigenvalues$\}$ and Os denote not necessarily idempotent observables. But that collection can be thought of as generated by $\{\mathrm{Def}(W)\}_{\mathcal{H}}$ if we adopt the natural condition that

$$O \in \underline{\mathrm{Def}}(W) \quad \text{if and only if} \quad \mathrm{SR}(O) \subseteq \mathrm{Def}(W),$$

where $SR(O)$ is the set of projection operators in the Spectral Resolution of O — a unique set of (not necessarily one-dimensional!) projections, by the spectral theorem. (In this paper, I shall always exclude from this set the projection, call it P_0, onto the trivial O-eigenspace of W, if W happens to have O as an eigenvalue.) The above condition is natural because one surely wants every observable that is a function of definite-valued observables to itself have a definite value, and O is (trivially) a function of its spectral projections while, conversely, each such projection is a characteristic function of O.[9] So given the above connection between what I have called the sets $\mathrm{Def}(W)$ and $\underline{\mathrm{Def}}(W)$ (of idempotent and general definite-valued observables, respectively), there is no loss of generality in focusing just on the sets $\{\mathrm{Def}(W)\}_{\mathcal{H}}$ and on projection observables in what follows.

Once $\mathrm{Def}(W)$ is specified for any W, all possible $0-1$ value assignments to its elements can be listed, and each will define a distinct 'value state' of the system in 'statistical state' W (note the two distinct notions of state being employed). The usual quantum statistical algorithm $\mathrm{Tr}(P_1P_2\ldots W)$, with $\{P_1, P_2, \ldots\}$ any pairwise commuting subset of $\mathrm{Def}(W)$, is then interpreted as specifying joint distributions of the form $\mathrm{Prob}([P_1]_W = [P_2]_W = \cdots = 1)$ from which all relevant marginal distributions can be recovered.

Once $\mathrm{Def}(W)$ is specified for any W, one can also go on and determine the relation between the sets $\mathrm{Def}(W_t)$ and $\mathrm{Def}(W_{t'})$ at different times, t and $t' > t$. Clearly, that relation will be parasitic on the time evolution of W, which itself depends on the overall Schrödinger-deterministic evolution of the composite state of which the system in state W is a part. If that system has no interaction with its environment, its W_t will evolve unitarily into $W_{t'}$. But in general

[9] All this talk of functional relations between observables in no way leads to any Bell-Kochen-Specker contradictions from the outset. We are simply discussing the issue of what set of observables should even be regarded as having definite values. This is separate from the issue of whether contextual value assignments should be introduced on that set once it has been chosen. (Such assignments can distinguish the particular definite value some projection has when considered as a function of one maximal observable O from the particular definite value it has as a function of another incompatible one O'.) Furthermore, as we shall see more clearly later, even when we bar contextual value assignments, Bell- Kochen-Specker contradictions need not arise if the definite-valued set of observables is appropriately restricted. (To avoid cumbersome notation, I have resisted introducing more subscripts on expressions like $[P]_W$ to indicate explicitly that such values can depend on context too; for it will not be long before we impose our requirement that rules this sort of contextualism out, making any added notation redundant.)

it will not, and we can see this already in, say, ideal measurements where the partially traced-out state of the object measured starts out pure $\left(W_t^2 = W_t\right)$ but (usually) ends up mixed $\left(W_{t'}^2 \neq W_{t'}\right)$, an evolution that cannot be unitarily implemented.[10]

Furthermore, given the various value assignments on $\mathrm{Def}(W_t)$ and $\mathrm{Def}(W_{t'})$, one can seek to define conditional ('transition') probabilities of realizing any value state on the latter set given any value state on the former. Of course, these conditional probabilities will have to come out consistent with the *unconditional* probabilities for value states at t and t' determined by applying the quantum statistical trace algorithm to W_t and $W_{t'}$ so they will in general depend on the overall state of the composite system in the same way that the statistical states W_t and $W_{t'}$ do.

Now these two dynamical aspects of KD-type modal interpretations are certainly worth delving into more deeply. But dynamics will play no further role in my considerations of what sets $\{\mathrm{Def}(W)\}_{\mathcal{H}}$ and valuations on those sets to opt for at any given time (though it is an interesting question as to whether plausible dynamic constraints can limit the options too).

All of the above defines the general framework within which KD's new rule for picking out definite-valued observables can be formulated. Clearly there is scope for a variety of 'KD-type' interpretations depending on what sets $\{\mathrm{Def}(W)\}_{\mathcal{H}}$ we choose.

So what sets $\{\mathrm{Def}(W)\}_{\mathcal{H}}$ do KD themselves opt for? Let $\mathrm{BAG}(\mathrm{SR}(W))$ denote the Boolean Algebra Generated by some compatible set of projections S on $\mathcal{H}$, where the lattice operations are the standard ones: $P \vee P' \equiv P \otimes P', P \wedge P' \equiv P \cap P'$ and $\neg P \equiv P^{\perp} = I - P$. For convenience, define $N(W) \equiv \{P|PW = 0\}$, which picks out the set of projections onto sub-spaces of the Null space of W.

[10]In such cases, we can in general expect the set $\mathrm{Def}(W_{t'})$ picked out at the later time t' to depend on the environment of the system during the interval (t, t') but no specific dependence on the environment will need to be assumed here (beyond the minimum necessary to solve the measurement problem — see condition 7(6) below). Some may find it natural to label such dependence 'contextualism', but then it needs to be sharply distinguished from the idea that value assignments defined on the set $\mathrm{Def}(W_{t'})$ are contextual in the sense of the Bell-Kochen-Specker theorem (which will be spelled out in a few sections).

Then, in effect, the KD proposal ('rule') is simply that for any W, pure or mixed (and possibly arising through partial tracing over some composite system's state):[11]

$$\begin{aligned}\mathrm{Def}(W) &= \{P \mid P = P_1 + P_2, P_1 \in \mathrm{BAG}(\mathrm{SR}(W)), P_2 \in N(W)\} \\ &\equiv \mathrm{Def}_{\mathrm{KD}}(W).\end{aligned}$$

Choosing $\mathrm{Def}_{\mathrm{KD}}(W)$ like this has at least one desirable consequence that can easily be checked, viz., it conforms to the following minimal 'reality' criterion: that any projection whose value is predictable with certainty in state W should get a definite value! Apart from that, it is not yet clear whether the choice $\mathrm{Def}_{\mathrm{KD}}(W)$ can be independently motivated.

Before seeking motivation, there are a few differences that need to be cleared up between what KD say and their proposal as I have formulated it above.

KD couch their proposal in terms of the definite-valued observables that get picked out on a system $\mathcal{H}$ correlated with another $\mathcal{H}'$ to form a larger system with some given pure state vector v on $\mathcal{H} \otimes \mathcal{H}'$. They use a consequence of the polar decomposition theorem that v can always be decomposed into the 'biorthonormal' form $v = \sum_i c_i v_i \otimes v_i'$, with $\{v_i\}(\{v_i'\})$ an orthonormal basis of $\mathcal{H}(\mathcal{H}')$. This type of decomposition, also often called the 'Schmidt' decomposition of v, is unique if and only if there is no degeneracy amongst the numbers $\{|c_i|^2\}$ (some of which may need to be zero in order for the sets $\{v_i\}$ and $\{v_i'\}$ to define bases). So, at least for nondegenerate states v, KD are able to latch onto this purely formal decomposition property of v and propose that system $\mathcal{H}$ possesses values for the projections in $\mathrm{BAG}(\{P_{v_i} \mid |c_i|^2 \neq 0\})$ (similarly here, and below, for $\mathcal{H}'$).[12]

---

[11]Caution: not every sum of a projection in $\mathrm{BAG}(\mathrm{SR}(W))$ and one in $N(W)$ defines a projection (e.g. the projection P_0 onto W's 0 eigenspace — if it has one — will be in both these sets, but $2P_0$ is not idempotent). That is why $\mathrm{Def}_{\mathrm{KD}}(W)$ is defined to include only those sums that do define projections!

[12]In fact, Kochen seems to have been the first to propose this, though Dieks has much more fully worked out the proposal's implications and, in some cases, extended it — hence the phrase 'Kochen-Dieks' interpretation.

How does this proposal relate to $\mathrm{Def}_{\mathrm{KD}}(W)$ as I have defined it above? In this non-degenerate case, the partially traced-out state of $\mathcal{H}$ as determined by the composite pure statistical state P_v will have the decomposition $W = \sum_i |c_i|^2 P_{v_i}$, so that $\mathrm{SR}(W) = \{P_{v_i} | |c_i|^2 \neq 0\}$. Furthermore, since at most one of the numbers $|c_i|^2$ — call it $|c_0|^2$ — is zero in the absence of degeneracy, $N(W) \subseteq \{0, P_{v_0}\} \subseteq \mathrm{BAG}(\{P_{v_i} | |c_i|^2 \neq 0\})$. Thus the set KD pick out for system $\mathcal{H}$ is precisely the set I have defined to be $\mathrm{Def}_{\mathrm{KD}}(W)$, since in this non–degenerate case that set reduces to just $\mathrm{BAG}(\mathrm{SR}(W))$. So far, then, the difference between my formulation and KD's is only apparent.

But what about the degenerate case? If degeneracy occurs, and the only distinct eigenvalues of W are $\{|c_j|^2\}$, then Dieks takes

$$\mathrm{Def}(W) = \mathrm{BAG}(\{P_j | |c_j|^2 \neq 0\}).$$

where, for all j, $P_j \equiv \sum \left\{ P_{v_i} | \, |c_i|^2 = |c_j|^2 \right\}$.[13] This again agrees with my formulation, but only in the case where there is no degeneracy arising due to repeated zeros amongst the numbers $\{|c_i|^2\}$ appearing in the state v's biorthonormal expansion. If there is, I have taken the liberty of enlarging the set $\mathrm{Def}_{\mathrm{KD}}(W)$ to include all the nontrivial projections that become predictable with certainty to have value 0 or 1 in state W. (That is why the set $N(W)$ appears in my formulation, but not KD's.) I am motivated to do this in response to what I believe is a valid criticism made by Arntzenius (1990, pp. 243–4), that the KD prescription does not always meet the minimal reality criterion mentioned above. Since the enlarged set does, I am confident they would not oppose it. My confidence is increased by the fact that the enlarged set, though no longer always a Boolean algebra, is still (as we shall see in the next section) an ortholattice, and that, despite that, the valuations on it with nonzero probability still behave in a classically truth-functional way (as we shall see in the section after that). So enlarging the set to alleviate Arntzenius' criticism does not lead to a set of definite-valued projections

[13]Kochen reserves judgment on the degenerate case for a later paper (not yet published). So this is one respect in which Dieks extends the Kochen proposal (see the previous note).

with properties that are against the 'classical' spirit of KD's original proposal.

I should also note that formulating KD's proposal as I have — with W, rather than the composite pure state v, seen as the entity that picks out the set of definite-valued projections for system $\mathcal{H}$ — not only captures everything their proposal aims to, but is more general. For my formulation applies even in situations where the composite system is not in a pure state, and so does not necessarily possess any biorthonormal decomposition.[14]

The definition I have given for $\mathrm{Def}_{\mathrm{KD}}(W)$ above has proven useful for comparing to what KD actually say about their proposal for what observables get definite values. But I now want to end this section by deriving an alternative, equivalent definition of the set $\mathrm{Def}_{\mathrm{KD}}(W)$ which will make it much easier to work with (in particular, easier to see that $\mathrm{Def}_{\mathrm{KD}}(W)$ does in fact satisfy the requirements I shall lay down to limit the choices of which sets $\{\mathrm{Def}(W)\}_{\mathcal{H}}$ one should opt for, and easier for the proof of the motivation theorem).

Recall that we have:

$$\mathrm{Def}_{\mathrm{KD}}(W) \equiv \{P | P = P_1 + P_2, P_1 \in \mathrm{BAG}(\mathrm{SR}(W)), P_2 \in N(W)\}\,.$$

But a projection P_1 is in $\mathrm{BAG}(\mathrm{SR}(W))$ if and only if P_1 can be written as a sum of distinct projections from the set $\mathrm{SR}(W)$, with the possible inclusion in the sum of 0 and the projection P_0 onto the zero eigenspace of W (projections which, you will recall, I have not included in the definition of $\mathrm{SR}(W)$). Why? Whatever set $\mathrm{BAG}(\mathrm{SR}(W))$ is, any projection that can be written as such a sum must certainly be in it (otherwise that set, which necessarily includes $\mathrm{SR}(W)$, would not be closed under the Boolean operations). And the set of all projections that are sums of distinct projections from $\mathrm{SR}(W) \cup \{0, P_0\}$ is quickly checked to be a Boolean algebra, so that set must in fact be the smallest Boolean algebra that contains the set $\mathrm{SR}(W)$, i.e., it must be $\mathrm{BAG}(\mathrm{SR}(W))$. Given this way

[14]For a discussion of this and, indeed, how to extend the notion of such a decomposition to mixed states, see Barnett and Phoenix (1992). Dieks himself has now taken to formulating the KD rule in terms of statistical operators as I have here (see Vermaas and Dieks 1995).

of characterizing projections in the set BAG(SR(W)), one alternative characterization of the full set $\mathrm{Def}_{\mathrm{KD}}(W)$ is to say that P is in $\mathrm{Def}_{\mathrm{KD}}(W)$ if and only if P can be written as a sum of distinct projections in SR(W), plus a projection in the set $N(W)$ (a set which includes $\{0, P_0\}$. So we have unpacked the 'Boolean algebra generated by' part of the definition of $\mathrm{Def}_{\mathrm{KD}}(W)$.

But there is an even simpler characterization of $\mathrm{Def}_{\mathrm{KD}}(W)$, which we can now quickly deduce, that requires no explicit reference to the set $N(W)$, only SR(W). This will be the characterization used in all that follows. Letting the set $\{P_i\}$ denote W's spectral projections, it is:

$$\mathrm{Def}_{\mathrm{KD}}(W) = \{P | PP_i = P_i \text{ or } 0, \text{ for every } P_i \in \mathrm{SR}(W)\}\,.$$

For the proof, first suppose P is in $\mathrm{Def}_{\mathrm{KD}}(W)$, so by the previous paragraph, P is a sum of distinct projections in SR(W) plus a projection in $N(W)$. For every P_i in SR(W), either P_i is in the sum for P, or it is not. If it is, then $PP_i = P_i$ if not, $PP_i = 0$. Conversely, suppose $PP_i = P_i$ or 0, for every P_i in SR(W). Let $P_1, P_2, \ldots$ be the spectral projections of W that P preserves, and $\underline{P}_1, \underline{P}_2, \ldots$ the ones P annihilates. Then $P[P_1 + P_2 + \cdots] = [P_1 + P_2 + \cdots]$ and $P[\underline{P}_1 + \underline{P}_2 + \cdots] = 0$. Adding these equations (and using the spectral theorem) yields $P[I - P_0] = P_1 + P_2 + \cdots$, or $P = P_1 + P_2 + \cdots + PP_0$, which exhibits P as a sum of distinct spectral projectors $P_1, P_2, \ldots$ of W plus the term PP_0. But if P can be so written, PP_0 must also be a projection operator, and thus $PP_0 = P_0P$ so that it is in fact a projection onto the null space of W. So P is indeed in the set $\mathrm{Def}_{\mathrm{KD}}(W)$, by the argument of the previous paragraph.

4 The independently natural conditions

A priori, $\mathrm{Def}_{\mathrm{KD}}(W)$ need not be the only choice one might make for Def(W). First of all, there are the choices made by the Naïve Realist and Orthodoxy, viz., $\mathrm{Def}_{\mathrm{NR}}(W) = \{P | P : \mathcal{H} \to \mathcal{H}\}$ (i.e., *everything* gets a definite value) and $\mathrm{Def}_{\mathrm{O}}(W) = \{P | PW = 0 \text{ or } W\}$ (i.e., the eigenstate-eigenvalue rule generalized to cover mixed states), which can also be viewed as KD-type interpretations fitting into the framework outlined in the previous section.

But although those choices too (trivially) conform to the minimal reality criterion, they generate the conceptual problems. So what other choices might there be that don't give rise to those problems? None, if we impose four natural conditions on the set $\text{Def}(W)$, including the requirement that it indeed be 'picked out' by W — and any further needed structure of $\mathcal{H}$ itself — in the spirit of a KD approach.

Let's start with the latter requirement. It amounts to the demand that the set $\text{Def}(W)$ be invariant under all automorphisms of $\mathcal{H}$ that preserve W, which are exhausted by the unitary operators U satisfying $UWU^{-1} = W$. For if the set $\text{Def}(W)$ itself were not also preserved under such Us, then structure in addition to that of W and $\mathcal{H}$ would need to be brought in to make the difference. We are thus led to impose:

For any W, and any U satisfying $UWU^{-1} = W$:

$$U[\text{Def}(W)]U^{-1} = \text{Def}(W). \tag{1}$$

(1) is trivially satisfied by $\text{Def}_{\text{NR}}(W)$, and is easily checked for $\text{Def}_{\text{O}}(W)$. But of most interest to us is that $\text{Def}_{\text{KD}}(W)$ satisfies (1). For any unitary transformation that preserves W must preserve its spectral projectors (by the spectral theorem for W); and we saw at the end of the last section how the set $\text{Def}_{\text{KD}}(W)$ could be defined solely in terms of $\text{SR}(W)$, so it too must be preserved.

I turn now to the three further conditions on $\text{Def}(W)$, not necessarily part and parcel of what I have been calling KD-type interpretations, but natural none the less. They too will be satisfied by the three particular choices for this set we have been looking at — again trivially for $\text{Def}_{\text{NR}}(W)$, and in a manner easily checked for $\text{Def}_{\text{O}}(W)$. So I shall focus just on $\text{Def}_{\text{KD}}(W)$'s satisfaction of these conditions.

The second condition is simply that $\text{Def}(W)$ conform to the minimal reality criterion:

$$\textit{For any } W: \ \{P | PW = 0 \textit{ or } W\} \subseteq \text{Def}(W), \tag{2}$$

which we have already noted $\mathrm{Def}_{\mathrm{KD}}(W)$ satisfies. (The proof is easy: If $PW = 0$ or W, then by the spectral theorem for W, $PP_i = 0$ for all P_i in $\mathrm{SR}(W)$, or $PP_i = P_i$ for all P_i in $\mathrm{SR}(W)$.) We want (2) because the existence of the value $[P]_W$ provides the best causal explanation of why we are bound to find 0 or 1 if P is measured in state W. (2) does *not* imply the stronger reality criterion used by Einstein-Podolsky-Rosen, which also allows inferences to values to be made on the basis of conditional predictions with certainty obtained via the joint probabilities prescribed by an entangled composite state out of which W is partially traced. (2) does not license such inferences, because those joint probabilities do not show up in W after partial tracing.

The third condition is that the set $\mathrm{Def}(W)$ form an orthocomplemented sublattice of the orthocomplemented lattice $\mathcal{L}(\mathcal{H})$ of all projections on $\mathcal{H}$. The reason is that we would like it if $\mathrm{Def}(W)$'s projections formed a logic of propositions within which conjunctions, disjunctions and negations can be formed. This can be a reasonable requirement on the possession of values even if the logic thereby obtained turns out not to be completely classically truth-functional (particularly with respect to propositions formed out of incompatible projections). So the requirement is:

For any W, and any $P, P' \in \mathrm{Def}(W)$:
$P \oplus P' \in \mathrm{Def}(W)$, $P \cap P' \in \mathrm{Def}(W)$, *and* $P^{\perp} \in \mathrm{Def}(W)$. (3)

Does $\mathrm{Def}_{\mathrm{KD}}(W)$ satisfy this? Yes (as I promised earlier). Closure under $\perp$ follows from the fact that $PP_i = P_i$ or 0 for all P_i in $\mathrm{SR}(W)$, then $(I - P)P_i = 0$ or P_i for all P_i in $\mathrm{SR}(W)$. For closure under $\oplus$, consider any projection $P_i \in \mathrm{SR}(W)$ and any two projections P and P' in $\mathrm{Def}_{\mathrm{KD}}(W)$. The subspace corresponding to P_i must be either contained in, or orthogonal to, the subspace corresponding to P; the same goes for P'. So the P_i-subspace must either be contained in, or orthogonal to, the $P \oplus P'$-subspace! Closure under $\cap$ now follows by de Morgan's law, or can be proved directly (exercise!). (To get a more geometrical picture of the ortholattice $\mathrm{Def}_{\mathrm{KD}}(W)$, its generating atoms consist of all elements of $\mathrm{SR}(W)$ and all projections onto rays in the subspace $N(W)$.)

The last natural condition pertains to Def(W) when W is mixed, i.e. $W^2 \neq W$. Let us first restrict to the case where W is a so-called 'proper' mixture, i.e. a statistical state for a system that is supposed to be prepared so that it is not entangled with any other system.

Such mixtures are standardly thought to admit of an 'ignorance interpretation' according to which each member of an ensemble of systems in state W can be said to actually be in one of the pure states represented by a (not necessarily orthogonal) set $\{P_k\}$ of one-dimensional projection operators satisfying $W = \sum_k \lambda_k P_k$ with $\sum_k \lambda_k = 1$. Here, each λ_k gives the probability that the system is actually in the corresponding state P_k. (We ignore projections occurring only trivially with zero coefficient in the expansion $W = \sum_k \lambda_k P_k$.)[15] There can be many different sets $\{P_k\}$ bearing this kind of relationship to W, so knowledge of W alone does not fix the possible candidates for the pure states which compose an ensemble in state W. It is standardly assumed that we also need to know the details of how that ensemble was prepared.

Now there is no reason to think that the standard ignorance interpretation of proper mixtures should straightforwardly carry over to a KD-type modal interpretation however such mixtures are prepared. *A priori*, we must leave room for such interpretations to tell their own story about what acquired definite values in the process of state preparation. But it is certainly desirable to be able to make room for at least *some vestige of* the ignorance interpretation of proper mixtures, if only to minimize our differences with the standard wisdom about them. So for any proper mixture W, there should exist *at least one* set $\{P_k\}$ of pure states (satisfying $W = \sum_k \lambda_k P_k$ and $\sum_k \lambda_k = 1$) with respect to which we can, if we want, say that in actual fact, each member of an ensemble of systems in state W is in one of the states $\{P_k\}$. Clearly, if a KD-type interpretation loses the ability to say even this, it loses a lot!

Given this very weak requirement (which I have intentionally formulated so as not to contain any *a priori* preference for one par-

[15]Such decompositions of W are to be sharply distinguished from its spectral decomposition, though W's spectral projectors can sometimes provide *one* choice for the set $\{P_k\}$ if W happens to be non-degenerate.

ticular decomposition of W into pure states $\{P_k\}$ over another), the next question is: What should the definite-valued observables in proper mixture W be? Well, since we want to be able to say that the true state could be any one of the pure states $\{P_k\}$, $\mathrm{Def}(W)$ had better not include any projections not in $\mathrm{Def}(P_k)$ for any such P_k, otherwise that inclusion would contradict what the set $\mathrm{Def}(P_k)$ says gets a definite value in state P_k! Put the other way around, increasing our knowledge about a system's true statistical state should only make us want to pick out more observables as having definite values, never less. This leads to the fourth natural requirement on KD-type interpretations (containing the $W^2 = W$ case trivially):

> *For any W, there should exist at least some (not necessarily orthogonal) set $\{P_k\}$ pure states satisfying $W = \sum_k \lambda_k P_k$ with $\sum_k \lambda_k = 1$, such that for every $P_k \in \{P_k\}$:* $\mathrm{Def}(W) \subseteq \mathrm{Def}(P_k)$. (4)

Why have I not restricted this last condition (4) to just proper mixtures W? After all, haven't my arguments so far only dealt with those? And isn't it true that the argument I have given to motivate (4) can only rest on the presumption that W is proper? For if W is improper, i.e., if the system with statistical state W is a component of a composite system in a pure entangled state, it is well known (see d'Espagnat 1976) that the ignorance interpretation of W leads to inconsistency: we cannot suppose that the composite's components are, in actual fact, in some unknown pure state, without also supposing that the composite was in a *mixture* of pure states from the start!

That last observation is of course completely correct. But the argument that (4) must hold also for improper mixtures (and so, any mixture whatsoever) is *not* based on any (necessarily dubious!) argument for the ignorance interpretation of improper mixtures. Rather, the argument now goes like this. The aim of KD-type interpretations, as I have construed them, is to use only the information contained in W (and perhaps $\mathcal{H}$ as well) to pick out the set $\mathrm{Def}(W)$. But given just any old statistical operator W, it is not possible to tell out of what circumstances it arose, and so impossible to distinguish it as proper or improper. Thus whatever formal condition on

$\mathrm{Def}(W)$ — such as condition (4) — we say should be satisfied at least when W is proper, must also be satisfied when W arises as an improper mixture, because the set $\mathrm{Def}(W)$ we pick out can only depend on W! Since any W one might entertain could have arisen as a proper mixture, condition (4) must then hold quite generally for any W.

With this argument, I have not attempted to slip an argument for the ignorance interpretation of improper mixtures through by the back door. Condition (4) is a necessary but nowhere near sufficient condition for securing the possibility of an ignorance interpretation of proper mixtures. Without extra assumptions, no condition on the set $\mathrm{Def}(W)$, including (4), could force us into an ignorance interpretation of even proper mixtures, let alone improper ones. So when we say (4) holds, not only for proper, but improper mixtures, we are not thereby committed to their ignorance interpretation. At most, we are committed to attributing definite values to observables in the same way we would if the systems in state W were each really in some pure state and W arose as a proper mixture, i.e., we are committed to attributing values 'as if' it were possible to ignorance interpret improper mixtures. But acting 'as if' something were true is a far cry from commitment to its really being true.

$\mathrm{Def}_{\mathrm{KD}}(W)$ satisfies (4) by choosing the set $\{P_k\}$ to be any (not necessarily orthogonal) set satisfying $W = \sum_k \lambda_k P_k$ and $\sum_k \lambda_k = 1$, where every P_k satisfies $P_i P_k$ for some P_i in $\mathrm{SR}(W)$. For let P be any projection in $\mathrm{Def}_{\mathrm{KD}}(W)$, and consider any P_k in the chosen set satisfying $P_i P_k = P_k$ for some P_i in $\mathrm{SR}(W)$. Either $PP_i = P_i$, in which case $PP_k = P_k$ or $PP_i = 0$, in which case $PP_k = 0$. In either case, P is in $\mathrm{Def}_{\mathrm{KD}}(P_k)$ and (4) follows.

Note that (4) would come out false for $\mathrm{Def}_{\mathrm{KD}}(W)$ if we had made the stronger demand that the set $\{P_k\}$ making (4) true specifically contain a projection onto a ray not in any of W's nontrivial eigenspaces (which is perfectly possible consistent with $W = \sum_k \lambda_k P_k$ and $\sum_k \lambda_k = 1$. So for those preparations of W which mix pure states from that kind of set, KD's value definiteness rule yields results in conflict with the standard ignorance interpretation of W.[16]

[16]The reasons for this have recently been elaborated in more detail by Vermaas

However, as I said before, we can't expect a KD-type interpretation to reproduce all that the standard wisdom says about proper mixtures.

5 Avoiding the conceptual problems

I now turn to the conditions expressing avoidance of contextualism and the problem of measurement, which is where we expect the choices $\mathrm{Def}_{\mathrm{NR}}(W)$ and $\mathrm{Def}_{\mathrm{O}}(W)$ to fall by the wayside.

A *noncontextual* value state, or truth value assignment, on $\mathrm{Def}(W)$ is a map $[]_W : \mathrm{Def}(W) \to \{0,1\}$ which, when restricted to any Boolean sublattice of $\mathrm{Def}(W)$, yields a homomorphism of that sublattice into the two-element Boolean lattice $\{0,1\}$. Requiring that a single map $[]_W$ do the job, rather than a collection of maps $\{[]_W\}$ each associated with a different Boolean sublattice in $\mathrm{Def}(W)$, ensures the noncontextuality of the values assigned. And restricting to homomorphisms on such sublattices at least guarantees that propositions built up from compatible ones in $\mathrm{Def}(W)$ are classically truth-functional (even if not for incompatible propositions — a possibility that we surely must at least be prepared for given our adoption of (3), the ortholattice condition).

By directly proving the relevant corollary to Gleason's theorem, Bell (1966) showed that no such $[]_W$ exists — for any $\mathcal{H}$ for which $\dim(\mathcal{H}) > 2$, and any W on $\mathcal{H}$ — if $\mathrm{Def}(W) = \mathcal{L}(\mathcal{H}) = \mathrm{Def}_{\mathrm{NR}}(W)$. Kochen and Specker (1967) strengthened this by showing there are even *finite* orthocomplemented sublattices of $\mathcal{L}(\mathcal{H})$ on which no noncontextual value states exist.[17] Of course that does not show all orthocomplemented sublattices of $\mathcal{L}(\mathcal{H})$ have that property, opening the door up to KD-type interpretations.

Kochen and Specker's (1967) treatment has another strength, viz., they explicitly show that one only needs to treat their finite set of definite-valued projections as forming a partial Boolean algebra, in which meets and joins of incompatible projections are left undefined, in order to prove that it admits no noncontextual value states

and Dieks (1995).

[17] In fact, Bell (1966) also establishes this (see Clifton 1993).

(which, in their framework, become simply homomorphisms of the partial Boolean algebra into $\{0,1\}$). We shall not be making use of this extra strength to constrain the set $\text{Def}(W)$ so that it avoids contextualism, since we have already imposed the ortholattice condition (3) on $\text{Def}(W)$. On the other hand, no extra restrictions on how the values of incompatible propositions work out will be needed beyond Kochen and Specker's requirements on compatible propositions.

Now it is not enough simply to require that noncontextual value states on $\text{Def}(W)$ exist; for there must be sufficiently many valuations to satisfy the quantum statistical algorithm.[18] The appropriate condition for circumventing the Bell-Kochen-Specker theorem is therefore:

> *For any W, there exist sufficiently many noncontextual value states $[]_W$ on $\text{Def}(W)$, and a measure μ_W on the set of all such $[]_W$s, such that for any pairwise commuting subset $\{P_1, P_2, \dots\}$ of $\text{Def}(W)$:*
>
> $$\text{Tr}(P_1 P_2 \cdots W) = \mu_W\{[]_W | [P_1]_W = [P_2]_W = \cdots = 1\}. \tag{5}$$

$\text{Def}_{\text{O}}(W)$ satisfies (5) easily, but what of $\text{Def}_{\text{KD}}(W)$? Let $W = \sum_i |c_i|^2 P_i$ be the spectral decomposition of W. Take the various $[]_W$s with nonzero measure to be in one-to-one correspondence with the P_is in $\text{SR}(W)$. For each such P_i, define $[P]^i_W = 1$ if $PP_i = P_i$, and $[P]^i_w = 0$ if $PP_i = 0$ which completely specifies the map $[]^i_W$ on the set $\text{Def}_{\text{KD}}(W)$. Now check that each such $[]^i_W$ defines a noncontextual value state on $\text{Def}_{\text{KD}}(W)$. Obviously $[I - P]^i_W = 1 - [P]^i_W$. And for $P, P' \in \text{Def}_{\text{KD}}(W)$, $[P \cap P']^i_W = [P]^i_W [P']^i_W$, since $(P \cap P')P_i = P_i$ implies that $PP_i = P_i$ *and* $P'P_i = P_i$, whereas $(P \cap P')P_i = 0$ implies $PP_i = 0$ *or* $P'P_i = 0$. $[P \oplus P']^i_W = [P]^i_W + [P']^i_W - [P \cap P']^i_W$ then follows by de Morgan's law (or can be proved directly). Finally, for each i, assign $\mu_W([]^i_W) = |c_i|^2 \dim(P_i)$. Then for commut-

[18]My 1993 gives examples where the first requirement, but not the second, is fulfilled so that contextual valuations are still needed.

ing $P_1, P_2, \dots$ in $\mathrm{Def}_{\mathrm{KD}}(W)$:

$$\mu_W\{[]_W | [P_1]_W = [P_2]_W = \cdots = 1\} = \sum_{P_1 P_i = P_2 P_i = \cdots = P_i} |c_i|^2 \dim(P_i),$$

which exactly mimics the results of applying the formula $\mathrm{Tr}(P_1 P_2 \cdots W)$.

Note that for the above argument that each $[]^i_W$ defines a non-contextual value state, we did not need to restrict to the case of commuting pairs of projections, $P, P' \in \mathrm{Def}_{\mathrm{KD}}(W)$! Thus each $[]^i_W$ defines a *lattice* homomorphism of $\mathrm{Def}_{\mathrm{KD}}(W)$ into $\{0, 1\}$ as well. So we might say that 'with probability 1' classical truth tables apply to the propositions in $\mathrm{Def}_{\mathrm{KD}}(W)$ — as promised a few sections back.[19]

To solve the measurement problem, $\mathrm{Def}(W)$ needs to be large enough to accommodate definiteness of apparatus pointer readings and the values of observables they are supposed to indicate, a test $\mathrm{Def}_{\mathrm{O}}(W)$ blatantly fails for mixtures arising out of entangled apparatus/object states. It is reasonable to require that:

For any W: $\mathrm{SR}(W) \subseteq \mathrm{Def}(W)$, (6)

which obviously $\mathrm{Def}_{\mathrm{KD}}(W)$ satisfies. Why?

Any state W can always be regarded as the partially traced-out state of an object appropriately entangled with some apparatus that has just performed an ideal measurement on that object of an observable O such that $\mathrm{SR}(O) = \mathrm{SR}(W)$. So the projections in $\mathrm{SR}(W)$ need to be in $\mathrm{Def}(W)$ if the apparatus is to fulfill its function of truly indicating O's value upon completing the measurement.[20] Even more so if the 'object' itself is sufficiently macroscopic to function as an apparatus that can (ideally) measure a third system using O as a pointer observable.

[19] The fact that the nonzero measure $[]_W$s I have introduced define lattice homomorphisms of $\mathrm{Def}_{\mathrm{KD}}(W)$ into $\{0, l\}$ does not conflict with the non-distributivity of $\mathrm{Def}_{\mathrm{KD}}(W)$ when the null space of W is more than one-dimensional, because none of these homomorphisms define Boolean embeddings.

[20] Strictly speaking, this assertion, which brings into the discussion observables which are not necessarily idempotent, only follows once we commit ourselves to the connection between the sets $\underline{\mathrm{Def}}(W)$ and $\mathrm{Def}(W)$ argued for earlier.

But what if the object's statistical state W did not in fact arise as a result of an $\mathcal{O}$ measurement on it or of its measuring another system with $\mathcal{O}$ as the pointer observable? Still (6) must hold of W, because KD-type interpretations allow only W, not how that state was brought about, to pick out the definite-valued observables. (The point here is the same as the one that was made in the previous section to justify condition (4) even for improper mixtures.) As a consequence, the ascription of values will not be limited to measurement interactions, which has been regarded by some (e.g. Bell) as also necessary for fully eradicating the measurement problem.

Pace KD's critics, I will take (6) to be sufficient for solving that problem as well. Thus, as signaled in section 2, I am assuming for present purposes that their worries about (6) have been satisfactorily dealt with elsewhere in the literature.[21]

6 The motivation theorem and its implications

Before giving a precise statement of the motivation theorem, it is worth briefly reviewing the requirements on Def(W) we have accumulated in the previous two sections. We have four independently natural requirements (1)–(4), and two requirements (5) and (6) which must hold for KD-type interpretations to avoid contextualism and overcome the measurement problem.

First, the independently natural requirements. Condition (1) required that Def(W) be invariant under any unitary transformation that preserves W, so that nothing else in addition to W and the Hilbert space $\mathcal{H}$ is 'smuggled in' to define the definite-valued observables. (2), which I've dubbed a 'minimal' reality criterion, required that Def(W) at least include all projections with values predictable with certainty in state W. With (3), we required that Def(W) form an ortholattice within which conjunctions, disjunctions, and negations of the propositions that correspond to Def(W)'s projec-

[21] To be explicit: worries about whether (6) is enough to cover cases of imperfect measurement, or even ideal cases where an apparatus, having measured a maximal observable, is nevertheless left with a highly degenerate statistical operator, e.g. one which is a multiple of the identity operator (see notes 4 and 7).

tions are defined and can be meaningfully entertained. Lastly, (4) was required so that Def(W) at least allows an ignorance interpretation of W, with respect to at least one of its decompositions into pure states, if W were to arise as a proper mixture.

To avoid the conceptual problems, we added two further conditions. (5) demanded that there be sufficiently many non-contextual value assignments on Def(W), so that the statistics prescribed by the standard quantum trace formula can be recovered. And (6) required that Def(W) at least contain the projections in the spectral resolution of W to answer the (ideal) measurement problem in case the system happens to get entangled with another that is measuring an observable sharing W's spectral resolution, or if such an observable acts as the pointer observable for the measurement of an observable on another system.

Keeping all this in mind, the result advertised in the abstract can now be stated with precision:

> *For any $\mathcal{H}$ with* $\dim(\mathcal{H}) > 2$, *the collection* $\mathrm{Def}(W)\}_{\mathcal{H}}$ *satisfies conditions (1)–(6) if and only if for all W:* $\mathrm{Def}(W) = \mathrm{Def}_{\mathrm{KD}}(W)$.

At this point, we need only prove necessity. Since the proof is somewhat involved, I have relegated it to an appendix, perhaps to be read only on a second reading of the paper. This is not to say that the proof is beyond understanding, and to aid in that it is very helpful to draw some diagrams (for the case $\dim(\mathcal{H}) = 3$ of course!) to get a geometric feel for what is going on behind all the algebraic arguments. It is also useful to commit to memory the above summary of each of the conditions (1)–(6), since I have not taken the trouble in the proof to restate each condition every time it is invoked. Finally, note that the motivation theorem above has precisely the same exceptions in the case $\dim(\mathcal{H}) = 2$ that Gleason's theorem does, e.g. nothing in that case prevents us from choosing $\mathrm{Def}(W) = \mathrm{Def}_{\mathrm{NR}}(W)$ for all W.

If there is a weak point in this motivation theorem, it may be thought to lie in the assumption that Def(W) should form an ortholattice, condition (3). That one cannot strengthen the theorem by replacing this requirement with one demanding merely that Def(W) form a partial Boolean algebra (while leaving the other conditions intact) is confirmed by the following counterexample. If $W^2 \neq W$, take Def(W) = Def$_{\text{KD}}$(W); but if $W^2 = W = P$, take Def(P) to be the partial Boolean algebra generated by P, all one-dimensional projections orthogonal to it, *and all projections onto rays making a 45° angle* with P. (Of course, the 'if' part of the theorem still goes through, since KD's ortholattices also define partial Boolean algebras.)

Note that this counterexample only disagrees with KD's rule in the case of pure states. It is presently not clear to me how far one can move away from KD's rule in the case of mixed states with the adoption of partial Boolean algebras instead of ortholattices — an issue that might be worth resolving, since it is only in those states where quantum measurement is a problem (if W is pure, the composite state out of which it arises cannot be entangled). In any case, I do not regard the failure of the motivation theorem to go through under the assumption that Def(W) just form a partial Boolean algebra as something which undercuts the motivation it provides. If anything, it just poses a challenge to improve on the theorem so that it also carries along those predisposed to believe that the attribution of definite truth values to a pair of incompatible propositions need not be accompanied by the attribution of any truth value (or meaning) to their conjunction or disjunction.

Before concluding, I should also re-emphasize that this motivation theorem cannot exonerate KD's modal interpretation against its critics, for I have not even attempted to deal with their points here. In fact, it seems to me that the most troublesome point for KD is one originally raised by Arntzenius which has nothing to do with measurement or contextualism or any of the other issues raised by conditions (1)–(6). Arntzenius points out that, according to KD's way of picking out definite-valued observables, when a projection P on $\mathcal{H}$ gets a value, that does not always mean that the corresponding projection $P \otimes I$ on the composite system described by $\mathcal{H} \otimes \mathcal{H}'$ does! Arntzenius remarks with incredulity (p. 244): 'in certain cases it will be truth valueless to claim of the object "the table" that it has the property that its left hand side is green, and nevertheless be true of the object "the left hand side of the table" that it has the property that it is green.' Indeed, one readily verifies that when the composite state on $\mathcal{H} \otimes \mathcal{H}'$ is a pure state P reducing to W on $\mathcal{H}$, $[\mathrm{Def}_{\mathrm{KD}}(W)] \otimes I \subseteq \mathrm{Def}_{\mathrm{KD}}(P)$ if and only if P is not entangled or W is a multiple of the identity operator. Thus the puzzling scenario Arntzenius relates will occur quite generically in KD's modal interpretation.[22]

Let me now end by summing up the situation for KD's modal interpretation in light of the above motivating theorem.

In the beginning there was Einstein versus Bohr. Today there is a whole panoply of competing interpretations of quantum mechanics vying for center stage. Sometimes it seems the odder an interpretation is, the more likely it is to gain acceptance. But there is a way to rein these interpretations in: demand that they yield satisfactory answers to the problems of measurement and contextualism. KD's interpretation does this, though not without stumbling into curious features of its own (see above!). But if we take on board KD's general understanding of what constitutes satisfactory answers,[23] then the specific features of their interpretation are not just artifacts of the particular formal rule they decide to use for picking out definite-valued observables, but can be seen as forced upon us by independent considerations through the motivation theorem.

Appendix: Proof of the motivation theorem

For any $\mathcal{H}$ *with* $\dim(\mathcal{H}) > 2$, *the collection* $\mathrm{Def}(W)$ *satisfies conditions (1)–(6) if and only if for all* W: $\mathrm{Def}(W) = \mathrm{Def}_{\mathrm{KD}}(W)$.

Fix an arbitrary Hilbert space $\mathcal{H}$ with $\dim(\mathcal{H}) > 2$.

The necessity proof starts from the simple observation that $\mathrm{Def}_{\mathrm{KD}}(W) \subseteq \mathrm{Def}(W)$ for all W on $\mathcal{H}$. For (2) and (6) automatically

[22] Van Fraassen (1991, pp. 290–4) seems to have been the only one to attempt to counter this objection, which can be raised against his own modal interpretation as well. But worries about it have recently resurfaced in Clifton (1995*b*) and Bacciagaluppi (1995).

[23] Of course, there is still considerable scope for disagreements about what general approach to take to the interpretation of quantum mechanics, as the interpretations of Bohm and GRW amply illustrate (see footnote 5).

require that $[\mathrm{SR}(W) \cup N(W)] \subseteq \mathrm{Def}(W)$, and so, by (3), $\mathrm{Def}(W)$ must at least include that ortholattice generated by $\mathrm{SR}(W) \cup N(W)$, which is precisely $\mathrm{Def}_{\mathrm{KD}}(W)$.

So all that's left to do is to show that $\mathrm{Def}(W) \subseteq \mathrm{Def}_{\mathrm{KD}}(W)$ for all W, the main part of the proof. That proof is a little longer(!). What makes it so is nothing particularly difficult, but merely the fact that certain conditions, particularly the ortholattice condition (3) and the ignorance interpretation condition (4), need to be invoked over and over again until the desired conclusion is reached. In any case, the main part of the proof is best divided into two stages. First, Stage 1 will prove $\mathrm{Def}(W) \subseteq \mathrm{Def}_{\mathrm{KD}}(W)$ for pure states $W = P$. Then Stage 2 will prove that for mixtures, the proof of which depends on the pure case having been established.

Stage 1: For all W satisfying $W^2 = W$: $\mathrm{Def}(W) \subseteq \mathrm{Def}_{\mathrm{KD}}(W)$.

Fix an arbitrary pure W on $\mathcal{H}$, which we might as well call P. Suppose for *reductio ad absurdum* that there is a $P' \in \mathrm{Def}(P)$ such that $P' \notin \mathrm{Def}_{\mathrm{KD}}(P)$, i.e. $P'P \neq P$ and $\neq 0$. It suffices to show from this that $\mathrm{Def}(P)$ must include all one-dimensional projections on $\mathcal{H}$, and therefore, by (3), the ortholattice those generate, $\mathcal{L}(\mathcal{H})$; for then, by Gleason's theorem, (5) can no longer be satisfied.

Clearly $0 < \mathrm{Tr}(P'P) < 1$; but without loss of generality, we may assume $\mathrm{Tr}(P'P) > \frac{1}{2}$. For if P' doesn't satisfy this inequality, $I - P'$ will. And $I - P'$ can always be substituted for P' in the argument to follow, since $I - P' \in \mathrm{Def}(P)$ if $P' \in \mathrm{Def}(P)$ by (3), and $(I = P')P \neq P$ and $\neq 0$ whenever $P'P \neq P$ and $\neq 0$.

Also clearly $0 < \dim(P') < \dim(\mathcal{H})$; but again without loss of generality, we may assume $\dim(P') > 1$. For if $\dim(P') = 1$, we can always (since $\dim(\mathcal{H}) > 2$) choose another projection P'' satisfying $P''P' = 0, P''P = 0$ and $0 < \dim(P'') < \dim(\mathcal{H}) - 1$. Then we have $P'' \oplus P' = P'' + P' \in \mathrm{Def}(P)$ by (3), since $P'' \in \mathrm{Def}_{\mathrm{KD}}(P) \subseteq \mathrm{Def}(P)$ and $P' \in \mathrm{Def}(P)$ by hypothesis. And by construction, the projection $P'' + P'$ satisfies $1 < \dim(P'' + P') < \dim(\mathcal{H})$, $(P'' + P')P = P'P \neq P$ and $\neq 0$, and $1 > \mathrm{Tr}((P'' + P')P) = \mathrm{Tr}(P'P) \geq \frac{1}{2}$. So $P'' + P$ could always be substituted for P' in the following argument.

Taking stock, then, there must be a P' with the properties that:

$$P' \in \mathrm{Def}(P),\ P'P \neq P \text{ and } \neq 0,\ 1 > \mathrm{Tr}(P'P) \geq \frac{1}{2}$$

and

$$\dim\mathcal{H} > \dim P' > 1.$$

For convenience, define $S \equiv \{P''|\dim P'' = 1 \text{ and } P'P'' = P''\}$. In words, S is the set of all projections onto rays in the subspace onto which P' projects.

Claim 1: $S \subseteq \mathrm{Def}(P)$. For if $P'' \in S$, $P'' \oplus P \in \mathrm{Def}_{\mathrm{KD}}(P) \subseteq \mathrm{Def}(P)$, and therefore $(P'' \oplus P) \cap P' = P'' \in \mathrm{Def}(P)$, using $P'P \neq P$ and (3).

Claim 2: For any real number $p \in [0, \frac{1}{2}]$, there exists a $P'' \in S$ such that $\mathrm{Tr}(P''P) = p$. The case $p = 0$ follows from the fact that $\dim(I - P) \cap P' = \dim P' - 1 > 0$. Furthermore, since $P'P \neq 0$, there must exist a $\underline{P}'' \in S$ such that $P' = (I - P) \cap P' + \underline{P}''$. And for this $\underline{P}''$, $\mathrm{Tr}(\underline{P}''P) = \mathrm{Tr}([(I - P) \cap P' + \underline{P}'']P) = \mathrm{Tr}(P'P) \geq \frac{1}{2}$. So, in sum, there exists a $P'' \in S$ such that $\mathrm{Tr}(P''P) = 0$, and another $\underline{P}'' \in S$ such that $\mathrm{Tr}(\underline{P}''P) \geq \frac{1}{2}$. Claim 2 then follows from the fact that the range of P', which contains all rays onto which the projections in S project, is a subspace, which allows us to find, for any p in the interval $(0, \mathrm{Tr}(\underline{P}''P)]$ as well, a $P'' \in S$ for which $\mathrm{Tr}(P''P) = p$.

Combining Claims 1 and 2, we can drop reference to P' and the set S and conclude: for any real number $p \in [0, \frac{1}{2}]$, there exists a one-dimensional projection $P'' \in \mathrm{Def}(P)$ such that $\mathrm{Tr}(P''P) = p$.

Since $\dim\mathcal{H} > 2$, it follows that whenever $\mathrm{Tr}(P_1''P) = \mathrm{Tr}(P_2''P) = p$ (with P_1'', P_2'' and P all one-dimensional), there exists a unitary operator U such that $P_2'' = UP_1''U^{-1}$ and $P = UPU^{-1}$. (Explicitly: if v, v_1'' and v_2'' are unit vectors in the ranges of P, P_1'' and P_2'' respectively, then U is the rotation about v which sends $(I - P)(v_1'')$ to $(I - P)(v_2'')$ and acts like the identity on the subspace orthogonal to the one in which this rotation takes place.) With this result and the conclusion reached above, (1) allows us to derive:

$$\{P''|\dim P'' = 1 \text{ and } \mathrm{Tr}(P''P) \in [0, 1/2]\} \subseteq \mathrm{Def}(P).$$

Finally, consider any one-dimensional P'' *not* satisfying $0 \leq \mathrm{Tr}(P''P) \leq \frac{1}{2}$. Nevertheless $I - P''$ will satisfy this inequality;

and therefore $I - P''$ can be decomposed into a sum of $\dim\mathcal{H} - 1$ orthogonal one-dimensional projections, each of which satisfy the inequality. By the above argument, each projection in that sum must be in $\mathrm{Def}(P)$, and therefore so must the sum, $I - P''$, by (3). So $P'' \in \mathrm{Def}(P)$, again using (3). We now have:

$$\{P'' | \dim P'' = 1\} = \{P'' | \dim P'' = 1 \text{ and } \mathrm{Tr}(P''P) \in [0,1]\} \subseteq \mathrm{Def}(P),$$

which is what we said at the beginning it sufficed to prove.

(Aside: Bub (1994) has recently shown that condition (1) is not necessary for this stage of the argument by reaching the same conclusion as above on the strength of the ortholattice condition (3). Unfortunately, Bub's observation does not lead to a strengthening of the motivation theorem by dropping its reliance on condition (1), since this condition is still necessary for the next stage of the proof.)

Stage 2: For all W satisfying $W^2 \neq W$: $\mathrm{Def}(W) \subseteq \mathrm{Def}_{\mathrm{KD}}(W)$.

Again fix an arbitrary statistical operator W on $\mathcal{H}$, this time mixed. The reason that this stage depends on the establishment of Stage 1 is that we cannot apply condition (4) unless we know what sets of observables must get picked out as having definite values in all pure states.

Let W have spectral projectors $\{P_i\}$, and let $\{P_k\}$ be any set of (not necessarily orthogonal) one-dimensional projections which makes (4) true for W. (There could be more than one such set, but recall that (4) requires only that there be at least one.) Claim: for all P_i, there must be a P_k such that $P_iP_k = P_k$. (In words: every nonzero eigenvalue eigenspace of W must contain at least one of the pure states in $\{P_k\}$.) For suppose we adopt the hypothesis to the contrary — that there is a P_i such that for all P_k, $P_iP_k \neq P_k$. By (6), $\{P_i\} \subseteq \mathrm{Def}(W)$, so using (4), $\{P_i\} \subseteq \mathrm{Def}(P_k)$ for all P_k, which by the previous Stage 1 implies that for all P_i, $P_iP_k = P_k$ or 0 for all P_k. In conjunction with our hypothesis, this implies that there is a P_i such that for all P_k, $P_iP_k = 0$. But then we could not have $W = \sum_k \lambda_k P_k$, since that would imply (for that very P_i) that $P_iW = 0$, contradicting the fact that P_i is one of the projections onto one of W's nonzero eigenvalue eigenspaces; i.e. that $P_iW \neq 0$.

The rest of the argument is just another *reductio ad absurdum* which turns on the *reductio* of Stage 1, i.e. we shall show, using the claim proved above, that if there is a $P \in \text{Def}(W)$ such that $P \notin \text{Def}_{\text{KD}}(W)$, then we must contradict the result proved in Stage 1. This then establishes the conclusion of the present stage, completing the whole proof.

If there is a $P \in \text{Def}(W)$ such that $P \notin \text{Def}_{\text{KD}}(W)$, there must exist a P_i such that $PP_i \neq P_i$ and $\neq 0$. But we have seen from the above claim that for that P_i, there must exist a P_k in the set $\{P_k\}$ (making (4) true) such that $P_i P_k = P_k$. And (4) demands that, since $P \in \text{Def}(W)$, $P \in \text{Def}(P_k)$ for that P_k, which means $PP_k = P_k$ or 0, by Stage 1. So, let's sum all this up. We have three projections P, P_i and P_k such that:

> $P \in \text{Def}(W)$, $P_i \in \text{SR}(W)$, $PP_i \neq P_i$ and $\neq 0$, $PP_k = P_k$ or 0, and $P_i P_k = P_k$.

We have derived that $PP_k = P_k$ or 0; but without loss of generality, we may in fact assume $PP_k = P_k$. For if $PP_k = 0$, $(I - P)P_k = P_k$. And since we have $PP_i \neq P_i$ and $\neq 0$, $(I - P)P_i \neq P_i$ and $\neq 0$. Since (3) implies $I - P \in \text{Def}(W)$ because $P \in \text{Def}(W)$, we could just as easily switch to using $I - P$ in place of P above, and in the argument to follow.

Since P_i is a nontrivial projection, $\dim P_i > 0$; but without loss of generality, we may also assume that $\dim P_i > 1$. For if $\dim P_i = 1$, $P_i P_k = P_k$ implies $P_i = P_k$ (since P_k is nontrivial too!), which already gives us a contradiction with the requirements that $PP_i \neq P_i$ and $PP_k = P_k$.

So we have three projections P, P_i and P_k such that:

$$P \in \text{Def}(W),\ P_i \in \text{SR}(W),\ \dim P_i > 1,$$
$$PP_i \neq P_i \text{ and } \neq 0,\ PP_k = P_k \text{ and } P_i P_k = P_k.$$

These relations between P, P_i and P_k automatically entail that the subspace corresponding to P must intersect that corresponding to P_i, but not contain it. By (3), the projection onto that intersection $P \cap P_i$ must be in $\text{Def}(W)$, since both P and P_i are (the latter, of course, because of (5)). So we now have a projection onto a proper subspace of the (nontrivial) eigenspace of W onto which P_i projects that has to be in Def(W).

(1) and (3) then force *all* projections onto subspaces of the P_i-subspace to be in Def(W). For any unitary transformation U on the P_i-subspace extends to a unitary transformation on $\mathcal{H}$ that preserves W. So $U(P \cap P_i)U^{-1}$ must be in Def(W) for any such U, and all these projections unitarily generated from $P \cap P_i$ 'inside' the P_i-subspace easily generate the set of all of P_i's subspaces under the ortholattice operations. Now since dim$P_i > 1$ and $P_i P_k = P_k$, we can choose a projection P' in the P_i-subspace such that $P'P_k \neq P_k$ *and* $P'P_k \neq 0$. Since $P' \in$ Def(W), (4) requires that $P' \in$ Def(P_k). But this contradicts Stage 1, because Def$(P_k) =$ Def$_{\mathrm{KD}}(P_k) = \{P | PP_k = P_k \text{ or } 0\}$.

Bibliography

Albert, D. & Loewer, B. (1990), Wanted dead or alive: Two attempts to solve Schrödinger's paradox, in A. Fine *et al.*, eds, 'PSA 1990, Volume 1', Philosophy of Science Association, East Lansing, MI, pp. 277–85.

Arntzenius, F. (1990), Kochen's interpretation of quantum mechanics, in A. Fine *et al.*, eds, 'PSA 1990, Volume 1', Philosophy of Science Association, East Lansing, MI, pp. 241–9.

Bacciagaluppi, G. (1995), 'Kochen-Specker theorem in the modal interpretation of quantum mechanics', *International Journal of Theoretical Physics* **34**, 1205–16.

Bacciagaluppi, G. & Dickson, W. M. (1999), 'Dynamics for modal interpretations', *Foundations of Physics* **29**, 1165–201.

Bacciagaluppi, G. & Hemmo, M. (1994), Making sense of approximate decoherence, in D. Hull *et al.*, eds, 'PSA 1994, Volume 1', Philosophy of Science Association, East Lansing, MI, pp. 345–54.

Barnett, S. M. & Phoenix, S. J. D. (1992), 'Bell's inequality and the Schmidt decomposition', *Physics Letters. A* **167**, 233–7.

Bell, J. S. (1966), 'On the problem of hidden variables in quantum mechanics', *Reviews of Modern Physics* **38**, 447–52.

Bohm, D. (1952*a*), 'A suggested interpretation of the quantum theory in terms of "hidden" variables. I', *Physical Review (2)* **85**, 166–79.

Bohm, D. (1952*b*), 'A suggested interpretation of the quantum theory in terms of "hidden" variables. II', *Physical Review (2)* **85**, 180–93.

Bohm, D. & Bub, J. (1966), 'A proposed solution of the measurement problem in quantum mechanics by a hidden variable theory', *Reviews of Modern Physics* **38**, 453–69.

Bub, J. (1991), 'Measurement and "beables" in quantum mechanics', *Foundations of Physics* **21**, 25–42.

Bub, J. (1994), 'On the structure of quantal proposition systems', *Foundations of Physics* **24**, 1261–79.

Clifton, R. (1993), 'Getting contextual and nonlocal elements-of-reality the easy way', *American Journal of Physics* **61**, 443–7.

Clifton, R. (1995*a*), Making sense of the Kochen-Dieks "no-collapse" interpretation of quantum mechanics independent of the measurement problem, in 'Fundamental problems in quantum theory', Vol. 755 of *Annals of the New York Academy of Science*, New York Acad. Sci., New York, pp. 570–8.

Clifton, R. (1995*b*), 'Why modal interpretations of quantum mechanics must abandon classical reasoning about physical properties', *International Journal of Theoretical Physics* **34**, 1303–12.

d'Espagnat, B. (1976), *Conceptual foundations of quantum mechanics*, W. A. Benjamin, Inc., Reading, MA.

Dickson, W. M. (1994), Wavefunction tails in the modal interpretation, in D. Hull *et al.*, eds, 'PSA 1994, Volume 1', Philosophy of Science Association, East Lansing, MI, pp. 366–76.

Dieks, D. (1989), 'Resolution of the measurement problem through decoherence of the quantum state', *Physics Letters A* **142**, 439–46.

Dieks, D. (1993), 'The modal interpretation of quantum mechanics, measurements and macroscopic behaviour', *Physical Review A* **49**, 2290–300.

Dirac, P. A. M. (1947), *The principles of quantum mechanics*, 3rd ed., Clarendon Press, Oxford.

Ghirardi, G. C., Rimini, A. & Weber, T. (1986), 'Unified dynamics for microscopic and macroscopic systems', *Physical Review D* **34**, 470–91.

Healey, R. (1989), *The philosophy of quantum mechanics*, Cambridge University Press, Cambridge.

Healey, R. (1993), 'Measurement and quantum indeterminateness', *Foundations of Physics Letters* **6**, 307–16.

Hemmo, M. (1993), 'Review of R. Healey's The philosophy of quantum mechanics: an interactive interpretation', *Foundations of Physics* **23**, 1137–45.

Kochen, S. (1985), A new interpretation of quantum mechanics, in 'Symposium on the foundations of modern physics', World Scientific Publishing, Singapore, pp. 151–69.

Kochen, S. & Specker, E. P. (1967), 'The problem of hidden variables in quantum mechanics', *Journal of Mathematics and Mechanics* **17**, 59–87.

van Fraassen, B. C. (1991), *Quantum mechanics: An empiricist view*, Clarendon Press, Oxford.

Vermaas, P. E. & Dieks, D. (1995), 'The modal interpretation of quantum mechanics and its generalization to density operators', *Foundations of Physics* **25**, 145–58.

von Neumann, J. (1968), *Mathematische Grundlagen der Quantenmechanik*, Springer-Verlag, Berlin.

Chapter 2

A uniqueness theorem for 'no collapse' interpretations of quantum mechanics

with Jeffrey Bub

1 The interpretation problem

On the orthodox (Dirac-von Neumann) interpretation[1] of quantum mechanics, an observable has a determinate (definite, sharp) value for a system in a given quantum state if and only if the state is an eigenstate of the observable. So, the orthodox interpretation selects a particular set of observables that have determinate values in a given quantum state; equivalently, a particular set of idempotent observables or propositions, represented by projection operators,

---

Thanks to Guido Bacciagaluppi and Michael Dickson for helpful critical comments on a draft version of the paper. R. C. would like to thank the Social Sciences and Humanities Research Council of Canada for research support.

[1] P. A. M. Dirac, *Quantum Mechanics* (Oxford: Clarendon Press, 1958), 4th ed., pp. 46–7; J. von Neumann, *Mathematical Foundations of Quantum Mechanics* (Princeton: Princeton University Press, 1955), p. 253.

that have determinate truth values. If the quantum state is represented by a ray or 1-dimensional projection operator e spanned by the unit vector $|e\rangle$, these are the propositions p such that $e \leq p$ or $e \leq p^{\perp}$ (where the relation '$\leq$' denotes subspace inclusion, or the corresponding relation for projection operators, and $p^{\perp}$ denotes the subspace orthogonal to p).

The orthodox interpretation involves a well-known measurement problem (see Section 3), which Dirac and von Neumann resolve formally by invoking a projection postulate[2] that characterizes the 'collapse' or projection of the quantum state of a system onto an eigenstate of the measured observable. Dynamical 'collapse' interpretations of quantum mechanics[3] modify the unitary, Schrödinger dynamics of the theory to achieve the required state evolution for both measurement and non-measurement interactions, while retaining the orthodox criterion for determinateness.

'No collapse' interpretations avoid the measurement problem by selecting other sets of observables as determinate for a system in a given quantum state. For example, certain versions of the 'modal' interpretation[4] exploit the polar decomposition theorem to select a preferred set of determinate observables for a system S as a subsys-

---

[2]See Dirac, *op. cit.*, p. 36; and von Neumann, *op. cit.*, pp. 351 and 418.

[3]e.g., D. Bohm and J. Bub, 'A Proposed Solution of the Measurement Problem in Quantum Mechanics by a Hidden Variable Theory', *Reviews of Modern Physics* 38 (1966), 453–69; G. C. Ghirardi, A. Rimini and T. Weber, 'Unified Dynamics for Microscopic and Macroscopic Systems', *Physical Review D* 34 (1966), 470–91.

[4]The idea of a modal interpretation of quantum mechanics was first introduced by van Fraassen. [B. van Fraassen, 'Hidden Variables and the Modal Interpretation of Quantum Statistics', *Synthèse* 42 (1979), 155–65; 'A Modal Interpretation of Quantum Mechanics', in E. Beltrametti and B. van Fraassen (eds), Current Issues in Quantum Logic (Singapore: World Scientific, 1981), pp. 229-58. See also B. van Fraassen, *Quantum Mechanics: An Empiricist View* (Oxford: Oxford University Press, 1991)]. We have in mind here specifically the interpretations developed by Kochen and by Dieks. [Kochen, 'A New Interpretation of Quantum Mechanics', in P. Lahti and P. Mittelstaedt (eds), *Symposium on the Foundations of Modern Physics* (Singapore: World Scientific, 1985), pp. 151–69; D. Dieks, 'Quantum Mechanics Without the Projection Postulate and Its Realistic Interpretation', *Foundations of Physics* 19 (1989), pp. 1397–423; D. Dieks, 'Modal Interpretation of Quantum Mechanics, Measurements, and Macroscopic Behavior', *Physical Review A* 49 (1994), pp. 2290–300.]

tem of a composite system $S + S^*$ in a state $e \in \mathcal{H} \otimes \mathcal{H}^*$. Bohm's 'causal' interpretation[5] selects position in configuration space as a preferred always determinate observable for any quantum state, and certain other observables are selected as inheriting determinate status at a given time from this preferred determinate observable and the state at that time. Bohr's complementarity interpretation[6] selects as determinate an observable associated with an individual quantum phenomenon manifested in a measurement interaction involving a specific classically describable experimental arrangement, and certain other observables inherit determinate status from this observable and the quantum state. We discuss these interpretations and the orthodox interpretation in Section 3.

There are restrictions on what sets of observables can be taken as simultaneously determinate without contradiction, if the attribution of determinate values to observables is required to satisfy certain constraints. The 'no-go' theorems for 'hidden variables' underlying the quantum statistics provide a series of such results. So, in fact, the options for a 'no collapse' interpretation of quantum mechanics in the sense we have in mind are rather limited. We begin with a brief review of some of these limiting results.

The constraint imposed by the Kochen and Specker theorem[7] requires that the values assigned to a set of mutually compatible observables, represented by pairwise commuting self-adjoint operators, should preserve the functional relations satisfied by these observables. For example, the constraint requires that the value assigned to an observable A should be the square of the value assigned to an observable B, if $A = B^2$. With sums and products defined for mutually compatible observables only, the observables of a quantum mechanical system form a partial algebra, and the idempotent observables form a partial Boolean algebra. Kochen

---

[5]D. Bohm, 'A Suggested Interpretation of Quantum Theory in Terms of "Hidden Variables": Parts I and II', *Physical Review* 85 (1952), pp. 166–79, 180–93. Different formulations of Bohm's theory treat observables such as spin differently. We discuss these differences in Section 3.

[6]N. Bohr, *Atomic Physics and Human Knowledge* (New York: Wiley, 1958).

[7]S. Kochen and E. P. Specker, 'The Problem of Hidden Variables in Quantum Mechanics', *Journal of Mathematics and Mechanics* 17 (1967), 59–87.

and Specker show that a necessary condition for the assignment of values to all the observables of a quantum mechanical system, in such a way as to satisfy the functional relationship constraint, is the existence of an embedding of the partial algebra of observables into a commutative algebra; equivalently, the embedding of the partial Boolean algebra of idempotent observables into a Boolean algebra. A necessary and sufficient condition for the existence of an embedding of a partial Boolean algebra into a Boolean algebra is that, for every pair of distinct elements p, q in the partial Boolean algebra, there exists a homomorphism h onto the 2-element Boolean algebra $\{0, 1\}$ such that $h(p) \neq h(q)$.

As Kochen and Specker show, there are no 2-valued homomorphisms on the partial Boolean algebra of projection operators on a Hilbert space of three or more dimensions (much less any embeddings). For the homomorphism condition implies that for every orthogonal n-tuple of 1-dimensional projection operators or corresponding rays in $\mathcal{H}_n$ one projection operator or ray is mapped onto 1 ('true') and the remaining $n-1$ projection operators or rays are mapped onto 0 ('false'). And this is shown to be impossible for the finite set of orthogonal triples of rays that can be constructed from 117 appropriately chosen rays in $\mathcal{H}_3$: any assignment of 1s and 0s to this set of orthogonal triples satisfying the homomorphism condition involves a contradiction.

Now, an observable A can be compatible with an observable B and also with an observable C, while B and C are incompatible (represented by non-commuting operators). A and B will be representable as functions of an observable X, while A and C will be representable as functions of an observable Y, incompatible with X. If A denotes an observable of a system S and B and C denote incompatible observables of a system S', space-like separated from S, then the functional relationship constraint, that the value of A as a function of X should be the same as the value of A as a function of Y, becomes a locality condition. Bell argued[8] that the general func-

[8]J. S. Bell, 'On the Problem of Hidden Variables in Quantum Mechanics', *Reviews of Modern Physics* 38 (1966), 447–75. Reprinted in *Speakable and Unspeakable in Quantum Mechanics* (Cambridge: Cambridge University Press, 1987).

tional relationship constraint cannot be justified physically, but this weaker locality condition can. Bell's theorem[9] shows that the locality condition cannot be satisfied in general in a Hilbert space of four or more dimensions: there are sets of observables for which value assignments satisfying the locality condition are inconsistent with the quantum statistics. [More recent versions of Bell's theorem, e.g., Greenberger, Horne and Zeilinger (GHZ)[10] do not require statistical arguments, or minimize the statistics needed, e.g., Hardy[11].]

Several authors have considered the problem of constructing the smallest set of observables that cannot be assigned values in such a way as to satisfy the functional relationship or locality constraints. Kochen and Conway have reduced the number of rays in $\mathcal{H}_3$ required to generate a contradiction from value assignments satisfying the Kochen-Specker homomorphism condition from 117 to 31.[12] Peres has shown how to derive a contradiction for a more symmetrical set of 33 rays in $\mathcal{H}_3$, and for 24 rays in $\mathcal{H}_4$.[13] Kernaghan has reduced Peres' 24 rays to 20.[14] Clifton has an eight-ray Kochen and Specker argument in $\mathcal{H}_3$ (but the proof requires quantum statistics to derive a contradiction).[15] Mermin proves a version of the Kochen and Specker theorem for nine observables in $\mathcal{H}_4$, and a version of Bell's theorem (the GHZ version) for ten observables in $\mathcal{H}_8$.[16] There

---

[9]J. S. Bell, 'On the Einstein-Podolsky-Rosen Paradox', *Physics* 1 (1964), 195–200. Reprinted in *Speakable and Unspeakable in Quantum Mechanics, op. cit.* Note that this article, published two years before the 1966 review article, was actually written after the review article.

[10]D. M. Greenberger, M. A. Horne and A. Zeilinger, 'Going Beyond Bell's Theorem', in M. Kafatos (ed.), *Bell's Theorem, Quantum Theory, and Conceptions of the Universe* (Dordrecht: Kluwer, 1989), pp. 73–6.

[11]L. Hardy, 'Quantum Mechanics, Local Realistic Theories, and Lorentz-Invariant Theories', *Physical Review Letters* 68 (1992), 2981–84.

[12]The proof is unpublished. For a reference, see A. Peres, *Quantum Theory: Concepts and Methods* (Dordrecht: Kluwer, 1993), p. 197.

[13]A. Peres, *op. cit.*, Chap. 7.

[14]M. Kernaghan, 'Bell-Kochen-Specker Theorem with 20 Vectors', *Journal of Physics A: Math. Gen.* 27 (1994), L829.

[15]R. Clifton, 'Getting Contextual and Nonlocal Elements of Reality the Easy Way', *American Journal of Physics* 61 (1993), 443–7.

[16]N. D. Mermin, 'Hidden Variables and the Two Theorems of John Bell', *Reviews of Modern Physics* 65 (1993), 803–815. Note that the Mermin proofs, which are con-

are related results by Penrose and others.[17]

The question of how small we can make the set of observables and still generate a Kochen-Specker contradiction is important in revealing structural features of Hilbert space, but of no immediate significance for a 'no collapse' interpretation of quantum mechanics. The relevant question in a sense concerns the converse issue. To provide a 'no collapse' interpretation of quantum mechanics in the sense we have in mind, we want to know how *large* we can take the set of determinate observables *without* generating a Kochen-Specker contradiction, i.e., we are interested in the maximal sets of observables that can be taken as having determinate (but perhaps unknown) values for a given quantum state, subject to the functional relationship constraint, or the maximal sets of propositions that can be taken as having determinate truth values, where a truth-value assignment is defined by a 2-valued homomorphism. (As we show below, even the orthodox interpretation selects such a maximal set.)

More precisely, the projection operators (or corresponding Hilbert space subspaces) of a quantum mechanical system form a lattice $\mathcal{L}$ that can be taken as representing the lattice of yes-no experiments or propositions pertaining to the system. We know that we cannot assign truth values to all the propositions in $\mathcal{L}$ in such a way as to satisfy the functional relationship constraint, or even the weaker locality condition, for all observables generated as spectral measures over these propositions. That is, we cannot take all the propositions in $\mathcal{L}$ as determinately true or false if truth values are assigned subject to these constraints. So the probabilities defined by the quantum state cannot be interpreted epistemically

ceptually very simple, are not nearly so economical as the Peres proofs in terms of the number of rays.

[17]See H. Brown, 'Bell's Other Theorem and its Connections with Nonlocality. Part I', in A. van der Merwe, F. Selleri and G. Tarozzi (eds), *Bell's Theorem and the Foundations of Modern Physics* (Singapore: World Scientific, 1992), pp. 104–16 for an interesting discussion of 'no-go' theorems, especially the Mermin proofs, and Peres, *op. cit.*, p. 212, for an account of Penrose's 40-ray proof for a spin-3/2 system. See also J. Zimba and R. Penrose, 'On Bell Non-Locality Without Probabilities: More Curious Geometry', *Studies in History and Philosophy of Modern Physics* 24 (1993), 697–720.

and represented as measures over the different possible truth-value assignments to all the propositions in $\mathcal{L}$. But we also know that any single observable can be taken as determinate for any quantum state (since the propositions associated with an observable generate a Boolean algebra), so we may suppose that fixing a quantum state represented by a ray e in $\mathcal{H}$ and an arbitrary preferred observable R as determinate places restrictions on what propositions can be taken as determinate for e in addition to R-propositions.

The natural question for a 'no collapse' interpretation of quantum mechanics would then appear to be: What are the *maximal* sublattices $\mathcal{D}(e, R)$ of $\mathcal{L}$ to which truth values can be assigned, where each assignment of truth values is defined by a 2-valued homomorphism on $\mathcal{D}(e, R)$, and the probabilities defined by e for mutually compatible sets of propositions in $\mathcal{D}(e, R)$ can be represented as measures over the different possible truth value assignments to $\mathcal{D}(e, R)$? As a further constraint, it seems appropriate to require that $\mathcal{D}(e, R)$ is invariant under lattice automorphisms that preserve e and R (so that $\mathcal{D}(e, R)$ is genuinely selected by e and R, and the lattice structure of $\mathcal{H}$), and that $\mathcal{D}(e, R)$ is unaffected if the quantum system S is regarded as a subsystem of a composite system $S + S'$, where S is not 'entangled' with S'. We shall refer to these sublattices as the 'determinate' sublattices of $\mathcal{L}$.

If the aim is to exclude determinate values for observables (perhaps fixed by underlying 'hidden variables'), i.e., to prove a 'no-go' theorem, then the constraints on values should be as weak as possible. For our problem, however, we are interested in characterizing the maximal sublattices of $\mathcal{L}$ that *allow* an interpretation of the associated observables as determinate for a given quantum state. The constraints we are thereby led to impose reflect, roughly, the strongest 'classicality' conditions we can get away with, consistent with such an interpretation. So we require that $\mathcal{D}(e, R)$ is a sublattice of $\mathcal{L}$ (rather than a partial Boolean subalgebra, in which operations corresponding to the conjunction and disjunction of propositions are defined only for compatible propositions represented by commuting projection operators), and that the possible truth value assignments are defined by 2-valued lattice homomorphisms (rather than 2-valued partial Boolean algebra homomorphisms, i.e., maps onto $\{0, 1\}$ that reduce to 2-valued lattice homomorphisms only on each Boolean sub-algebra of $\mathcal{D}(e, R)$).

Note that the problem of characterizing the maximal sets of observables or propositions that can be taken as (simultaneously) determinate, without generating a Kochen-Specker contradiction, is not well-defined *unless* we impose constraints on the sets — there are clearly many different infinite sets of propositions that can be assigned determinate truth values without contradiction in this sense. The problem is only interesting relative to the requirement that a maximal set of determinate propositions is an extension of some physically significant algebraic structure of determinate propositions. Since the dynamical variables in classical mechanics are all simultaneously determinate, and any single observable in quantum mechanics can be taken as determinate, our proposal is to consider what part of the non-Boolean lattice of quantum propositions can be added to the Boolean algebra of propositions defined by the spectral measure of a particular quantum mechanical observable R, for a given quantum state e, before this extended structure becomes too 'large' to support sufficiently many truth value assignments, defined by 2-valued homomorphisms, to generate the quantum statistics for the propositions in the extended structure as measures over these different truth value assignments.

In the following section, we prove that the determinate sublattices $\mathcal{D}(e, R)$ are uniquely characterized as follows:[18] In an n-dimensional Hilbert space $\mathcal{H}_n$, suppose R has $m \leq n$ distinct eigenspaces r_i and the rays $e_{r_i} = (e \vee r_i^{\perp}) \wedge r_i$, $i = 1, \ldots, k \leq m$, are the non-zero projections of the state e onto these eigenspaces. The determinate sublattice $\mathcal{D}(e, R)$ of $\mathcal{L}$ is then the sublattice $\mathcal{L}_{e_{r_1} e_{r_2} \ldots e_{r_k}}$ generated by the k orthogonal rays e_{r_i} and all the rays in the subspace $(e_{r_1} \vee e_{r_2} \vee \cdots e_{r_k})^{\perp}$ orthogonal to the k-dimensional subspace

[18] The uniqueness theorem supersedes earlier related results by one of us. See J. Bub, 'Quantum Mechanics Without the Projection Postulate', *Foundations of Physics* 22 (1992), 737–54; J. Bub, 'On the Structure of Quantal Proposition Systems', *Foundations of Physics* 24 (1994), 1261–80; J. Bub, 'How to Interpret Quantum Mechanics', *Erkenntnis* 41 (1994), 253–73.

spanned by the e_{r_i}. $\mathcal{L}_{e_{r_1}e_{r_2}\ldots e_{r_k}}$ can also be characterized as $\{e_{r_i} : i = 1,\ldots,k\}'$, the commutant in $\mathcal{L}$ of e_{r_i}, $i = 1,\ldots,k$, or as $\{p : e_{r_i} \leq p \quad \text{or} \quad e_{r_i} \leq p^{\perp}, i = 1,\ldots,k\}$.

Physically, $\mathcal{L}_{e_{r_1}e_{r_2}\ldots e_{r_k}}$ contains all those projections with values strictly correlated to the values of R when the system is in the state e. We note that the full set of (not necessarily idempotent) observables associated with $\mathcal{L}_{e_{r_1}e_{r_2}\ldots e_{r_k}}$ includes any observable whose eigenspaces are spanned by rays in $\mathcal{L}_{e_{r_1}e_{r_2}\ldots e_{r_k}}$. The set of maximal observables includes any maximal observable with k eigenvectors in the directions e_{r_i}, $i = 1,\ldots,k$.

Bell,[19] and also Bohm and Bub,[20] objected to the Jauch and Piron 'no-go' theorem,[21] which required that a truth value assignment h to the lattice of quantum propositions (equivalently, the probabilities assigned by dispersion-free states) should satisfy the constraint (a consequence of axiom 4° in their numbering) that

$$h(p \wedge q) = 1 \text{ if } h(p) = h(q) = 1,$$

for any propositions p, q (even incompatible propositions represented by non-commuting projection operators). To reproduce the quantum statistics, the constraint should be required to hold only for expectation values generated by distributions over the hidden variables corresponding to quantum states, but not necessarily for arbitrary hidden variable distributions (in particular, not for the truth value assignments themselves).

Now, there exist 2-valued lattice homomorphisms on our determinate sublattices $\mathcal{L}_{e_{r_1}e_{r_2}\ldots e_{r_k}}$ so axiom 4° holds for these sublattices. The assumption that fails is Jauch and Piron's axiom 5, which requires that any proposition not equal to the null proposition is assigned the value 1 (i.e., 'true') by some dispersion-free state. This is certainly a reasonable requirement on the full lattice $\mathcal{L}$: if a proposition is assigned the value 0 by every dispersion-free state, it will

[19] J. S. Bell, 'On the Problem of Hidden Variables in Quantum Mechanics', *op. cit.*

[20] D. Bohm and J. Bub, 'A Refutation of the Proof by Jauch and Piron that Hidden Variables Can Be Excluded in Quantum Mechanics', *Reviews of Modern Physics* 38 (1966), 470–5.

[21] J. M. Jauch and C. Piron, 'Can Hidden Variables be Excluded in Quantum Mechanics?', *Helvetica Physica Acta* 36 (1963), 827–37.

have zero probability for every quantum state (represented as a measure over dispersion-free states), hence will be orthogonal to every quantum state, and so can only be the null proposition. However, a similar argument does not apply to interpretations that select proper sublattices of $\mathcal{L}$. In fact, there are propositions in $\mathcal{L}_{e_{r_1}e_{r_2}\ldots e_{r_k}}$ not equal to the null proposition (propositions in $\left(e_{r_1} \vee \cdots \vee e_{r_k}\right)^{\perp}$) that have zero probability in the state e and are assigned the value 0 by all 2-valued homomorphisms on $\mathcal{L}_{e_{r_1}e_{r_2}\ldots e_{r_k}}$. The reason this is possible is that $\mathcal{L}_{e_{r_1}e_{r_2}\ldots e_{r_k}}$ is always a proper sublattice of $\mathcal{L}$ and is constructed on the basis of what the particular state e of the system is. So the Jauch and Piron argument for axiom 5 does not apply to the sublattices $\mathcal{L}_{e_{r_1}e_{r_2}\ldots e_{r_k}}$.

Similarly, the Kochen and Specker argument that the existence of hidden variables requires a Boolean embedding of the full partial Boolean algebra of idempotent observables of a quantum mechanical system breaks down. If the lattice operations in $\mathcal{L}_{e_{r_1}e_{r_2}\ldots e_{r_k}}$ are confined to compatible elements (corresponding to commuting projection operators), then $\mathcal{L}_{e_{r_1}e_{r_2}\ldots e_{r_k}}$ can be regarded as a partial Boolean algebra. There exist 2-valued partial Boolean homomorphisms on $\mathcal{L}_{e_{r_1}e_{r_2}\ldots e_{r_k}}$, in fact sufficiently many to generate the probabilities defined by the quantum state for the propositions in $\mathcal{L}_{e_{r_1}e_{r_2}\ldots e_{r_k}}$, but insufficiently many to provide an embedding of $\mathcal{L}_{e_{r_1}e_{r_2}\ldots e_{r_k}}$ into a Boolean algebra. Our proposal is that the determinate sublattices $\mathcal{L}_{e_{r_1}e_{r_2}\ldots e_{r_k}}$ for all R, provide a class of perfectly viable 'no collapse' interpretations of quantum mechanics, in which the functional relationship constraint is satisfied without a Boolean embedding.

In the final section, we show that the determinate sublattice of a composite system $S + M$ (representing a system and a measuring instrument), for the 'entangled' state of $S + M$ arising dynamically from a unitary evolution representing a quantum mechanical measurement interaction, contains the propositions corresponding to the S-observable correlated with the pointer observable of M, if the pointer observable (or some observable correlated with the pointer observable) is taken as the preferred determinate observable R. Note that this determinate sublattice of $S + M$ is derived *without any reference to measurement as an unanalyzed operation*, i.e., this natural description of the measurement process falls out as just a special instance of the linear Schrödinger quantum dynamics, without requiring the projection or 'collapse' of the quantum state to validate the determinateness of measured values, as in the orthodox interpretation. This is perhaps evident from the statement of the uniqueness theorem: $\mathcal{L}_{e_{r_1}e_{r_2}...e_{r_k}}$ is generated from the non-zero Lüders projections of the quantum state e onto the eigenspaces of R. Different choices for the preferred determinate observable R correspond to different 'no collapse' interpretations of quantum mechanics. We shall illustrate this with the orthodox interpretation (without the projection postulate), the modal interpretation (in the versions developed by Kochen and by Dieks), Bohm's 'causal' interpretation (one natural way to develop an Einsteinian realism within quantum mechanics), and Bohr's complementarity interpretation. But first we turn to the theorem.

2 The uniqueness theorem

We begin with some definitions: Consider a composite quantum mechanical system $S^{**} = S + S^*$ represented on a Hilbert space $\mathcal{H}^{**} = \mathcal{H} \otimes \mathcal{H}^*$. Suppose the state of S^{**} is represented by a ray $e^{**} \in \mathcal{H}^{**}$. Let R, R^* and R^{**} denote (preferred) observables of S, S^*, and S^{**} represented by self-adjoint operators defined on the Hilbert spaces $\mathcal{H}, \mathcal{H}^*$ and $\mathcal{H}^{**}$ respectively.

Definition 1. We define a *compound* observable, denoted by $R \,\&\, R^*$, as any S^{**}-observable on $\mathcal{H}^{**}$ whose eigenspaces are the tensor products of the eigenspaces of R and R^*. So the number of distinct eigenvalues of $R \,\&\, R^*$ is equal to the product of the number of distinct eigenvalues of R and the number of distinct eigenvalues of R^*. (For example, if $\mathcal{H}$ and $\mathcal{H}^*$ are both 2-dimensional and R and R^* each have two distinct eigenvalues, ± 1, then $R \,\&\, R^*$ is an observable in $\mathcal{H}^{**} = \mathcal{H} \otimes \mathcal{H}^*$ with four distinct eigenvalues corresponding to four 1-dimensional eigenspaces, while the tensor product $R \otimes R^*$ has only two distinct eigenvalues, ± 1, corresponding to two 2-dimensional eigenspaces.)

Definition 2. We define an observable *induced by* R^{**} on the subsystem S as an observable R on $\mathcal{H}$ that exists if and only if there is an observable R^* defined on $\mathcal{H}^*$ such that $R^{**} = R\&R^*$. (It follows that R is unique up to a choice of eigenvalues, i.e. different induced observables share the same set of eigenspaces.)

Definition 3. We define *the state induced by* e^{**} *on the subsystem* S as the state represented by the ray $e \in \mathcal{H}$ that exists if and only if there is a ray $e^* \in \mathcal{H}^*$ such that $e^{**} = e \otimes e^*$.

Definition 4. We define *the restriction of* $\mathcal{D}^{**}(e^{**}, R^{**})$ *to* $\mathcal{H}$, denoted by $\mathcal{D}(e^{**}, R^{**})$, as the set of all projection operators p on $\mathcal{H}$ such that $p \otimes I^* \in \mathcal{D}^{**}(e^{**}, R^{**})$, where I^* is the identity operator on $\mathcal{H}$.

Theorem 1 (Uniqueness Theorem)

Consider a (pure) quantum state represented by a ray e in an n-dimensional Hilbert space $\mathcal{H}$ and an observable R with $m \leq n$ distinct eigenspaces r_i. Let $e_{r_i} = (e \vee r_i^{\perp}) \wedge r_i,\ i = 1, \ldots, k \leq m$, denote the non-zero projections of the ray e onto the eigenspaces r_i. Then $\mathcal{D}(e, R) = \mathcal{L}_{e_{r_1} e_{r_2} \ldots e_{r_k}}$ is the unique maximal sublattice of $\mathcal{L}(\mathcal{H})$ satisfying the following four conditions:

1. Truth and probability (TP): $\mathcal{D}(e, R)$ is an ortholattice admitting sufficiently many 2-valued homomorphisms, $h : \mathcal{D}(e, R) \rightarrow \{0, 1\}$, to recover the joint probabilities assigned by the state e to mutually compatible sets of elements $\{p_i\}_{i \in I}, p_i \in \mathcal{D}(e, R)$, as measures on a Kolmogorov probability space (X, F, μ), where X is the set of 2-valued homomorphisms on $\mathcal{D}(e, R)$, F is a field of subsets of X, and

$$\mu(\{h : h(p_i) = 1, \text{ for all } i \in I\}) = \mathrm{Tr}(e \prod_{i \in I} p_i).$$

2. R-preferred (R-PREF): The eigenspaces r_i of R belong to $\mathcal{D}(e, R)$.

3. e, R-definability (DEF): For any $e \in \mathcal{H}$ and observable R of S defined on $\mathcal{H}$, $\mathcal{D}(e, R)$ is invariant under lattice automorphisms that preserve e and R.

4. Weak separability (WEAK SEP): If $\mathcal{H}$ is a factor space of a tensor product Hilbert space $\mathcal{H}^{**} = \mathcal{H} \otimes \mathcal{H}^{*}$, and the state e^{**} and preferred observable R^{**} on $\mathcal{H}^{**}$ induce the state e and preferred observable R on $\mathcal{H}$, then $\mathcal{D}(e, R) = \mathcal{D}(e^{**}, R^{**})$.

Remarks: The motivation for the conditions TP and R-PREF is clear from the preceding discussion. The condition DEF requires that the determinate sub-lattice is selected by the state e and a preferred observable R. The condition WEAK SEP is introduced to avoid a dimensionality constraint in the theorem: without this condition, the eigenspaces of R are required to be more than 2-dimensional. The idea behind WEAK SEP is simply that we want two systems that are not 'entangled' (and are endowed with their own preferred observables) to be separable, in the sense that each system is independently characterized by its own determinate sublattice of properties, where the determinate sublattice of a component system is the restriction of the determinate sublattice of the composite system to the component system. Put differently, the determinate sublattice of a model quantum mechanical universe should be unaffected if we add a system to the universe, and there is no entanglement arising from any interaction between the universe and the added system. The qualification 'weak' here is added to contrast the condition with Einstein's stronger separability requirement (see Section 3), that the determinate properties of two spatially separated systems should be independent of each other, even if the quantum state of the composite system is an 'entangled' state (a linear superposition of product states) resulting from a past interaction between the systems (as in the Einstein-Podolsky-Rosen argument).

The strategy of the proof of the theorem proceeds by showing that, if $p \in \mathcal{D}(e, R)$, then for any e_{r_i}, $i = 1, \ldots, k$, either $e_{r_i} \leq p$ or $e_{r_i} \leq p^{\perp}$. Since $\mathcal{L}_{e_{r_1} e_{r_2} \ldots e_{r_k}} = \{p : e_{ri} \leq p \text{ or } e_{r_i} \leq p^{\perp},\ i, \ldots, k\}$ will be shown to satisfy the conditions of the theorem, maximality requires that $\mathcal{D}(e, R) = \mathcal{L}_{e_{r_1} e_{r_2} \ldots e_{r_k}}$.

Proof: For any system S represented on a Hilbert space $\mathcal{H}$, and any state $e \in \mathcal{H}$ and preferred observable R on $\mathcal{H}$, we can always choose a Hilbert space $\mathcal{H}^*$, a state $e^* \in \mathcal{H}^*$, and observable R^* on $\mathcal{H}^*$, so that the composite system, $S + S^*$, is in the state $e \otimes e^*$ and the composite preferred observable is $R\&R^*$. Furthermore, we can always choose an R^* with at least one eigenspace, r, of dimensionality greater than two. By WEAK SEP, if $p \in \mathcal{D}(e, R)$, then $p \otimes I^* \in \mathcal{D}^{**}(e \otimes e^{**}, R\&R^*)$, since $\mathcal{D}(e, R)$ must be the restriction of $D^{**}(e \otimes e^*, R\&R^*)$ to $\mathcal{H}$. Even if an eigenspace r_1 of R is 1-dimensional or 2-dimensional, the dimension of the eigenspace $r_1 \otimes r$ in $\mathcal{H} \otimes \mathcal{H}^*$ is greater than two.

We shall show that if $p \in \mathcal{D}(e, R)$, then for any e_{r_i}, $i = 1, \ldots, k$ either $e_{r_i} \leq p$ or $e_{r_i} \leq p^\perp$, if $\dim r_1 > 2$. But the proof applies also to the determinate sub-lattice $\mathcal{D}^{**}(e^{**}, R^{**})$. In this sublattice, $\dim r_i \otimes r_j^* > 2$ even if $\dim r_1 < 2$, so $e_{r_i} \otimes e^*_{r_j} \leq p \otimes I^*$ or ${}_{r_i} \otimes e^*_{r_j} \leq p^\perp \otimes I^*$ (where $e_{r_i} \otimes e^*_{r_j}$ is the orthogonal projection of $e \otimes e^*$ onto the ijth eigenspace of $R \,\&\, R^*$). It follows that $e_{r_i} \leq p$ or $e_{r_i} \leq p^\perp$, even if $\dim r_1 \leq 2$.

So suppose $p \in \mathcal{D}(e, R)$. Consider the k eigenspaces r_i of R on which e has a nonzero projection. Since these eigenspaces are in $\mathcal{D}(e, R)$ (by R-PREF) and are assigned nonzero probability by the state e, for each such r_i there exists at least one 2-valued lattice homomorphism h_i, with nonzero measure μ, such that $h_i(r_i) = 1$ (by TP). It follows that, for each $i = l, \ldots, k$, either $r_i \wedge p \neq 0$ or $r_i \wedge p^\perp \neq 0$ (or else $h_i(p) = h_i(p^\perp) = 0$, contradicting TP).

Suppose, then, without loss of generality (as shown above), that $\dim r_i > 2$. If $r_i \wedge p = r_i$, then $r_i \leq p$ and $e_{r_i} \leq p$. If $r_i \wedge p^\perp = r_i$, then $r_i \leq p^\perp$ and $e_{r_i} \leq p^\perp$. So, if either $r_i \wedge p = r_i$ or $r_i \wedge p^\perp = r_i$, then either $e_{r_i} \leq p$ or $e_{r_i} \leq p^\perp$.

If $r_i \wedge p \neq r_i$ and $r_i \wedge p^\perp \neq r_i$, then either $0 \neq r_i \wedge p < r_i$ or $0 \neq r_i \wedge p^\perp < r_i$ (for either $r_i \wedge p \neq 0$ or $r_i \wedge p^\perp \neq 0$). Suppose $0 \neq r_i \wedge p < r_i$ (a similar argument applies if $0 \neq r_i \wedge p^\perp < r_i$). We write $r_i \wedge p = b$ for convenience. Clearly, $b \in \mathcal{D}(e, R)$, since $r_i \in \mathcal{D}(e, R)$ by R-PREF and $p \in \mathcal{D}(e, R)$ by hypothesis, and TP applies. There are two cases to consider: (i) b is orthogonal to e_{r_i} in r_i, and (ii) b is not orthogonal to e_{r_i} in r_i. We first show that in either case we must have $e_{r_i} \in \mathcal{D}(e, R)$.

(i) Suppose b is orthogonal to e_{r_i} in r_i, i.e., $b \leq e'_{r_i}$, where here the $'$ denotes the orthogonal complement in r_i. Consider all lattice automorphisms U that are rotations about e_{r_i} in r_i and the identity in $r_i^{\perp}$ [such rotations are possible because $\dim r_i > 2$ by hypothesis]. Clearly, U preserves e and R, because U preserves the eigenspaces of R, and U preserves the projections of e onto the eigenspaces of R. Then, by DEF, $U(b) \in \mathcal{D}(e, R)$, for all such rotations, and there are clearly sufficiently many rotations (regardless of b's dimension) to generate a set of elements $\{U(b)\}$ in $\mathcal{D}(e, R)$ whose span is e'_{r_i}. It follows that $e'_{r_i} \in \mathcal{D}(e, R)$, by lattice closure (i.e. by TP), and since $r_i \in \mathcal{D}(e, R)$ by R-PREF, lattice closure further requires that $e_{r_i} \in \mathcal{D}(e, R)$.

(ii) Suppose b is not orthogonal to e_{r_i} in r_i. We may suppose that b is not 1-dimensional [for if it is, we can instead consider b', the orthocomplement of b in r_i, since $\dim r_i > 2$ by hypothesis]. Since b is skew to e_{r_i}, $b = c \vee d$, where $c \leq e'_{r_i}$ (i.e., c is orthogonal to e_{r_i} in r_i) and d is a ray skew to e_{r_i} — the projection of the ray e_{r_i} onto the subspace b. Consider a lattice automorphism U that is a 'reflection' through $e_{r_i} \vee c$ in r_i (thus U preserves e_{r_i} and c, but not d) and the identity in $r_i^{\perp}$. As in (i) above, U preserves e and R, so $U(b) \in \mathcal{D}(e, R)$ and $b \wedge U(b) = c \in \mathcal{D}(e, R)$ by lattice closure. As in (i), since c is orthogonal to e_{r_i} in r_i, we can further consider rotations of c about e_{r_i} in r_i to show that $e'_{r_i} \in \mathcal{D}(e, R)$ and hence that $e_{r_i} \in \mathcal{D}(e, R)$.

Now, since $e_{r_i} \in \mathcal{D}(e, R)$ and $e_{r_i} \neq 0$, there exists a 2-valued homomorphism h_i with nonzero measure μ such that $h_i(e_{r_i}) = 1$. Suppose $p \in \mathcal{D}(e, R)$, and neither $e_{r_i} \leq p$ nor $e_{r_i} \leq p^{\perp}$. Then $e_{r_i} \wedge p = 0$ and $e_{r_i} \wedge p^{\perp} = 0$ (because e_{r_i} is a ray). It follows that for this homomorphism h_i, $h_i(p) = 0$ and $h_i(p^{\perp}) = 0$ contradicting TP. So either $e_{r_i} \leq p$ or $e_{r_i} \leq p^{\perp}$.

We conclude that if $p \in \mathcal{D}(e, R)$, then either $e_{r_i} \leq p$ or $e_{r_i} \leq p^{\perp}$, for all $i = 1, \ldots, k$. Maximality now requires that $\mathcal{D}(e, R)$ contain the set of all such elements p, i.e. $\mathcal{D}(e, R) = \{p : e_{r_i} \leq p \text{ or } e_{r_i} \leq p^{\perp}, i = 1, \ldots, k\} = \mathcal{L}_{e_{r_1} e_{r_2} \ldots e_{r_k}}$, because $\mathcal{L}_{e_{r_1} e_{r_2} \ldots e_{r_k}}$ satisfies TP, R-PREF, DEF, and WEAK SEP.

To see that TP is satisfied, we need to show that $\mathcal{L}_{e_{r_1}e_{r_2}\ldots e_{r_k}}$ is closed under the ortholattice operations, that there exist 2-valued homomorphisms on $\mathcal{L}_{e_{r_1}e_{r_2}\ldots e_{r_k}}$, and that the joint probabilities assigned by e to mutually compatible sets of elements $\{p_i\}_{i\in I}$ in $\mathcal{L}_{e_{r_1}e_{r_2}\ldots e_{r_k}}$ can be recovered as measures on a Kolmogorov probability space (X, F, μ), where X is the set of 2-valued homomorphisms on $\mathcal{L}_{e_{r_1}e_{r_2}\ldots e_{r_k}}$.

We consider closure first. If $p \in \mathcal{L}_{e_{r_1}e_{r_2}\ldots e_{r_k}}$, then either $e_{r_i} \leq p$ or $e_{r_i} \leq p^{\perp}$, for all $i = 1, \ldots, k$. So, $p^{\perp} \in \mathcal{L}_{e_{r_1}e_{r_2}\ldots e_{r_k}}$, because either $e_{r_i} \leq p^{\perp}$ or $e_{r_i} \leq (p^{\perp})^{\perp} = p$. To show that $p \wedge q \in \mathcal{L}_{e_{r_1}e_{r_2}\ldots e_{r_k}}$ and $p \vee q \in \mathcal{L}_{e_{r_1}e_{r_2}\ldots e_{r_k}}$, if $p \in \mathcal{L}_{e_{r_1}e_{r_2}\ldots e_{r_k}}$ and $q \in \mathcal{L}_{e_{r_1}e_{r_2}\ldots e_{r_k}}$, we need to show that $e_{r_i} \leq p \wedge q$ or $e_{r_i} \leq (p \wedge q)^{\perp}$, and $e_{r_i} \leq p \vee q$ or $e_{r_i} \leq (p \vee q)^{\perp}$, for all $i = 1, \ldots, k$. If $p \in \mathcal{L}_{e_{r_1}e_{r_2}\ldots e_{r_k}}$ and $q \in \mathcal{L}_{e_{r_1}e_{r_2}\ldots e_{r_k}}$, then, for each $i = 1, \ldots, k$, $e_{r_i} \leq p$ or $e_{r_i} \leq p^{\perp}$, and $e_{r_i} \leq q$ or $e_{r_i} \leq q^{\perp}$. So, either (i) $e_{r_i} \leq p$ and $e_{r_i} \leq q$, in which case $e_{r_i} \leq p \wedge q$ and $e_{r_i} \leq p \leq p \vee q$, or (ii) $e_{r_i} \leq p$ and $e_{r_i} \leq q^{\perp}$, in which case $e_{r_i} \leq p \leq p \vee q$ and $e_{r_i} \leq q^{\perp} \leq p^{\perp} \vee q^{\perp} = (p \wedge q)^{\perp}$, or (iii) $e_{r_i} \leq p^{\perp}$ and $e_{r_i} \leq q$, in which case $e_{r_i} \leq q \leq p \vee q$ and $e_{r_i} \leq p^{\perp} \leq p^{\perp} \vee q^{\perp} = (p \wedge q)^{\perp}$, or (iv) $e_{r_i} \leq p^{\perp}$ and $e_{r_i} \leq q^{\perp}$, in which case $e_{r_i} \leq p^{\perp} \wedge q^{\perp} = (p \vee q)^{\perp}$ and $e_{r_i} \leq p^{\perp} \leq p^{\perp} \vee q^{\perp} = (p \wedge q)^{\perp}$.

To show the existence of a 2-valued homomorphism on $\mathcal{L}_{e_{r_1}e_{r_2}\ldots e_{r_k}}$, consider k maps $h_i : \mathcal{L}_{e_{r_1}e_{r_2}\ldots e_{r_k}} \rightarrow \{0, 1\}$ defined as follows: For any element $p \in \mathcal{L}_{e_{r_1}e_{r_2}\ldots e_{r_k}}$, $h_i(p) = 1$, if $e_{r_i} \leq p$ and $h_i(p) = 0$ if $e_{r_i} \leq p^{\perp}$. Each h_i is a 2-valued homomorphism on $\mathcal{L}_{e_{r_1}e_{r_2}\ldots e_{r_k}}$. Clearly $h_i(0) = 0$ and $h_i(I) = 1$, because $e_{r_i} \leq I = 0^{\perp}$. If $e_{r_i} \leq p$, $h_i(p) = 1$ and $h_i(p^{\perp}) = 0$, because $p = (p^{\perp})^{\perp}$. Similarly, if $e_{r_i} \leq p^{\perp}$, then $h_i(p^{\perp}) = 1$ and $h_i(p) = 0$. So $h_i(p) = 1 - h_i(p^{\perp})$. If $e_{r_i} \leq p$ and $e_{r_i} \leq q$, so that $h_i(p) = 1$ and $h_i(q) = 1$, then $e_{r_i} \leq p \wedge q$, so $h_i(p \wedge q) = 1$. If $e_{r_i} \leq p$ and $e_{r_i} \leq q^{\perp}$, so that $h_i(p) = 1$ and $h_i(q) = 0$, or $e_{r_i} \leq p^{\perp}$ and $e_{r_i} \leq q$, so that $h_i(p) = 0$ and $h_i(q) = 1$, then $e_{r_i} \leq p^{\perp} \vee q^{\perp} = (p \wedge q)^{\perp}$, so that $h_i(p \wedge q) = 0$. If $e_{r_i} \leq p^{\perp}$ and $e_{r_i} \leq q^{\perp}$, so that $h_i(p) = h_i(q) = 0$, then $e_{r_i} \leq p^{\perp} \vee q^{\perp} = (p \wedge q)^{\perp}$, so that $h_i(p \wedge q) = 0$. So $h_i(p \wedge q) = h_i(p)h_i(q)$. Since $p \vee q = (p^{\perp} \wedge q^{\perp})^{\perp}$, it follows that $h_i(p \vee q) = h_i(p) \vee h_i(q) = h_i(p) + h_i(q) - h_i(p)h_i(q)$. So each h_i is a homomorphism.

[We note that the k 2-valued homomorphisms h_i on $\mathcal{L}_{e_{r_1}e_{r_2}\ldots e_{r_k}}$ are in general the *only* 2-valued homomorphisms on $\mathcal{L}_{e_{r_1}e_{r_2}\ldots e_{r_k}}$. Since any 2-valued homomorphism that maps any ray in $(e_{r_1} \vee e_{r_2} \vee \cdots \vee e_{r_k})^{\perp}$ onto 1 must map each of the rays e_{r_i}, $i = 1, \ldots, k$, onto 0, it follows that any such homomorphism must map one of each orthogonal $(n-k)$-tuple of rays in $(e_{r_1} \vee e_{r_2} \vee \cdots \vee e_{r_k})^{\perp}$ onto 1 and the remaining members of the $(n-k)$-tuple onto 0. The infimum of any two rays that are mapped onto 1 must also be mapped onto 1, but this is the zero element. So no 2-valued homomorphism can map any of the rays in $(e_{r_1} \vee e_{r_2} \vee \cdots e_{r_k})^{\perp}$ onto 1 — except, of course, when $\dim(e_{r_1} \vee e_{r_2} \vee \cdots e_{r_k})^{\perp} = 1$, i.e., when $n = k+1$. In all other cases ($n > k+1$), it follows that the only 2-valued homomorphisms of $\mathcal{L}_{e_{r_1}e_{r_2}\ldots e_{r_k}}$ are the homomorphisms h_i, $i = 1, \ldots, k$, where each h_i maps the ray e_{r_i} onto 1 and every other ray in $\mathcal{L}_{e_{r_1}e_{r_2}\ldots e_{r_k}}$ onto 0 (since all the other rays in $\mathcal{L}_{e_{r_1}e_{r_2}\ldots e_{r_k}}$ are orthogonal to e_{r_i}).]

To generate the probabilities assigned by e to mutually compatible sets of elements $\{p_i\}_{i \in I}$, $p_i \in \mathcal{L}_{e_{r_1}e_{r_2}\ldots e_{r_k}}$, on the Kolmogorov probability space, the 2-valued homomorphism that maps e_{r_i} onto 1 (more precisely, the corresponding singleton subset) is assigned the measure $\mathrm{Tr}(ee_{r_i})$, for $i = 1, \ldots, k$. (The homomorphism that maps $(e_{r_1} \vee e_{r_2} \vee \cdots \vee e_{r_k})^{\perp}$ onto 1, assuming $\dim(e_{r_1} \vee e_{r_2} \vee \cdots \vee e_{r_k})^{\perp} = 1$, is assigned the measure 0.) To see that $\mu(\{h : h(p_1) = h(p_2) = \cdots = 1\}) = \mathrm{Tr}(ep_1p_2\cdots)$, first note that $p_1p_2\cdots = p_1 \wedge p_2 \wedge \cdots$ for compatible projections $p_1, p_2, \ldots$, and so the product $p_1p_2\cdots$ defines an element, call it p, that also belongs to $\mathcal{L}_{e_{r_1}e_{r_2}\ldots e_{r_k}}$, since $\mathcal{L}_{e_{r_1}e_{r_2}\ldots e_{r_k}}$ is closed under the ortholattice operations. Furthermore, $h(p_1) = h(p_2) = \cdots = 1$ if and only if $h(p_1 \wedge p_2 \wedge \cdots) = h(p) = 1$, since h is a lattice homomorphism. So it suffices to show that, for a general element $p \in \mathcal{L}_{e_{r_1}e_{r_2}\ldots e_{r_k}}$, $\mu(\{h : h(p) = 1\}) = \mathrm{Tr}(ep)$.

Now any $p \in \mathcal{L}_{e_{r_1}e_{r_2}\ldots e_{r_k}}$ can be expressed as $p = e_{r_i} \vee e_{r_j} \vee \cdots \vee q$, for some $i, j, \ldots$, where q is the orthogonal projection operator onto a subspace (possibly the zero subspace) orthogonal to *all* the e_{r_i}, i.e., $q \in (e_{r_1} \vee e_{r_2} \vee \cdots \vee e_{r_k})^{\perp}$. If $h(p) = 1$, $h(e_{r_i} \vee e_{r_j} \vee \cdots \vee q) = h(e_{r_i}) \vee$

$h(e_{r_j}) \vee \cdots \vee h(q) = 1$. So:

$$\mu(\{h : h(p) = 1\}) = \mu\left(\{h : h(e_{r_i}) \vee h(e_{r_j}) \vee \cdots \vee h(q) = 1\}\right)$$
$$= \mu(\{h : h(e_{r_i}) = 1\}) + \mu\left(\{h : h(e_{r_j}) = 1\}\right) + \cdots,$$

because $\mu(\{h : h(q) = 1\}) = 0$ and $h(e_{r_i}) \neq h(e_{r_j})$ if $i \neq j$. It follows that:

$$\begin{aligned}\mu(\{h : h(p) = 1\}) &= \text{Tr}(ee_{r_i}) + \text{Tr}(e_{r_j}) + \cdots \\ &= \text{Tr}(e(e_{r_i} + e_{r_j} + \cdots + q))\end{aligned}$$

because $\mu(\{h : h(e_{r_i}) = 1\}) = \text{Tr}(ee_{r_i})$ and $\text{Tr}(eq) = 0$ and hence:

$$\begin{aligned}\mu(\{h : h(e_{r_i}) = 1\}) &= \text{Tr}(e(e_{r_i} \vee e_{r_j} \vee \cdots \vee q)) \\ &= \text{Tr}(ep)\end{aligned}$$

since $e_{r_i}, e_{r_j}, \ldots$ and q are mutually orthogonal.

R-PREF is satisfied by construction. Since $e_{r_i} \in \mathcal{L}_{e_{r_1}e_{r_2}\ldots e_{r_k}}$, $i = 1, \ldots, k$, and every ray in $(e_{r_1} \vee e_{r_2} \vee \cdots \vee e_{r_k})^{\perp}$ belongs to $\mathcal{L}_{e_{r_1}e_{r_2}\ldots e_{r_k}}$, it follows that every ray in e'_{r_i}, $i = 1, \ldots, k$, (the orthocomplement of e_{r_i} in r_i) belongs to $\mathcal{L}_{e_{r_1}e_{r_2}\ldots e_{r_k}}$, so $r_i \in \mathcal{L}_{e_{r_1}e_{r_2}\ldots e_{r_k}}$, $i = 1, \ldots, k$ by lattice closure (established above). The remaining $m - k$ elements r_i also belong to $\mathcal{L}_{e_{r_1}e_{r_2}\ldots e_{r_k}}$, because each of these is represented by a subspace orthogonal to e and hence by a subspace in $(e_{r_1} \vee e_{r_2} \vee \cdots \vee e_{r_k})^{\perp}$. But since every ray in $(e_{r_1} \vee e_{r_2} \vee \cdots \vee e_{r_k})^{\perp}$ belongs to $\mathcal{L}_{e_{r_1}e_{r_2}\ldots e_{r_k}}$, every subspace of $(e_{r_1} \vee e_{r_2} \vee \cdots \vee e_{r_k})^{\perp}$ belongs to $\mathcal{L}_{e_{r_1}e_{r_2}\ldots e_{r_k}}$.

DEF is satisfied because lattice automorphisms that preserve e and R automatically preserve the nonzero projections of e onto the eigenspaces r_i of R, i.e. the rays e_{r_i}, $i = 1, \ldots, k$, that generate $\mathcal{L}_{e_{r_1}e_{r_2}\ldots e_{r_k}}$.

To see that WEAK SEP is satisfied, consider a composite system, $S + S^*$, represented on a Hilbert space $\mathcal{H} \otimes \mathcal{H}^*$. Suppose the state of $S + S^*$ is $e \otimes e^*$ and the privileged observable of $S + S^*$ is an observable $R\,\&\,R^*$. Then, for all i, $e_{r_i} \leq p$ or $e_{r_i} \leq p^{\perp}$ if and only if, for all i, j, $e_{r_i} \otimes e^*_{r^*_j} \leq p \otimes I^*$ or $e_{r_i} \otimes e^*_{r^*_j} \leq p^{\perp} \otimes I^*$. It follows that $\mathcal{D}(e \otimes e^*, R\&R^*) = \mathcal{D}(e, R)$. QED

3 Interpretations

The above analysis shows that the determinate sublattices of the lattice of projection operators or subspaces, $\mathcal{L}(\mathcal{H})$, of a Hilbert space $\mathcal{H}$, representing the propositions ('yes-no' experiments) of a quantum mechanical system are just the lattices $\mathcal{D}(e, R) = \mathcal{L}_{e_{r_1}e_{r_2}\ldots e_{r_k}} = \{p : e_{r_i} \leq p \text{ or } e_{r_i} \leq p^{\perp},\ i = 1, \ldots, k\}$. These are the maximal subcollections of quantum propositions that can be determinately true or false, given the quantum state e and a preferred observable R, subject to certain constraints that essentially require these subcollections to be lattices determined by e and R on which sufficiently many truth valuations exist to recover the usual quantum statistics. The set of observables associated with $\mathcal{L}_{e_{r_1}e_{r_2}\ldots e_{r_k}}$ includes any observable whose eigenspaces are spanned by rays in $\mathcal{L}_{e_{r_1}e_{r_2}\ldots e_{r_k}}$. The set of maximal observables includes any maximal observable with k eigenvectors in the directions e_{r_i}, $i = 1, \ldots, k$.

The uniqueness theorem characterizes a class of 'no collapse' interpretations of quantum mechanics, where each interpretation involves the selection of a particular preferred determinate observable, and hence the selection, via the quantum state at a particular time, of a particular determinate sublattice with respect to which the probabilities defined by the quantum state have the usual epistemic significance. So the quantum probabilities defined on the determinate sublattice can be understood as measures of ignorance about the actual values of observables associated with propositions in the determinate sublattice, i.e. the actual values of the preferred determinate observable and observables that inherit determinate status via the quantum state and the preferred observable.

We note that Bell concluded his seminal critique of impossibility proofs for hidden variables in quantum mechanics[22] by observing that the equations of motion in Bohm's theory 'have in general a grossly non-local character', so that 'in this theory an explicit causal mechanism exists whereby the disposition of one piece of apparatus affects the results obtained with a distant piece'. He remarked that

[22] J. S. Bell, 'On the Problem of Hidden Variables in Quantum Mechanics', *op. cit.*, p. 452.

'the Einstein-Podolsky-Rosen paradox is resolved in the way which Einstein would have liked least' and raised the question whether one could prove 'that *any* hidden variable account of quantum mechanics *must* have this extraordinary character'. Bell subsequently proved[23] that no hidden variable theory satisfying a locality constraint could reproduce the quantum statistics. Our theorem goes beyond this result by characterizing all possible 'completions' of quantum mechanics ('no collapse' interpretations), subject to certain natural constraints.

3.1 The orthodox (Dirac-von Neumann) interpretation

Without the projection postulate, the orthodox interpretation is a 'no collapse' interpretation of quantum mechanics in our sense. On the orthodox interpretation, an observable has a determinate value if and only if the state of the system is an eigenstate of the observable. Equivalently, the propositions that are determinately true or false of a system are the propositions represented by subspaces that either contain the state of the system, or are orthogonal to the state, i.e., the propositions assigned probability 1 or 0 by the state. The orthodox interpretation can therefore be formulated as the proposal that the preferred determinate observable is the unit observable I, and that $\mathcal{D}(e, I) = \mathcal{L}_e$ is the determinate sublattice of a system in the state e, where $\mathcal{L}_e = \{p : e \leq p \quad \text{or} \quad e \leq p^{\perp}\}$.

The choice of the preferred determinate observable as the unit observable leads to the measurement problem. For a composite system $S + M$ in an entangled state of the form $|e\rangle = \sum_i c_i|a_i\rangle|r_i\rangle$, neither R-propositions nor A-propositions belong to the sublattice $\mathcal{L}_e$. In order to avoid the problem, Dirac and von Neumann assume the projection postulate,[24] that unitary evolution is suspended in the case of a measurement interaction, and the state of $S + M$ is projected onto the ray spanned by one of the unit vectors $|a_i\rangle|r_i\rangle$ with probability $|c_i|^2$, because only in such a state, according to the orthodox interpretation, do the observables A and R have determinate values.

---

[23] J. S. Bell, 'On the Einstein-Podolsky-Rosen Paradox', *op. cit.*

[24] Dirac, *op. cit.*, p. 36; and von Neumann, *op. cit.*, pp. 351 and 418.

3.2 Resolution of the measurement problem

There is nothing in the mathematical structure of quantum mechanics that forces the choice of the preferred determinate observable R as the unit observable I, and indeed there is every reason to avoid this choice because it leads to the measurement problem. 'No collapse' interpretations that seek to solve this problem represent alternative proposals for choosing R. For such interpretations, there is no measurement problem if R plays the role of a pointer observable in all measurement interactions, or an observable correlated with the pointer observable.

To see this, consider a model quantum mechanical universe consisting of two systems, S and M, associated with a Hilbert space $\mathcal{H}_S \otimes \mathcal{H}_M$. A measurement interaction between S and M, say a dynamical evolution of the quantum state of the composite system $S + M$ described by a unitary transformation that correlates eigenstates $|a_i\rangle$ of an S-observable A with eigenstates $|r_i\rangle$ of an M-observable R, will result in a state represented by a unit vector of the form $|e\rangle = \sum_i c_i |a_i\rangle |r_i\rangle$ (assuming initial pure states for S and M). If we take $I \otimes R$ as the preferred determinate observable, the projections of the ray e onto the eigenspaces $\mathcal{H}_S \otimes r_i$, of $I \otimes R$ are the rays e_{r_i}, spanned by the unit vectors $|a_i\rangle|r_i\rangle$. So for this state, the determinate sublattice contains propositions represented by the projection operators $a_i \otimes I_M$, where a_i here represents the projection operator onto the subspace in $\mathcal{H}_S$ spanned by the unit vector $|a_i\rangle$, i.e., propositions corresponding to the eigenvalues of A (and also, of course, propositions corresponding to pointer positions, represented by the projection operators $I_S \otimes r_i$).

Evidently, the same conclusion follows if we take some observable $I \otimes I \otimes T$ as the preferred observable, where R is correlated with T via the dynamical evolution of the quantum state in a measurement interaction, so that the state after the measurement takes the form $|e\rangle = \sum c_i |a_i\rangle |r_i\rangle |t_i\rangle$. In this case, the projections of the ray e onto the eigenspaces $\mathcal{H}_S \otimes \mathcal{H}_M \otimes t_i$ of $I \otimes I \otimes T$ are the rays e_{t_i} spanned by the unit vectors $|a_i\rangle|r_i\rangle|t_i\rangle$.

It follows that, without introducing any measurement constraints on the determinate sublattices (apart from the choice of R, which could still be justified on grounds independent of measurement), we derive that the propositions corresponding to the observable correlated with an appropriate pointer observable in the 'entangled' state arising from a unitary transformation representing a quantum mechanical measurement interaction are determinately true or false. So we derive the interpretation of the probabilities defined by a quantum state for the eigenvalues of an observable A as 'the probabilities of finding the different possible eigenvalues of A in a measurement of A', where a measurement is represented as a dynamical process that yields determinate values for A.

3.3 The modal interpretations of Kochen and of Dieks

The idea behind a 'modal' interpretation of quantum mechanics is that quantum states, unlike classical states, constrain possibilities rather than actualities — which leaves open the question of whether one can introduce 'value states'[25] that assign values to the observables of the theory, or equivalently, truth values to the corresponding propositions. As van Fraassen puts it:[26]

> In other words, the [quantum] state delimits what can and cannot occur, and how likely it is — it delimits possibility, impossibility, and probability of occurrence — but it does not say what actually occurs. The transition from the possible to the actual is not a transition *of* state, but a transition *described by* the state.

Apart from Van Fraassen's original version of the modal interpretation,[27] there are now a variety of other 'no collapse' interpretations of quantum mechanics that can be seen as modal in this sense, for example, the interpretations of Krips, Kochen, Healey,

[25]The terminology is van Fraassen's. See B. van Fraassen, *Quantum Mechanics: An Empiricist View*, *op. cit.*, p. 275.

[26]B. van Fraassen, *ibid.*, p. 279.

[27]First formulated in B. van Fraassen, 'Hidden Variables and the Modal Interpretation of Quantum Statistics', *op. cit.*

Dieks, and Bub.[28] All these modal interpretations share with van Fraassen's interpretation the feature that an observable can have a determinate value even if the quantum state is not an eigenstate of the observable, so they preserve the linear, unitary dynamics for quantum states without requiring the projection postulate to validate the determinateness of pointer readings and measured observable values in quantum measurement processes. The modal interpretations of Kochen and Healey exploit the polar decomposition theorem to define value states. By that theorem, any pure quantum state $|e\rangle$ of a system $S + S^*$ can be expressed in the form:

$$|e\rangle = \sum_i c_i |u_i\rangle |v_i\rangle,$$

for some orthonormal set of vectors $\{|u_i\rangle\}$ in $\mathcal{H}(S)$ and some orthonormal set $\{|v_i\rangle\}$ in $\mathcal{H}(S^*)$. The decomposition is unique if and only if $|c_i|^2 \neq |c_j|^2$ for any $i \neq j$. In the non-degenerate case, the basic idea is to take the propositions that are determinately true or false for S in the quantum state $|e\rangle$ as the propositions represented by the Boolean algebra of projection operators generated by the set $\{P_{|u_i\rangle}\}$. (Similar remarks apply to S^*, of course.) There are alternative proposals for the degenerate case. Clifton's proposal,[29] closely related to Dieks', is to take the set of determinate propositions of S as the propositions represented by projection operators of the form $P_1 + P_2$, where P_1 belongs to the spectral measure of the density operator W representing the reduced state of $|e\rangle$ for S, and P_2 belongs to the null space of W (i.e. $P_2 W = 0$). All S-propositions assigned probability 1 or 0 are included in this set.

---

[28]S. Kochen, 'A New Interpretation of Quantum Mechanics', *op. cit.*; H. Krips, *The Metaphysics of Quantum Theory* (Oxford: Clarendon Press, 1987); R. Healey, *The Philosophy of Quantum Mechanics* (Cambridge: Cambridge University Press, 1989); D. Dieks, 'Quantum Mechanics Without the Projection Postulate and Its Realistic Interpretation', *op. cit.*; 'Modal Interpretation of Quantum Mechanics, Measurements, and Macroscopic Behavior', *op. cit.*; J. Bub, 'Quantum Mechanics Without the Projection Postulate', *op. cit.*; 'How to Interpret Quantum Mechanics', *op. cit.*

[29]R. Clifton, 'Independently Motivating the Kochen-Dieks Modal Interpretation of Quantum Mechanics', *British Journal for the Philosophy of Science* 46 (1995), 33–57 (Chapter 1 of this volume).

The interpretations of Kochen and of Dieks can therefore be understood as the proposal that for any quantum state W of a system S, pure or mixed [where W arises from partial tracing over $\mathcal{H}(S^*)$ if S is a subsystem of system $S + S^*$], the determinate propositions of S are the propositions represented by the projection operators in the set: $\mathrm{Def_{KD}}(S) = \{P : P = P_1 + P_2, P_1 \in$ spectral measure of W, $P_2 \in$ null space of $W\}$. As Clifton has shown,[30] an equivalent formulation is:

$$\mathrm{Def_{KD}}(S) = \left\{P : PP_{W_i} = P_{W_i} \text{ or } 0,\ \forall P_{W_i} \in \mathrm{SR}(W)\right\},$$

where $\mathrm{SR}(W)$ is the set of projection operators in the spectral representation of W (i.e. the projection operators onto the eigenspaces corresponding to the nonzero eigenvalues of W).[31] It is then easy to see that $\mathrm{Def_{KD}}(S)$ forms a sublattice of $\mathcal{H}(S)$.

We can now prove a recovery theorem for the interpretations of Kochen and of Dieks, that these versions of the modal interpretation are 'no collapse' interpretations, in the sense of Theorem 1, for the preferred determinate observable $R = W \otimes I^*$, i.e. the determinate sublattices of S, as a subsystem the system $S + S^*$, are the sublattices $\mathcal{D}(e, W \otimes I^*)|_S$, where '$|_S$' denotes the restriction of the sublattice to the Hilbert space $\mathcal{H}$ of the subsystem S, e is the ray representing the quantum state of $S + S^*$, and W is the reduced state of e for the system S. (So the preferred determinate observable is not fixed, but changes with time as the state e evolves.)

Theorem 2 (Recovery Theorem)

If S is a subsystem of a quantum mechanical universe $S + S^*$ represented on a Hilbert space $\mathcal{H} \otimes \mathcal{H}^*$, and the state of $S + S^*$ is represented by a ray $e \in \mathcal{H} \otimes \mathcal{H}^*$, then

$$\mathcal{D}(e, W \otimes I^*)|_S = \mathrm{Def_{KD}}(S).$$

Proof: We introduce the following notation to capture the distinctions required to formulate the proof precisely:

[30] ibid.

[31] The spectral measure of W is the Boolean algebra generated by $\mathrm{SR}(W)$.

W_i:	The ith eigenspace of W (in $\mathcal{H}$).
P_{W_i}:	the projection operator onto W_i (in $\mathcal{H}$).
$(W \otimes I^*)_i$:	the ith eigenspace of $W \otimes I^*$ (in $\mathcal{H} \otimes \mathcal{H}^*$).
$P_{(W\otimes I^*)_i}$:	the projection operator onto $(W \otimes I^*)_i$ (in $\mathcal{H} \otimes \mathcal{H}^*$)
$e_{(W\otimes I^*)_i}$:	the nonzero projection of the ray e onto $(W \otimes I^*)_i$ (in $\mathcal{H} \otimes \mathcal{H}^*$).
$P_{e_{(W\otimes I^*)_i}}$:	the (nonzero) projection operator onto the ray $e_{(W\otimes I^*)_i}$ (in $\mathcal{H} \otimes \mathcal{H}^*$).

Note that there is a 1–1 correspondence between the eigenspaces W_i in $\mathcal{H}$ and $(W \otimes I^*)_i$ in $\mathcal{H} \otimes \mathcal{H}^*$, because w_i is an eigenvalue of W if and only if w_i is an eigenvalue of $W \otimes I^*$. In our formulation of the uniqueness theorem in Section 2, we showed that

$$\mathcal{D}(e,R) = \mathcal{L}_{e_{r_1}e_{r_2}\ldots e_{r_k}} = \{p : e_{r_i} \leq p \quad \text{or} \quad e_{r_i} \leq p^{\perp},\ i = 1,\ldots,k\},$$

where the e_{r_i} are the nonzero projections of e onto the k eigenspaces of R. Writing P for the projection operator corresponding to the lattice element p, and $P_{e_{r_i}}$ for the projection operator corresponding to the lattice element e_{r_i}, the determinate sublattice of projection operators can be expressed equivalently as:

$$\{P : PP_{e_{r_i}} = P_{e_{r_i}} \quad \text{or} \quad 0,\ \forall i\},$$

where the quantifier over the index i ranges over all nonzero projections of e onto the eigenspaces of R. So what we want to show is that

$$\begin{aligned} \mathcal{D}(e, W \otimes I^*)|_S &\equiv \{P : (P \otimes I^*)P_{e_{(W\otimes I^*)_i}} = P_{e_{(W\otimes I^*)_i}} \text{ or } 0,\ \forall i\} \\ &= \mathrm{Def}_{\mathrm{KD}}(S) \\ &\equiv \left\{P : PP_{W_i} = P_{W_i} \text{ or } 0,\ \forall i\right\}. \end{aligned}$$

The quantifier over the index i in the expression for $\mathrm{Def}_{\mathrm{KD}}(S)$ ranges over all projection operators P_{W_i} in $\mathrm{SR}(W)$. The quantifier over the index i in the expression for $\mathcal{D}(e, W \otimes I^*)|_S$ ranges over all nonzero projections of e onto the eigenspaces of $W \otimes I^*$, including in principle the eigenspace corresponding to the zero eigenvalue (the null space of $W \otimes I^*$), if $W \otimes I^*$ has a zero eigenvalue. Since W is the reduced state of e for the system S, e lies entirely in the span of the nonzero eigenvalue eigenspaces of $W \otimes I^*$, and hence has zero projection onto the zero eigenvalue eigenspace, if such an eigenspace exists. So the quantifier in the expression for $\mathcal{D}(e, W \otimes I^*)|_S$ in fact ranges over the eigenspaces with projection operators in the set $\mathrm{SR}(W \otimes I^*)$, which are in 1–1 correspondence with projection operators in the set $\mathrm{SR}(W)$. In the following, quantification over the index i refers to a set of projection operators or subspaces in 1–1 correspondence with the set $\mathrm{SR}(W)$.

(i) We first prove that

$$\mathrm{Def}_{\mathrm{KD}}(S) \subseteq \mathcal{D}(e, W \otimes I^*)|_S.$$

It suffices to show that the generators of $\mathrm{Def}_{\mathrm{KD}}(S)$ are in $\mathcal{D}(e, W \otimes I^*)|_S$, because $\mathcal{D}(e, W \otimes I^*)|_S$ is an ortholattice and so must include the lattice $\mathrm{Def}_{\mathrm{KD}}(S)$ as a sublattice if it includes the generators of $\mathrm{Def}_{\mathrm{KD}}(S)$. The generators of $\mathrm{Def}_{\mathrm{KD}}(S)$ are the projection operators in the set:

$$\{P_{W_i} : \forall i\} \cup \{P : PP_{W_i} - 0, \forall i\}.$$

To see that $\{P_{W_i} : \forall i\} \subseteq \mathcal{D}(e, W \otimes I^*)|_S$, note that

$$P_{(W \otimes I^*)_i} \in \mathcal{D}(e, W \otimes I^*),\ \forall i,$$

because these are just the projection operators onto the nonzero eigenvalue eigenspaces of the preferred determinate observable $W \otimes I^*$, which always belong to $\mathcal{D}(e, W \otimes I^*)$ by assumption. But since

$$P_{(W \otimes I^*)_i} = P_{W_i} \otimes I^*,\ \forall i,$$

it follows that

$$P_{W_i} \in \mathcal{D}(e, W \otimes I^*)|_S,\ \forall i.$$

To see that $\{P : PP_{W_i} = 0, \forall i\} \subseteq \mathcal{D}(e, W \otimes I^*)|_S$, note that if $PP_{W_i} = 0$ for all $P_{W_i} \in \mathrm{SR}(W)$, then

$$(P \otimes I^*)(P_{W_i} \otimes I^*) = (P \otimes I^*)P_{(W \otimes I^*)_i} = 0,\ \forall i,$$

and so

$$(P \otimes I^*)P_{e_{(W \otimes I^*)_i}} = 0,\ \forall i.$$

But if $(P \otimes I^*)P_{e_{(W\otimes I^*)_i}} = 0$ for all nonzero projections of the ray e onto the nonzero eigenvalue eigenspaces of $W \otimes I^*$, then, since e has zero projection onto the zero eigenvalue eigenspace of $W \otimes I^*$ (if such an eigenspace exists), $(P \otimes I^*)P_{e_{(W\otimes I^*)_i}} = 0$ for all nonzero projections of e onto all the eigenspaces of $W \otimes I^*$, and so, by definition:

$$P \in \mathcal{D}(e, W \otimes I^*)|_S.$$

(ii) We now prove that

$$\mathcal{D}(e, W \otimes I)|_S \subseteq \mathrm{Def}_{\mathrm{KD}}(S).$$

Suppose the state vector $|e\rangle$ can be represented in *one* of its decompositions with respect to $\mathcal{H} \otimes \mathcal{H}^*$ as:

$$|e\rangle = \sum_{ij} c_{ij}|u_{ij}\rangle|v_{ij}\rangle.$$

We can express this vector as

$$|e\rangle = \sum_i \left(\sum_j c_{ij}|u_{ij}\rangle|v_{ij}\rangle \right),$$

where for fixed i, $|c_{ij_1}|^2 = |c_{ij_2}|^2 = \cdots \equiv |c_i|^2$ and the index i ranges over the *distinct* numbers $|c_i|^2$. These are just the nonzero eigenvalues of W that index $\mathrm{SR}(W)$, and so:

$$P_{W_i} = \sum_j |u_{ij}\rangle\langle u_{ij}|.$$

We now have, for all i:

$$|e\rangle_{(W\otimes I^*)_i} = P_{(W\otimes I^*)_i}|e\rangle = P_{W_i} \otimes I^*|e\rangle = \sum_j c_{ij}|u_{ij}\rangle|v_{ij}\rangle.$$

Suppose $P \in \mathcal{D}(e, W \otimes I^*)|_S$, then:

$$(P \otimes I^*)P_{e_{(W\otimes I^*)_i}} = P_{e_{(W\otimes I^*)_i}} \text{ or } 0,\ \forall i.$$

It follows that

$$(P \otimes I^*)|e_{(W \otimes I^*)_i}\rangle = |e_{(W \otimes I^*)_i}\rangle \text{ or } |0\rangle, \ \forall i,$$

and so

$$\sum_j c_{ij} P|u_{ij}\rangle|v_{ij}\rangle = \sum_j c_{ij}|u_{ij}\rangle|v_{ij}\rangle \text{ or } |0\rangle, \ \forall i.$$

For a fixed i, this implies that either $P|u_{ij}\rangle = |u_{ij}\rangle$ for all j, or $P|u_{ij}\rangle = |0\rangle$ for all j, because $c_{ij} \neq 0$ for any j. And since $\{|u_{ij}\rangle, \forall j\}$ is a basis for the eigenspace W_i:

$$PP_{W_i} = P_{W_i} \text{ or } 0, \ \forall i,$$

i.e.,

$$P \in \mathrm{Def}_{\mathrm{KD}}(S).$$

3.4 Bohmian mechanics

Bohm's 1952 hidden variable theory or 'causal' interpretation[32] can be understood as a proposal for implementing an interpretation in which the preferred determinate observable R is fixed once and for all as position in configuration space, instead of being defined by the time-evolving quantum state as in the modal interpretations of Kochen and Dieks. (Alternative versions of Bohm's theory taking R as momentum or some other observable instead of position are considered by Epstein[33] and Stone.[34]) Bohmian dynamics arises as the dynamics of 'value states' on the determinate sublattice $\mathcal{L}_{e_{r_1} e_{r_2} \ldots e_{r_k}}$ (i.e., states defined by 2-valued homomorphisms on $\mathcal{L}_{e_{r_1} e_{r_2} \ldots e_{r_k}}$) as the quantum state e evolves in time. As we have seen, value states defined by 2-valued homomorphisms (with nonzero measure) are

[32] D. Bohm, 'A Suggested Interpretation of Quantum Theory in Terms of "Hidden Variables": Parts I and II', *op. cit.*

[33] T. Epstein, 'The Causal Interpretation of Quantum Mechanics', *Physical Review* 89 (1952), 319 and *Physical Review* 91 (1953), 985.

[34] A. Stone, 'Does the Bohm Theory Solve the Measurement Problem?', *Philosophy of Science* 61 (1994), 250-66.

in 1–1 correspondence with the rays e_{r_i}, hence with the eigenspaces r_i of R, and assign the same value to e_{r_i}, and r_i, for all i. Since the evolution of such states is completely determined by the evolution of e and of R, we want an equation of motion for the determinate values of R that will preserve the distribution of R-values specified by e, as e evolves in time according to Schrödinger's time-dependent equation of motion. (Several authors have proposed dynamical evolution laws for value states in the modal interpretations of Kochen and of Dieks, where the preferred determinate observable is not fixed but is defined by the quantum state and so changes with time as the state evolves.)[35] Recall that Bohm extracts two real equations from Schrödinger's time-dependent complex equation of motion for the wave function of a single particle of mass M,

$$i\hbar\frac{\partial\psi}{\partial t} = -\frac{\hbar}{2M}\nabla^2\psi + V\psi,$$

by substituting $\psi = R\exp\left(\frac{iS}{\hbar}\right)$:[36]

$$\frac{\partial S}{\partial t} + \frac{(\nabla S)^2}{2M} + V - \frac{\hbar^2}{2m}\frac{\nabla^2 R}{R} = 0,$$
$$\frac{\partial R^2}{\partial t} + \nabla\cdot\left(\frac{R^2\nabla S}{M}\right) = 0.$$

The first equation (derived from the real part of the Schrödinger equation) can be interpreted as a Hamilton-Jacobi equation for the motion of the particle under the influence of a potential function V and an additional 'quantum potential' $\frac{\hbar}{2m} = \frac{\nabla^2 R}{R}$. The trajectories of these particles are given by the solutions to the equation:

$$\frac{\mathrm{dx}}{\mathrm{d}t} = \frac{\mathbf{J}}{\rho} = \frac{\nabla S}{M} = \frac{\hbar}{M}\mathrm{Im}\left(\frac{\nabla\psi}{\psi}\right) = \frac{\hbar}{2iM}\frac{\psi^*\nabla\psi - \psi\nabla\psi^*}{|\psi|^2},$$

[35]D. Dieks, 'Modal Interpretations of Quantum Mechanics, Measurements, and Macroscopic Behavior'; *op. cit.*; P. E. Vermaas, 'Unique Transition Probabilities in the Modal Interpretation', *Studies in History and Philosophy of Modern Physics* 27 (1996) 133–59; G. Bacciagaluppi and M. Dickson, 'Dynamics for modal interpretations', *Foundations of Physics* 29 (1999) 1165–201.

[36]The symbol R here should not, of course, be confused with the symbol R for the preferred determinate observable.

where Im( ) denotes the imaginary part of ( ), $\rho = R^2 = |\psi|^2$, and $\mathbf{J} = \frac{R^2 \nabla S}{M} = \frac{\hbar}{M}\mathrm{Im}(\psi^* \nabla \psi)$, i.e., the particle trajectories are given by the integral curves of a velocity field defined by the gradient of the phase S. So the trajectories $\mathbf{x}(t)$ depend on the wave function ψ. The second equation (derived from the imaginary part of the Schrödinger equation) can be written as a continuity equation for an ensemble density $\rho = |\psi|^2$, and a probability current $\mathbf{J}$:

$$\frac{\partial \rho}{\partial t} + \nabla \cdot \mathbf{J} = 0.$$

The continuity equation guarantees that if $\rho = |\psi|^2$ initially, ρ will remain equal to $|\psi|^2$ at all times.

Vink[37] has shown how to formulate a dynamics for any always determinate observable by generalizing a proposal by Bell[38] for constructing stochastic Bohm-type trajectories for fermion number density regarded as an always determinate observable or 'beable' for quantum field theory. Vink considers an arbitrary complete set of commuting observables R^i $(i - l, 2, \ldots, I)$, with simultaneous eigenvectors $|r^1_{n^1}, r^2_{n^2}, \ldots, r^I_{n^I}\rangle$, where the $n^i = 1, 2, \ldots, N^i$ label the finite and discrete eigenvalues of R^i. Suppressing the index i, these are written as $|r_n\rangle$. (Equivalently, one can take the $|r_n\rangle$ to be the different eigenvectors of a maximal observable R of which each of the R^i is a function. Then supplying the dynamics for R automatically induces a dynamics on each R^i.) The time evolution of the state vector is given by the equation of motion:

$$i\hbar \frac{\mathrm{d}|\psi(t)\rangle}{\mathrm{d}t} = H|\psi(t)\rangle,$$

or

$$i\hbar \frac{\mathrm{d}\langle r_n|\psi\rangle}{\mathrm{d}t} = \langle r_n|H\psi\rangle = \sum_m \langle r_n|H|r_m\rangle\langle r_m|\psi\rangle$$

[37] J. C. Vink, 'Quantum Mechanics in Terms of Discrete Beables', *Physical Review A* 48 (1993), 1808–18.

[38] J. S. Bell, 'Beables for Quantum Field Theory', in *Speakable and Unspeakable in Quantum Mechanics, op. cit.*, pp. 173–80. See pp. 176–7.

in the R-representation. The imaginary part of this equation yields the continuity equation:

$$\frac{\mathrm{d}P_n}{\mathrm{d}t} = \frac{1}{\hbar}\sum_m J_{nm},$$

where the probability density P_n and the current matrix J_{nm} are defined by:

$$\begin{aligned} P_n(t) &= |\langle r_n|\psi(t)\rangle|^2, \\ J_{nm}(t) &= 2\mathrm{Im}\left(\langle\psi(t)|r_n\rangle\langle r_n|H|r_m\rangle\langle r_m|\psi(t)\rangle\right). \end{aligned}$$

For the non-maximal (degenerate) observables R^i, the probability density and current matrices are defined by summing over the remaining indices, e.g.

$$P^i_n = \sum_r |\langle r^i_{n^i}, r|\psi\rangle|^2,$$

where $r = r^j_{m^j}$ for $j \neq i$, and similarly for J^i_{nm}.

We want a stochastic dynamics for the discrete observable R consistent with the continuity equation. Suppose the jumps in R-values are governed by transition probabilities $T_{mn}\mathrm{d}t$, where $T_{mn}\mathrm{d}t$ denotes the probability of a jump from value r_n to value r_m in time $\mathrm{d}t$.

The transition matrix gives rise to time-dependent probability distributions of R-values. It follows that the rate of change of the probability density $P_n(t)$ for r_n must satisfy the master equation:

$$\frac{\mathrm{d}P_n(t)}{\mathrm{d}t} = \sum_m \left(T_{nm}P_m - T_{mn}P_n\right).$$

But, from the Schrödinger equation, $\frac{\mathrm{d}P_n(t)}{\mathrm{d}t}$ also satisfies the continuity equation

$$\frac{\mathrm{d}P_n(t)}{\mathrm{d}t} = \frac{1}{\hbar}\sum_m J_{nm},$$

so we require

$$\frac{J_{nm}}{\hbar} = T_{nm}P_m - T_{mn}P_n.$$

We want solutions for the transition matrix T, given P and J, with $T_{mn} \geq 0$. Since $J_{mn} = -J_{nm}$ (hence $J_{nn} = 0$), the above equation yields $n(n-1)/2$ equations for the n^2 elements of T. So there are many solutions. Bell's choice[39] was:

$$\begin{aligned} T_{nm} &= \frac{J_{nm}}{\hbar P_m}, \text{ if } J_{nm} \geq 0, \\ T_{nm} &= 0, \text{ if } J_{nm} \leq 0. \end{aligned}$$

For $n = m$, T_{nn} is fixed by the normalization $\sum_m T_{mn}\mathrm{d}t = 1$.

Vink shows that Bell's solution for the transition matrix leads to Bohm's theory in the continuum limit, when R is position in configuration space. For example, consider a single particle of mass M on a 1-dimensional lattice. Let $x = nd$, with $n = 1, 2, \ldots, N$ and d the lattice distance. Vink shows that to first order in d:

$$J_{nm} = \frac{\hbar}{Md}\left[S'(md)P_m\delta_{m,n-1} - S'(md)P_m\delta_{m,n+1}\right],$$

where S is the phase, and the derivative, F', of a function F on the lattice is defined as $F'(x) = \frac{F(x+d)-F(x)}{d}$. Thus Bell's solution yields:

$$\begin{aligned} T_{nm} &= \frac{S'(md)}{Md}\delta_{m,n-1}, \; S'(md) \geq 0, \\ T_{nm} &= \frac{S'(md)}{Md}\delta_{m,n+1}, \; S'(md) \leq 0. \end{aligned}$$

For positive $S'(md)$ the particle can jump from site m to site $m+1$ with probability $\frac{|S'(md)|\mathrm{d}t}{Md}$ and for negative $S'(md)$ the particle can jump from site m to $m-1$ with the same probability. So the nearest neighbour interactions in the Hamiltonian induce jumps only between neighbouring lattice sites for this transition matrix. Since each jump is over a distance d, the average displacement in a time interval $\mathrm{d}t$ is:

$$\mathrm{d}x = \frac{S'(x)\mathrm{d}t}{M},$$

i.e.,

$$\frac{\mathrm{d}x}{\mathrm{d}t} = \frac{S'(x)}{M}.$$

[39] In J. S. Bell, 'Beables for Quantum Field Theory', *op. cit.*

As $d \rightarrow 0$, $S' \rightarrow \partial_x S$, and so in the continuum limit:

$$\frac{\mathrm{d}x}{\mathrm{d}t} = \frac{\partial_x S}{M},$$

as for the continuous trajectories in Bohm's theory. Vink shows that the dispersion vanishes in the limit as $d \rightarrow 0$, and so the trajectories become smooth and identical to the trajectories in Bohm's theory as $d \rightarrow 0$. Other solutions for the transition matrix induce jumps between distant sites, so differentiable deterministic trajectories are not recovered even in the continuum limit. Nelson's stochastic dynamics[40] is characterized by one such solution. Bohmian dynamics, as the continuum limit of a stochastic dynamics, appears to be the unique deterministic dynamics for a 'no collapse' interpretation of quantum mechanics based on position in configuration space as an always determinate observable.

Vink himself proposes to take *all* observables as simultaneously determinate, with their determinate values all evolving independently in accordance with the above stochastic dynamics. These determinate values do not satisfy the Kochen and Specker functional relationship constraint in general, but it turns out that this constraint is satisfied for the measured values of simultaneously measured observables. This follows because a measurement, represented as an evolution of the quantum state to a linear superposition of product eigenstates of a pointer observable and measured observable, leads to an effective collapse of the state to one of the product eigenstates, as in the analysis of measurement in Bohm's theory discussed below. The determinate values of observables in eigenstates of these observables are just the corresponding eigenvalues, and the functional relationship constraint is satisfied for common eigenstates of a set of observables.[41]

There are two empirically equivalent proposals in the literature for handling observables other than position in configuration space

---

[40] E. Nelson, 'Derivation of the Schrödinger Equation from Newtonian Mechanics', *Physical Review* 150 (1966), 1079–85.

[41] For a discussion, see J. Bub, 'Interference, Noncommutativity, and Determinateness in Quantum Mechanics', *Topoi* 14 (1995), 39–43.

in Bohm's theory. Bohm, Schiller, and Tiomno,[42] and also Dewdney,[43] Holland and Vigier,[44] and Bohm and Hiley[45] treat spin as well as position in configuration space as an always determinate observable. So a quantum particle always has a determinate spin property as well as a determinate position in configuration space. The spin observable can take determinate values that are not eigenvalues of spin in states that are not eigenstates of spin, but the measured value of spin is always an eigenvalue of spin. This treatment of spin could, in principle, be extended to all observables.

The alternative way of handling observables other than position in configuration space, favoured by authors such as Bell,[46] Dürr, Goldstein and Zanghì,[47] Albert,[48] and Cushing,[49] conforms to an interpretation in our sense. The propositions in a Bohmian determinate sublattice are selected by the quantum state and position in configuration space as the only preferred determinate observable R. While position in configuration space is always determinate, other observables are sometimes determinate and sometimes indeterminate, depending on the quantum state. These other observables, with temporary determinate status inherited from R and the quantum state, can be associated with dispositions of the system. The observable spin, for example, is determinate in some quantum states and indeterminate in other quantum states. The possibility that the state can evolve to a form in which a spin component has

[42]D. Bohm, R. Schiller and J. Tiomno, 'A Causal Interpretation of the Pauli Equation', *Suppi. Nuovo Cimento* 1 (1955), 48–66.

[43]C. Dewdney, 'Constraints on Quantum Hidden-Variables and the Bohm Theory', *Journal of Physics A: Math. Gen.* 25 (1992), 3615–26.

[44]P. R. Holland and J. P. Vigier, 'The Quantum Potential and Signalling in the Einstein-Podolsky-Rosen Experiment', *Foundations of Physics* 18 (1988), 741–9.

[45]D. Bohm and B. J. Hiley, *The Undivided Universe: An Ontological Interpretation of Quantum Theory* (London: Routledge, 1993).

[46]J. S. Bell, 'Quantum Mechanics for Cosmologists'; and 'On the Impossible Pilot Wave', in *Speakable and Unspeakable in Quantum Mechanics, op. cit.*

[47]D. Dürr, S. Goldstein and N. Zanghì, 'Quantum Equilibrium and the Origin of Absolute Uncertainty', *Journal of Statistical Physics* 67 (1992), 843–907.

[48]D. Albert, *Quantum Mechanics and Experience* (Cambridge, MA: Harvard University Press, 1992).

[49]J. T. Cushing, *Quantum Mechanics: Historical Contingency and the Copenhagen Hegemony* (Chicago: University of Chicago Press, 1994).

a determinate value reflects a feature of the dynamics of the quantum state understood as a new kind of field in R-space, that this field evolves in a certain way in the presence of magnetic fields, i.e., it reflects a disposition of the system to undergo a certain kind of change of R-value under certain physical conditions. The different eigenvalues of the spin component mark the different possible changes in R associated with this evolution of the state. As we have seen, when the state takes a form in which a spin component has a determinate value (where this value is to be understood dispositionally), the probabilities assigned to the different eigenvalues of the spin component can be interpreted epistemically, as measures over a range of possible spin properties, one of which is actual. But these spin properties play no role in the dynamical evolution of the R-trajectories, which depends entirely on the initial value of R and the quantum state. From this point of view, the only real change in a Bohmian universe is the change in the quantum state and the change in R, and this suffices to account for all quantum phenomena. The propositions in the determinate sublattice for a given quantum state are the propositions that can be taken as determinately true or false consistently with R-propositions, and these will be associated with observables that we can *interpret* as measured via R as the pointer observable when the quantum state takes an appropriate form correlating values of R with values of these observables. A 'measurement' in this sense reveals dispositions of the system, grounded in the value of R and the quantum state, not pre-existing real occurrent values of any measured observable. So there is no interpretative advantage in taking any observables other than R as always determinate.

To illustrate, we consider the measurement of spin-related observables on Bohm's theory (following an analysis by Pagonis and Clifton).[50]

Let $S_x^2, S_{x'}^2, S_y^2, S_{y'}^2, S_z^2$ represent the squared components of spin in the x, x', y, y' and z directions of a spin-1 particle, respectively, where x, y, z and x', y', z form two orthogonal triples of directions

[50]C. Pagonis and R. Clifton, 'Unremarkable Contextualism: Dispositions in the Bohm Theory', *Foundations of Physics* 25 (1995), 281–96.

with the z-direction in common. Each of these observables has eigenvalues 1 and 0 (taking units in which $\hbar = 1$ and a spin component has eigenvalues $-1, 0$ and $+1$), corresponding respectively to a plane and a ray in $\mathcal{H}_3$. The three 0-eigenrays of S_x^2, S_y^2, S_z^2 form an orthogonal triple in $\mathcal{H}_3$, and the three 0-eigenrays of $S_{x'}^2, S_{y'}^2, S_z$ form another orthogonal triple in $\mathcal{H}_3$, with the 0-eigenray of S_z in common. Define the observables H and H' as:

$$\begin{aligned} H &= S_x^2 - S_y^2 \\ H' &= S_{x'}^2 - S_{y'}^2. \end{aligned}$$

The observables H and H' are maximal, with three eigenvalues, $-1, 0$ and $+1$, and incompatible (i.e., the corresponding operators do not commute). The eigenvalues $-1, 0$ and $+1$ of H correspond to the orthogonal triple of eigenrays defined by the 0-eigenrays of S_x^2, S_y^2 and S_z^2 respectively. The eigenvalues $-1, 0$ and $+1$ of H' correspond to the orthogonal triple of eigenrays defined by the 0-eigenrays of $S_{x'}^2, S_{y'}^2$ and S_z^2 respectively.

The non-maximal observable S_z^2 can be represented as a function of H and also as a function of H':

$$S_z^2 = H^2 = H'^2.$$

So S_z^2 can be measured via a measurement of H or of H'. The observable H can be measured, in principle, by passing the particle through a suitable inhomogeneous electromagnetic field, which functions much like a Stern-Gerlach magnet for the measurement of spin (see Swift and Wright[51]). The interaction of the particle with the field will be governed by a Hamiltonian of the form:

$$H_{\text{int}} = g i^{-1} \frac{\partial}{\partial q} H,$$

where g is a positive coupling constant that is nonzero only during the interaction, and q is a component of the particle's position. We

[51] A. R. Swift and R. Wright, 'Generalized Stern-Gerlach Experiments and the Observability of Arbitrary Spin Operators', *Journal of Mathematical Physics* 21 (1980), 77–82.

make the usual assumption that the measurement is impulsive, so that the Schrödinger equation reduces to

$$\frac{\partial \psi}{\partial t} = -iH_{\text{int}}\psi$$

during the interaction, and has the solution

$$\psi(q,t) = \exp\left(-g\frac{\partial}{\partial q}Ht\right)\psi(q,0).$$

Taking the initial quantum state as

$$\psi(q,0) = \phi(q)\sum_j c_j|H = j\rangle,$$

where $\phi(q)$ is a narrow wave packet symmetric about $q = 0$, and $|H = j\rangle$ is an eigenstate of H, the state at any time t during the interaction is:

$$\psi(q,t) = \sum_j c_j\phi(q - gjt)|H = j\rangle.$$

With a suitable choice for the coupling constant g, after a time $t \geq T$ (at the end of the interaction), gt will be significantly larger than the width of the packet $\phi(q)$, so that the overlap between adjacent wave packets $\phi(q - gjt)$ is negligible. As a result of the interaction, the particle's H-value will therefore become correlated with the particle's q-position. The Bohmian particle trajectories during the interaction are governed by the equation of motion for q:

$$\frac{\mathrm{d}q}{\mathrm{d}t} = \frac{J_q}{\rho},$$

where $\rho(q) = |\psi(q,t)|^2$ is the probability density and

$$J_q = \psi^*(q,t)gH\psi(q,t)$$

is the q-component of the probability current. So

$$\frac{\mathrm{d}q}{\mathrm{d}t} = \frac{g\sum_j j|c_j|^2|\phi(q - gjt)|^2}{\sum_j |c_j|^2|\phi(q - gjt)|^2}.$$

This equation can be solved to yield different trajectories for the different initial positions of the particle in the initial wave packet. By the continuity equation, $\frac{\partial \rho}{\partial t} + \frac{\partial J_q}{\partial q} = 0$, the particle trajectories at time $t \geq T$ will be distributed over the positions $q = gjt$ with probabilities $|c_j|^2$, for $j = -1, 0, +1$. So q acts as a measurement pointer for H-values. A displacement of the particle from its initial position in the narrow wave packet $\phi(q)$ centred about $q = 0$ by an amount gjt as the particle leaves the field can be understood as a measurement of H with the outcome $H = j$. Now, the same analysis for H' instead of H will yield an equation of motion for the particle trajectories in terms of a position coordinate q':

$$\frac{\mathrm{d}q'}{\mathrm{d}t} = \frac{Jq'}{\rho} = \frac{g \sum_j j|c'_j|^2 |\phi(q' - gjt)|^2}{\sum_j |c'_j|^2 |\phi(q' - gjt)|^2},$$

where the initial state of the particle is

$$\psi(q', 0) = \phi(q') \sum_j c'_j |H' = j\rangle,$$

and q' is the position coordinate of the particle that becomes correlated with the value of H' during the interaction (in which the particle is passed through an electromagnetic field oriented in a direction suitable for a measurement of $H' = S^2_{x'} - S^2_{y'}$ rather than $H = S^2_x - S^2_y$). We assume that the form of the initial position wave function is the same for q' as for q, and that the strength, g, and duration, T, of the interaction is the same for H' as for H.

Distinct possible particle trajectories cannot cross along the q-axis in an H-measurement, or the q'-axis in an H'-measurement, because the equation of motion is deterministic and gives velocity only as a function of position. This means that in an H-measurement, trajectories that end up, after time $t \geq T$, at one of the three possible final q-positions, $-gt$, 0 or gt, in that order, begin in one of three possible q-regions in the initial wave packet $\phi(q)$, ordered from negative to positive values of q. The relative sizes of the q-regions in an H-measurement will differ from the relative sizes of the q'-regions in an H'-measurement, because fractions $|c_{-1}|^2, |c_0|^2, |c_{+1}|^2$ of the initial q-positions (in an ensemble distributed according to the distribution function $|\phi(q)|^2$) end up at the positions $q = -gt, q = 0, q = gt$ respectively (corresponding to the results $H = -1, 0, +1$ in an H'-measurement), while *different* fractions $|c'_{-1}|^2, |c'_0|^2, |c'_{+1}|^2$ of the initial positions (in an ensemble distributed according to the same distribution function $|\phi(q')|^2$) end up at the positions $q' = -gt, q' = 0, q' = gt$, respectively (corresponding to the results $H' = -1, 0, +1$ in an H'-measurement). Note that $|c_i|^2 \neq |c'_i|^2$, unless $H = H'$. It follows that a given initial position of the particle will be affected differently by different interaction Hamiltonians, and so a measurement of S_z^2 via an H-measurement need not yield the same result as a measurement of S_z^2 via an H'-measurement, for the same initial position and quantum state of the particle.

In this sense, Bohm's theory violates the functional relationship constraint. If values are assigned to all observables of a spin-1 particle as the values that would be obtained on measurement (where a measurement of an observable is understood as an evolution of the quantum state of the particle to a form that correlates position values with values of the observable in the above sense) then, for some initial positions of the particle, the value assigned to S_z^2 will *not* be equal to the value assigned to H^2, and also equal to the value assigned to H'^2. For example, suppose the initial position of the particle is such that a measurement of H would yield the value $+1$, while a measurement of H' would yield the value 0. If these values are assigned to H and H', then the functional relationship constraint requires that $S_z^2 = 1$ and also that $S_z^2 = 0$, since $S_z^2 = H^2 = H'^2$. Of course, this contradiction does not show any inconsistency in Bohm's interpretation of quantum mechanics. Rather, it shows that observables like H, H' and S_z — observables associated with dispositions of the system — are 'contextual', in the sense that no determinate values can be attributed to them, except in the context of a specific measurement process, understood as the dynamical evolution of the quantum state to a certain form.

In a similar sense, Bohm's theory is non-local. Consider two spin-1/2 particles, S_1 and S_2, in the singlet spin state at time $t = 0$:

$$\psi(q_1, q_2, 0) = \phi(q_1)\phi(q_2)\left(\frac{1}{\sqrt{2}}|+\rangle_1|-\rangle_2 - \frac{1}{\sqrt{2}}|-\rangle_1|+\rangle_2\right)$$

as in Bohm's version of the Einstein-Podolsky-Rosen experiment.[52] A measurement of spin in the z-direction on S_1 (via a Stern-Gerlach interaction at S_1 that correlates the position of S_1 to the z-spin of S_1) induces an evolution of the quantum state of the 2-particle system to a form:

$$\psi(q_1, q_2, 0) = \phi(q_2)\frac{1}{\sqrt{2}}\left[\phi(q_1 + gt)|+\rangle_1|-\rangle_2 - \phi(q_1 - gt)|-\rangle_1|+\rangle_2\right]$$

where the wave packets $\phi(q_1 - gt)$ and $\phi(q_1 + gt)$ for the relevant position coordinates of S_1 are separated with negligible overlap. The q_1 position coordinate of the particle S_1 is effectively associated with just one of these wave packets. This means that the position of the 2-particle system $S_1 + S_2$ in configuration space, i.e., the q_1, q_2 position (ignoring other position coordinates not affected by the Stern-Gerlach interaction at S_1), can be associated with just one of the wave packets $\phi(q_2)\phi(q_1 + gt)$ or $\phi(q_2)\phi(q_1 - gt)$. So in a subsequent measurement of the spin of S_2 via a Stern-Gerlach interaction at S_2, the evolution of q_2 will depend on the configuration space position of $S_1 + S_2$, which is effectively correlated either with the spin state $|+\rangle_1|-\rangle_2$ or with the spin state $|-\rangle_1|+\rangle_2$. It follows that $S_1 + S_2$ will behave in the S_2-interaction as if its quantum state has effectively collapsed to $\phi(q_2)\phi(q_1 + gt)|+\rangle_1|-\rangle_2$ or $\phi(q_2)\phi(q_1 - gt)|-\rangle_1|+\rangle_2$ and so q_2 will evolve to a position corresponding to a z-spin value for S_2 opposite to the z-spin value correlated with the final position of q_1. Evidently, then, the outcome of a spin measurement at S_2 will depend on the type of spin measurement at S_1, i.e. on the orientation of the Stern-Gerlach magnet at S_1. The quantum state of $S_1 + S_2$ will evolve differently for a measurement of x-spin at S_1 rather than z-spin, say, and so will affect the evolution of q_2 differently in a subsequent spin measurement at S_2, as in the measurement of H and H' on the spin-1 system.

[52] D. Bohm, *Quantum Theory* (Englewood Cliffs, NJ: Prentice-Hall, 1951).

There is another sense, though, in which both the functional relationship constraint and the locality condition are satisfied by Bohm's causal interpretation. Consider the spin-1 system again, and the quantum state:

$$\psi(q,t) = \sum_j c_j \phi(q - gjt)|H = j\rangle$$

at times $t \geq T$ after the interaction. If q is a preferred determinate observable then, *in this state*, we can take H as determinate and also S_z^2 as determinate for the system, i.e., H and S_z^2 inherit determinate status from q and the quantum state, to the extent that the measurement can be regarded as ideal, and the wave functions $\phi(q - gjt)$ approximate orthogonal eigenfunctions. Of course, the wave function tails will always overlap to some small extent, so even if we discretize q to three relevant values (representing three non-overlapping ranges of q-values associated with the three peaks of the wave functions), the projections of the state $e(= \psi(q,t))$ onto the three corresponding eigenspaces will not yield the rays e_{q_j} spanned by the vectors $\phi(q - gjt)|H = j\rangle$, for $j = -1, 0, +1$, but only rays e'_{q_j} that are arbitrarily close to the rays e_{q_j} (for sufficiently large t). Here we simply treat the wave functions $\phi(q - gjt)$ as non-overlapping and effectively eigenfunctions of q, as Bohm does in his analysis of measurement in quantum mechanics.[53] Then H-propositions and S_z^2-propositions (more precisely, propositions arbitrarily 'close to' H-propositions and S_z^2-propositions) belong to the lattice $\mathcal{L}_{e_{r_1}e_{r_2}\dots e_{r_k}}$. By the theorem of Section 2, $\mathcal{L}_{e_{q_0}e_{q_{-1}}e_{q_{+1}}}$ contains the maximal set of propositions that we can take as determinately true or false for the state $\psi(q,t)$, given that q has a determinate value — the maximal set of propositions associated with observables that are measured via the pointer q in the Bohmian sense. For the state $\psi(q,t)$, observables like H and S_z^2 associated with $\mathcal{L}_{e_{q_0}e_{q_{-1}}e_{q_{+1}}}$ satisfy the functional relationship constraint. The value assigned S_z^2 is the square of the value assigned to H, for any position of the pointer q, and similarly for

[53]See Chapter 6, 'Measurement as a Special Case of Quantum Process', in D. Bohm and B. J. Hiley, *The Undivided Universe: An Ontological Interpretation of Quantum Theory, op. cit.*, pp. 97–133.

other observables associated with $\mathcal{L}_{e_{q0}e_{q-1}e_{q+1}}$ (H'-propositions do not, of course, belong to $\mathcal{L}_{e_{q0}e_{q-1}e_{q+1}}$). So, if we consider the set of observables that are measured in the Bohmian sense, by the evolution of the quantum state to a specific form that correlates the preferred determinate observable to values of these other observables, then for these observables the functional relationship constraint is satisfied. It follows that the locality condition is satisfied in *this* sense, for composite systems associated with tensor product Hilbert spaces, because locality is a special case of the functional relationship constraint.

Bohm's interpretation is a 'beable' interpretation of quantum mechanics in the spirit of Bell's notion. As Bell put it:[54]

> It would be foolish to expect that the next basic development in theoretical physics will yield an accurate and final theory. But it is interesting to speculate on the possibility that a future theory will not be *intrinsically* ambiguous and approximate. Such a theory could not be fundamentally about 'measurements', for that would again imply incompleteness of the system and unanalyzed interventions from outside. Rather it should again become possible to say of a system not that such and such may be *observed* to be so but that such and such *be so*. The theory would not be about '*observ*ables' but about '*be*ables'. [...]
>
> Many people must have thought along the following lines. Could one not just promote *some* of the 'observables' of the present quantum theory to the status of beables? The beables would then be represented by linear operators in the state space. The values which they are allowed to *be* would be the eigenvalues of those operators. For the general state the probability of a beable *being* a particular value would be calculated just as was formerly calculated the probability of *observing* that value.

[54] J. S. Bell, *Speakable and Unspeakable in Quantum Mechanics, op. cit.*, p. 41.

Bell's notion of a 'beable', referring to an objective property of a physical system, is equivalent to Einstein's 'element of reality', the terminology employed in the Einstein-Podolsky-Rosen argument.[55] Bell writes:[56]

> In particular we will exclude the notion of 'observable' in favour of that of 'beable'. The beables of the theory are those elements which might correspond to elements of reality, to things which exist. Their existence does not depend on 'observation'. Indeed observation and observers must be made out of beables.

The 'causal' interpretation, as Bohm and Hiley characterize it,[57] is a reformulation of quantum mechanics in terms of beables:

> This theory is formulated basically in terms of what Bell has called 'beables' rather than of 'observables.' These beables are assumed to have a reality that is independent of being observed or known in any other way. The observables therefore do not have a fundamental significance in our theory but rather are treated as statistical functions of the beables that are involved in what is currently called a measurement.

The interpretation of quantum mechanics in terms of beables is motivated by certain realist principles formulated by Einstein, a separability principle and a locality principle. These principles are implicit in the Einstein-Podolsky-Rosen argument (which appears to have been largely written by Podolsky) and explicit in various reformulations of the argument by Einstein. For example, referring to the Einstein-Podolsky-Rosen argument, Einstein writes as follows:[58]

---

[55] A. Einstein, B. Podolsky and N. Rosen, 'Can Quantum-Mechanical Description of Physical Reality be Considered Complete?', *Physical Review* 47 (1935), 777–80.

[56] J. S. Bell, *Speakable and Unspeakable in Quantum Mechanics, op. cit.*, p. 174.

[57] D. Bohm and B. J. Hiley, *The Undivided Universe: An Ontological Interpretation of Quantum Theory, op. cit.*, p. 41.

[58] A. Einstein, in *Albert Einstein: Philosopher-Scientist*, P. A. Schilpp (ed.) (Chicago: Open Court, 1991), 3rd ed., pp. 681–2.

> Of the 'orthodox' quantum theoreticians whose position I know, Niels Bohr's seems to me to come nearest to doing justice to the problem. Translated into my own way of putting it, he argues as follows:
>
> If the partial systems A and B form a total system which is described by its ψ-function $\psi|(AB)$, there is no reason why any mutually independent existence (state of reality) should be ascribed to the partial systems A and B viewed separately, *not even if the partial systems are spatially separated from each other at the particular time under consideration*. The assertion that, in this latter case, the real situation of B could not be (directly) influenced by any measurement taken on A is therefore, within the framework of quantum theory, unfounded and (as the paradox shows) unacceptable.
>
> By this way of looking at the matter it becomes evident that the paradox forces us to relinquish one of the following two assertions:
>
> 1. The description by means of the ψ-function is complete.
> 2. The real states of spatially separated objects are independent of each other.
>
> On the other hand, it is possible to adhere to (2), if one regards the ψ-function as the description of a (statistical) ensemble of systems (and therefore relinquishes (1)). However, this view blasts the framework of the 'orthodox quantum theory.'

Einstein's 'real state' or 'real situation' of a physical system (elsewhere[59] he speaks of the 'being-thus' of a system) corresponds in our formulation to a 2-valued homomorphism on the determinate

[59] A. Einstein, *Dialectica* 2 (1948), 320–24. Translated as 'Quantum Mechanics and Reality', in M. Born (ed.), *The Born-Einstein Letters* (London: Walker & Co., 1971), p. 192.

sublattice defined by the quantum state of the system and the preferred determinate observable. Each 2-valued homomorphism selects a particular determinate value for the preferred observable, and also particular determinate values for other observables associated with the determinate sublattice, i.e., each 2-valued homomorphism selects a set of preferred and derived determinate properties for the system. Since a determinate sublattice is uniquely defined by the quantum state and a preferred determinate observable, and the 2-valued homomorphisms on a determinate sublattice are in 1–1 correspondence with the values of the preferred observable, Einstein's 'real state' can be characterized equivalently as the specification of the quantum state of the system and the value of the preferred observable, which constitute the only genuine beables of the system (other observables associated with a determinate sublattice represent dispositions in the sense discussed above). The separability and locality principles can therefore be formulated as follows:

Separability: The determinate properties ('real states') of spatially separated systems are independent of each other.

Locality: If two systems are spatially separated, then the determinate properties ('real state') of one system cannot be directly influenced by any measurement on the other system.

As Einstein presents it, the issue of the completeness of quantum mechanics — the heart of the dispute between Einstein and Bohr — concerns the separability principle. What our determinate sublattices preserve is only a weak separability principle: the determinate properties ('real states') of spatially separated systems are independent of each other, i.e., each system is independently characterized by its own determinate sublattice, if and only if the quantum state of the composite system is not an 'entangled' state (linear superposition of product states) arising from past interaction between the systems.

Several authors, notably Fine[60] and Jammer,[61] have suggested

[60] A. Fine, *The Shaky Game: Einstein, Realism, and the Quantum Theory* (Chicago: University of Chicago Press, 1986).

[61] M. Jammer, *The Philosophy of Quantum Mechanics* (New York: Wiley, 1974).

that Einstein had something other than hidden variables in mind when he argued that quantum mechanics is incomplete. Einstein's negative reaction to Bohm's hidden variables theory in correspondence with Renninger, Born and others is often cited in support of this view. (In a letter to Born dated 12 May 1952, Einstein dismissed the theory as 'too cheap for me.'[62]) However, we agree with Bell's endorsement[63] of Shimony's characterization of Einstein as 'the most profound advocate of hidden variables', in the sense of a 'beable' interpretation of quantum mechanics. Einstein's lack of enthusiasm for Bohm's theory should not be construed as a rejection of the hidden variables program *per se*, but only a particular way of developing this program. What our theorem shows is that the possible 'completions' of quantum mechanics in Einstein's sense can be uniquely characterized and reduced to the choice of a fixed preferred determinate observable, i.e., a fixed beable. So, in fact, the option of preserving separability in the strong sense is excluded in a beable interpretation.

We have noted that the determinate sublattices satisfy the Kochen and Specker functional relationship constraint in what might be termed an 'ontological' sense, i.e., the values of the determinate observables associated with a determinate sublattice, as assigned by all 2-valued homomorphisms on the lattice, preserve the functional relationships satisfied by these observables, and hence preserve locality as a special case of the functional relationship constraint. But in what might be termed a 'dynamical' sense, the interpretations associated with determinate sublattices are nonlocal. If we understand a measurement as an interaction that induces an evolution of the quantum state to a form that correlates values of a preferred determinate observable with values of other observables, in virtue of which these other observables achieve derived determinate status in the time-evolved state, then a measurement on a system S_1 can make determinate a dispositional property of a system S_2, spatially separated from S_1, that was not determinate before the

[62] A. Einstein, in M. Born (ed.), *The Born-Einstein Letters* (London: Walker & Co., 1971), p. 192.

[63] J. S. Bell, *Speakable and Unspeakable in Quantum Mechanics, op. cit.*, p. 89.

measurement. And that will translate into a different value for the preferred observable when the disposition is actualized in a measurement. In this sense, Einstein's locality principle is violated. The 'real states' of one system can be directly influenced by a measurement on another spatially separated system.

3.5 Bohr's complementarity interpretation

While beable interpretations of quantum mechanics like Bohm's interpretation take a fixed preferred observable as determinate once and for all, on Bohr's complementarity interpretation an observable can be said to have a determinate value only in the context of a specific, classically describable experimental arrangement suitable for measuring the observable. For Bohr, a quantum 'phenomenon' is an individual process that occurs under conditions defined by a specific experimental arrangement. The experimental arrangements suitable for locating an atomic object in space and time, and for a determination of momentum-energy values, are mutually exclusive. We can choose to investigate either of these 'complementary' phenomena at the expense of the other, so there is no unique description of the object in terms of determinate properties.

Summing up a discussion on causality and complementarity, Bohr writes:[64]

> Recapitulating, the impossibility of subdividing the individual quantum effects and of separating a behaviour of the objects from their interaction with the measuring instruments serving to define the conditions under which the phenomena appear implies an ambiguity in assigning conventional attributes to atomic objects which calls for a reconsideration of our attitude towards the problem of physical explanation. In this novel situation, even the old question of an ultimate determinacy of natural phenomena has lost its conceptual basis, and it is against this background that the viewpoint of comple-

[64]N. Bohr, *Dialectica* 2 (1948), 312–19.

> mentarity presents itself as a rational generalization of the very ideal of causality.

Pauli characterizes Bohr's position this way:[65]

> While the means of observation (experimental arrangements and apparatus, records such as spots on photographic plates) have still to be described in the usual 'common language supplemented with the terminology of classical physics', the atomic 'objects' used in the theoretical interpretation of the 'phenomena' cannot any longer be described 'in a unique way by conventional physical attributes'. Those 'ambiguous' objects used in the description of nature have an obviously symbolic character.

We can understand the complementarity interpretation as the proposal to take the classically describable experimental arrangement (suitable for either a space-time or a momentum-energy determination) as defining the preferred determinate observable in what Bohr calls a quantum 'phenomenon'. So the preferred determinate observable is not fixed for a quantum system, but is defined by the classically described 'means of observation'. On this view, the determinate sublattice of a quantum system depends partly on what and how we choose to measure, not on objective features of the system itself. To echo Pauli, the properties we attribute to a quantum object in a measurement are 'ambiguous', or merely 'symbolic'. The complementarity interpretation, unlike a beable interpretation, which selects a fixed preferred determinate observable, is not a realist interpretation.[66]

It is instructive to consider the application of the complementarity interpretation to the Einstein-Podolsky-Rosen experiment. Re-

[65] W. Pauli, *Dialectica* 2 (1948), 307–11.

[66] It could also be argued that the complementarity interpretation does not really solve the measurement problem, unlike the other 'no collapse' interpretations we have been discussing (excluding the orthodox Dirac-von Neumann interpretation), since Bohr gives no measurement-independent prescription for what R should be taken to be. We concentrate on recovering other aspects of this interpretation, once Bohr's choice of R is granted.

ferring to this experiment, Bohr remarks[67] that the difference between the position coordinates of two particles, $Q_1 - Q_2$, and the sum of their corresponding momentum components, $P_1 + P_2$, are compatible observables, i.e., they are represented by commuting operators. (This follows immediately from the commutation relation for position and momentum, $QP - PQ = i\hbar I$.) So we can prepare a quantum state in which both these observables have determinate values. It follows that a measurement of either Q_1 or P_1 on the first particle will allow the prediction of the outcome of a subsequent measurement of either Q_2 or P_2, respectively, on the second particle. Or putting it another way, the assignment of a determinate value to Q_1 or P_1 will fix a determinate value for Q_2 or P_2, respectively. But now it would appear, if the two particles are separated and no longer interacting, that the second particle must have both a determinate Q_2-value and a determinate P_2-value prior to the Q_1 or P_1 measurements, which contradicts the assumption that the quantum state is a complete description of the system (since no quantum state assigns determinate values to two incompatible observables). What this argument fails to note, says Bohr, is that the experimental arrangements that allow accurate measurements of Q_1 and P_1 are mutually exclusive, so the predictions concerning Q_2 and P_2 refer to complementary phenomena.

It is unclear from Bohr's discussion how the attribution of a determinate value to an observable Q_1 (or P_1) of a system via a measurement on that system can make determinate an observable Q_2 (or P_2) of a second system spatially separated from and not interacting with the first system — how, that is, Q_2 (or P_2) can inherit determinate status from the selection of Q_1 (or P_1) as a preferred determinate observable. This puzzle is resolved if we take Bohr as proposing that the determinate sublattice for the 2-particle system is the determinate sublattice defined by a simultaneous eigenstate of the observables $Q_1 - Q_2$ and $P_1 + P_2$, and the observable Q_1 (or P_1) as the preferred determinate observable (on the basis of the particular experimental arrangement introduced for the measurement on the first particle). For then the determinate sublattice will contain

[67]See Bohr, *op. cit.*, p. 316.

propositions associated with the observable Q_2 (or P_2) of the second particle, in virtue of the form of the quantum state of the composite system as a strictly correlated linear superposition of product states. The nonlocality here is analogous to the nonlocality of dispositions discussed above for Bohm's causal interpretation.

Thus, the framework for interpretations of quantum mechanics presented here accommodates Bohr's complementarity interpretation as well as Einstein's realism, in the beable sense (stripped of separability/locality requirements, as in Bohm's interpretation). The opposing positions appear as two quite different proposals for selecting the preferred determinate observable — either fixed, once and for all, as the realist would require, or settled pragmatically by what we choose to observe.

Chapter 3

Revised proof for the uniqueness theorem for 'no collapse' interpretations of quantum mechanics

with Jeffrey Bub and Sheldon Goldstein

Bub and Clifton (1996) proved a uniqueness theorem for 'no collapse' interpretations of quantum mechanics. The proof was repeated, with minor modifications, in Bub (1997). The original proof involved a 'weak separability' assumption (introduced to avoid a dimensionality constraint) that required several preliminary definitions and considerably complicated the formulation of the theorem. Sheldon Goldstein has pointed out that the proof goes through without this assumption.

The question at issue is this: Consider an arbitrary 'pure' quantum state represented by a ray e on a Hilbert space $\mathcal{H}$ and the Boolean algebra or lattice, $\mathcal{B}(R)$, generated by the eigenspaces of a single observable R on $\mathcal{H}$. The probabilities defined by e for the ranges of values of R can be represented by a probability measure over the 2-valued homomorphisms on $\mathcal{B}(R)$. What is the maximal lattice extension $\mathcal{D}(e,R)$ of $\mathcal{B}(R)$, generated by eigenspaces of observables other than R, on which there exist sufficiently many 2-valued homomorphisms so that we can represent the probabilities defined by e, for ranges of values of R and these additional observables, in terms of a measure over the 2-valued homomorphisms on $\mathcal{D}(e,R)$? The theorem provides an answer to this question on the assumption that $\mathcal{D}(e,R)$ is invariant under automorphisms of the lattice $\mathcal{L}(\mathcal{H})$ of all subspaces of Hilbert space that preserve the state e and the 'preferred observable' R, without the further assumption of 'weak separability'.

We sketch a revised proof of the theorem here.

Theorem: Consider a quantum system S in a (pure) quantum state represented by a ray e in an n-dimensional Hilbert space $\mathcal{H}$ ($n < \infty$), and an observable R with $m \leq n$ distinct eigenspaces r_i of $\mathcal{H}$. Let $e_{r_i} = (e \vee r_i^{\perp}) \wedge r_i, i = 1, 2, \ldots, k \leq m$, denote the nonzero projections of the ray e onto the eigenspaces r_i. Then $\mathcal{L}_{e_{r_1}e_{r_2}\ldots e_{r_k}} = \{p : e_{r_i} \leq p$ or $e_{r_i} \leq p^{\perp}$, for all $i = 1, \ldots, k\}$ is the unique maximal sublattice $\mathcal{D}(e,R)$ of $\mathcal{L}(\mathcal{H})$ satisfying the following three conditions:

1. *Truth and probability (TP):* $\mathcal{D}(e,R)$ is an ortholattice admitting sufficiently many 2-valued homomorphisms, $h : \mathcal{D}(e,R) \rightarrow \{0,1\}$, to recover all the (single and joint) probabilities assigned by the state e to mutually compatible sets of elements $\{p_i\}_{i \in I}, p_i \in \mathcal{D}(e,R)$, as measures on a Kolmogorov probability space $(X, \mathcal{F}, \mu)$, where X is the set of 2-valued homomorphisms on $\mathcal{D}(e,R)$, $\mathcal{F}$ is a field of subsets of X, and
$$\mu(\{h : h(p_i) = 1, \text{ for all } i \in I\}) = tr(e \prod_{i \in I} p_i).$$

2. *R-preferred (R-PREF):* the eigenspaces r_i of R belong to $\mathcal{D}(e,R)$.

3. *e, R-definability (DEF):* for any $e \in \mathcal{H}$ and observable R of S defined on $\mathcal{H}$, $\mathcal{D}(e,R)$ is invariant under lattice homomorphisms that preserve e and R.

Proof: The strategy of the proof is to show, firstly, that if $p \in \mathcal{D}(e,R)$, then for any $e_{r_i}, i = 1, \ldots, k$, either $e_{r_i} \leq p$ or $e_{r_i} \leq p^{\perp}$, and secondly, that the sublattice $\mathcal{L}_{e_{r_1}e_{r_2}\ldots e_{r_k}} = \{p : e_{r_i} \leq p \text{ or } e_{r_i} \leq p^{\perp}, i = 1, \ldots, k\}$ satisfies the conditions of the theorem. Maximality then requires that $\mathcal{D}(e, R) = \mathcal{L}_{e_{r_1}e_{r_2}\ldots e_{r_k}}$.

Consider an arbitrary subspace p of $\mathcal{H}$, and suppose $p \in \mathcal{D}(e, R)$. Clearly, if r_i is included in p or in $p^{\perp}$, for any i, then $e_{r_i} \leq p$ or $e_{r_i} \leq p^{\perp}$.

Suppose this is not the case. That is, suppose some r_i is neither included in p nor in $p^{\perp}$. Then either the intersection of r_i with p or the intersection of r_i with $p^{\perp}$ is nonzero. (For, by the conditions TP and R-PREF, there must be a 2-valued homomorphism mapping r_i onto 1, and this homomorphism would have to map both p and $p^{\perp}$ onto 0 if $r_i \wedge p = r_i \wedge p^{\perp} = 0$, which contradicts TP.) Assume that the intersection of r_i with p is nonzero (a similar argument applies if we assume that the intersection of r_i with $p^{\perp}$ is nonzero). Call this intersection b.

If p belongs to $\mathcal{D}(e, R)$, then $b = r_i \wedge p$ also belongs to $\mathcal{D}(e, R)$ (by the lattice closure assumption of TP, since r_i belongs to $\mathcal{D}(e, R)$ by R-PREF). Now, either b is orthogonal to e_{r_i} in r_i, or b is skew to e_{r_i} in r_i. We shall show that in either case e_{r_i} must belong to $\mathcal{D}(e, R)$. It follows that $e_{r_i} \leq p$ or $e_{r_i} \leq p^{\perp}$, because if $e_{r_i} \not\leq p$ and $e_{r_i} \not\leq p^{\perp}$, then — since e_{r_i} is a ray — e_{r_i} must be skew to both p and $p^{\perp}$; that is, $e_{r_i} \wedge p = 0$ and $e_{r_i} \wedge p^{\perp} = 0$. But this contradicts TP, because there must be a 2-valued homomorphism mapping e_{r_i} onto 1, and this homomorphism would have to map both p and $p^{\perp}$ onto 0.

So if r_i is included in p or in $p^{\perp}$ then $e_{r_i} \leq p$ or $e_{r_i} \leq p^{\perp}$, and if r_i is not included in p or in $p^{\perp}$, then e_{r_i} must belong to $\mathcal{D}(e, R)$, from which it follows that $e_{r_i} \leq p$ or $e_{r_i} \leq p^{\perp}$.

To see that e_{r_i} must belong to $\mathcal{D}(e, R)$ if (i) $b = r_i \wedge p$ is orthogonal to e_{r_i} in r_i, or (ii) b is skew to e_{r_i} in r_i, consider each of these cases in turn.

(i) Suppose b is orthogonal to e_{r_i} in r_i; that is, $b \leq e'_{r_i}$ (where the $'$ here denotes the orthogonal complement in r_i). If r_i is 1-dimensional, $e_{r_i} = r_i$ and it follows immediately from R-PREF that $e_{r_i} \in \mathcal{D}(e, R)$. If r_i is 2-dimensional, $b = e'_{r_i}$ and $e_{r_i} = r \wedge b^{\perp}$, so $e_{r_i} \in \mathcal{D}(e, R)$ by lattice closure (that is, by TP), because both r_i and $b^{\perp}$ belong to $\mathcal{D}(e, R)$. If r_i is more than 2-dimensional, consider all lattice automorphisms U that are rotations about e_{r_i} in r_i and the identity in $r_i^{\perp}$. Such rotations preserve e and R, because they preserve the eigenspaces of R and the projections of e onto the eigenspaces of R. It follows that $U(b) \in \mathcal{D}(e, R)$, by DEF. There are clearly sufficiently many rotations to generate a set of elements $U(b)$ whose span is e'_{r_i}. So $e_{r_i} \in \mathcal{D}(e, R)$, and hence $e_{r_i} \in \mathcal{D}(e, R)$ by lattice closure.

(ii) Suppose b is skew to e_{r_i} in r_i. Consider first the case that the eigenspace r_i is more than 2-dimensional. In that case, we may suppose that b is not a ray (or else we apply the following argument to b'). The subspace b can therefore be represented as the span of a subspace c orthogonal to e_{r_i} in r_i, and the ray d, the projection of the ray e_{r_i} onto b. Consider a lattice automorphism U that is a reflection through the hyperplane $e_{r_i} \vee c$ in r_i and the identity in $r_i^{\perp}$. (So U preserves e_{r_i} and c, but not d.) As in (i) above, $U(b) \in \mathcal{D}(e, R)$ because U preserves e and R, and $b \wedge U(b) = c$, so $c \in \mathcal{D}(e, R)$. We can now consider rotations of c about e_{r_i} in r_i as in (i) to show that $e_{r_i}{}' \in \mathcal{D}(e, R)$, and hence that $e_{r_i} \in \mathcal{D}(e, R)$.

The argument fails if r_i is 2-dimensional. If b is skew to e_{r_i} in r_i and r_i is 2-dimensional, then b must be a ray in r_i; that is, c is the null subspace and $d = b$. The automorphism U reduces to a reflection through e_{r_i} in r_i. Suppose b is not at a 45° angle to e_{r_i}. The four rays $b, b', U(b), U(b)'$ would all have to belong to $\mathcal{D}(e, R)$ if b belongs to $\mathcal{D}(e, R)$, but this contradicts TP. A 2-valued homomorphism would have to map one of the rays b or b' onto 1, and one of the rays $U(b)$ or $U(b)'$ onto 1, and hence the intersection of these two rays — the null subspace — would also have to be mapped onto 1. So b cannot be skew to e_{r_i} in r_i, unless b is at a 45° angle to e_{r_i}. In this case, reflecting b through e_{r_i} yields b', and there is no contradiction with TP.

So, if r_i is 2-dimensional (and r_i is not included in p or $p^{\perp}$), we could conclude from this argument only that $e_{r_i} \leq p$ or $e_{r_i} \leq p^{\perp}$, or that $b \leq p$ or $b \leq p^{\perp}$, where b is a ray at a 45° angle to e_{r_i} in r_i. In the original formulation of the proof, the 'weak separability' condition was introduced to exclude this one anomalous possibility, that the determinate sublattice $\mathcal{D}(e, R)$ might contain propositions

p such that $b \leq p$ or $b \leq p^{\perp}$, where b is at a 45° angle to e_{r_i} in r_i.

To see that such a condition is not required, suppose r_i is 2-dimensional. Consider a lattice automorphism U that is the identity in $r_i^{\perp}$, preserves e_{r_i} in r_i, and maps b onto $U(b) \neq b$. For example, suppose we represent e_{r_i} by the vector $\begin{pmatrix} 1 \\ 0 \end{pmatrix}$ in r_i and take the action of U on r_i as represented by the matrix $\begin{pmatrix} i & 0 \\ 0 & 1 \end{pmatrix}$. If b is represented by a vector $\begin{pmatrix} b_1 \\ b_2 \end{pmatrix}$ in r_i, then:

$$\begin{pmatrix} i & 0 \\ 0 & 1 \end{pmatrix}\begin{pmatrix} 1 \\ 0 \end{pmatrix} = \begin{pmatrix} i \\ 0 \end{pmatrix} = i\begin{pmatrix} 1 \\ 0 \end{pmatrix}$$

$$\begin{pmatrix} i & 0 \\ 0 & 1 \end{pmatrix}\begin{pmatrix} b_1 \\ b_2 \end{pmatrix} = \begin{pmatrix} ib_1 \\ b_2 \end{pmatrix} \neq k\begin{pmatrix} b_1 \\ b_2 \end{pmatrix}$$

for any constant k. So U maps e_{r_i} onto the same ray but maps b onto a different ray, whether or not b is at a 45° angle to e_{r_i}. The four rays $b, b', U(b), U(b)'$ would all have to belong to $\mathcal{D}(e, R)$ if b belongs to $\mathcal{D}(e, R)$, which contradicts TP, by the argument above. So, if r_i is 2-dimensional, b cannot be skew to e_{r_i} in r_i.

We have now established that if $p \in \mathcal{D}(e, R)$ then either $e_{r_i} \leq p$ or $e_{r_i} \leq p^{\perp}$, for all $i = 1, \ldots, k$. The final stage of the proof involves showing that the set of *all* such elements $\{p : e_{r_i} \leq p$ or $e_{r_i} \leq p^{\perp}$, for all $i = 1, \ldots, k\} = \mathcal{L}_{e_{r_1}e_{r_2}\ldots e_{r_k}}$ satisfies the three conditions TP, R-PREF, and DEF. Maximality then requires that $\mathcal{D}(e, R) = \mathcal{L}_{e_{r_1}e_{r_2}\ldots e_{r_k}}$. (For details, see Bub (1997).)

Bibliography

Bub, J. (1997), *Interpreting the Quantum World*, Cambridge University Press, Cambridge.

Bub, J. & Clifton, R. (1996), 'A uniqueness theorem for "No collapse" interpretations of quantum mechanics', *Studies in the History and Philosophy of Modern Physics* **27**, 181–219. Chapter 2 of this volume.

Chapter 4

Lorentz-invariance in modal interpretations

with Michael Dickson

Although Bell's theorem tells us that any interpretation of quantum mechanics must fail to satisfy Bell's locality condition, the question of Lorentz-invariance is apparently left open by Bell's theorem. Indeed, it has been argued by many that nonlocality is compatible with Lorentz-invariance, given certain interpretations of the latter, or of the former. Some give a 'weak' interpretation of Lorentz-invariance, compatible with a 'strong' violation of locality. For example, Bohm's (1952) theory violates locality in a way that requires a preferred frame; but empirical Lorentz-invariance is preserved in the Bohm theory — there is no experiment that could determine the preferred frame. (See, for example, Valentini (1991*a*,*b*), Bohm and Hiley (1993), and Cushing (1994).) On the other hand, 'orthodox' interpretations of special relativity maintain a notion of 'fundamen-

---

We thank Tim Budden and F. A. Muller for considerable correspondence on an earlier draft of the paper. We thank Laura Ruetsche for rooting out an error in an earlier version of the argument (in section 2) that Lorentz-transformations for compound systems whose components are not interacting are represented by tensor-products of the Lorentz-transformations of the constituent systems. Thanks also to Dennis Dieks and Pieter Vermaas for comments and to Rivka Kfia for alerting us to an error in Appendix B.

tal' Lorentz-invariance — which, in particular, denies the existence of a preferred frame — and advocates of this view try to interpret the violation of locality in a way that is compatible with fundamental Lorentz-invariance. (See, for example, Jarrett (1984), Shimony (1986), and Redhead (1986).)

In any case, for a given hidden-variables theory, it is evidently an open question whether the theory is Lorentz-invariant, and if so, in what sense. Hardy (1992) has tried to close this gap with a general argument that *any* hidden-variables theory must fail to be Lorentz-invariant. However, as we show in detail in Appendix B, Hardy's assumptions about hidden-variables theories are too strong. Neither modal interpretations, nor the Bohm theory, fall under the rubric of 'hidden-variables theories' in Hardy's sense.

In this paper, we investigate the prospects for fundamental Lorentz-invariance in modal interpretations of quantum mechanics, and find that in general they *cannot* be fundamentally Lorentz-invariant, in a sense to be made precise. The nonlocality of modal interpretations is truly in conflict with the orthodox interpretation of relativity theory.

By 'modal interpretations' we mean, roughly, the interpretations due to van Fraassen (1979, 1991, Chap. 9), Kochen (1985), Dieks (1988, 1989), Healey (1989), Bub (1992, 1995), Clifton (1995), Vermaas and Dieks (1995), and Bacciagaluppi & Dickson (1999). All of these interpretations deny the eigenstate-eigenvalue link (a term coined by Fine (1970)), allowing that a system can possess a property even when it is not in the relevant eigenstate. These interpretations fall into two rough classes: those that employ density operators to constrain system properties, and (just) one that does not carry that commitment, namely Bub's (see also Bub and Clifton (1996)). For the bulk of this paper we will concentrate on the former class, and take the modal interpretation of Vermaas and Dieks (1995) as our representative. Though their interpretation differs in certain respects from the other density operator interpretations, our argument against fundamental Lorentz invariance impinges on those other interpretations, as we explain below. In Appendix A, we examine the scope of our argument in the context of Bub's modal interpretation, which selects the definite-valued properties in quite a different way.

Given the (unique) spectral decomposition,

$$W = \sum_i w_i P_i, \tag{1}$$

of a system's density operator, W, the Vermaas-Dieks modal interpretation says that the system possesses a property corresponding to one of the P_i, each with probability $\mathrm{Tr}[P_i]w_i$. For a given system, call the P_i the 'definite properties'. Which of the definite properties is actually possessed by a system is a hidden variable in this theory, in the sense that the quantum-mechanical state does not in general determine which of the definite properties is possessed. (We refer to the latter as the 'possessed property' of a system.) Finally, to treat compound systems, we say that for a compound system composed of the non-overlapping subsystems, $\alpha, \beta, \ldots, \omega$, with definite properties $P^{\alpha}_{i_\alpha}, \ldots, P^{\omega}_{i_\omega}$ determined by the (reduced) density operators for the subsystems, the joint probabilities are given by

$$\Pr(P^{\alpha}_{i_\alpha}, \ldots, P^{\omega}_{i_\omega}) = \mathrm{Tr}[W P^{\alpha}_{i_\alpha} \otimes \ldots \otimes P^{\omega}_{i_\omega}], \tag{2}$$

where W is the density operator for the compound system composed of $\alpha, \ldots, \omega$. (We are assuming that the reader is generally familiar with these ideas. See the bibliography for references to more complete discussions.)

Among density-operator interpretations, there are different treatments of compound systems. For example, some attribute non-supervenient properties to compound systems, while others attribute to compound systems only those properties that they inherit from their subsystems.[1] However, such differences are unimportant here.[2] All we will need is what we have said thus far: our argument against fundamental Lorentz invariance will apply to any

[1]Clifton (1996) and Dickson (1997) provide accounts of how various versions of the modal interpretation differ in their treatment of compound systems.

[2]This claim might be too quick. Our argument will require the existence of joint probabilities for one system, α, to possess a property, P^{α}, and another system, β, to possess a property, P^{β}. If one denies 'property composition' — the condition that if α possesses P^{α} and β possesses P^{β}, then the compound system $\alpha\&\beta$ pos-

density-operator modal interpretation that agrees to what is said in the previous paragraph, and then only in the context of the EPR-Bohm experiment, as discussed in the next section.

Notice that we have said nothing about the dynamics of the possessed properties. That is, we do not yet know how to answer questions of the form: Given that a system possesses P_1 at time t_1, what is the probability that it possesses P_2 at the later time, t_2? A general framework has been developed (by Bacciagaluppi & Dickson (1999); cf. Dickson (1997), and Vermaas (1996)) to answer such questions, but we do not need that framework here. We will show that *no* dynamics can be made fundamentally Lorentz-invariant.

In the next section, we review the EPR-Bohm experiment in a suitably relativistic setting, and we give the definite properties of the systems as determined by the Vermaas-Dieks version of the modal interpretation. In Section 3 we say more precisely what we mean by 'fundamental Lorentz-invariance' and we argue that it is a reasonable condition for an 'orthodox' relativistic theory to obey. In Section 4, we argue for a condition on transition probabilities, dubbed 'stability'. The meat of our argument is in Section 5, where we prove the following theorem:

> The conditions of fundamental Lorentz-invariance and stability (plus the condition that the modal rule for determining the possible properties of a system takes the same form in every frame) entail a contradiction.

On the other hand, in Section 6 we define 'empirical Lorentz-invariance' and show that it *does* hold in a stable dynamics. For

---

sesses the compound property $P^\alpha \& P^\beta$ — then one *might* have a basis for denying the existence of such joint probabilities. (Note the (proposed) distinction between 'α possesses P^α and β possesses P^β' and '$\alpha\&\beta$ possesses $P^\alpha \& P^\beta$'.) However, even in the absence of property composition, it remains unclear why these joint probabilities should fail to exist. After all, they need not be interpreted as probabilities for the compound system to possess a compound property, but may instead be interpreted as the probability for α to possess P^α and β to possess P^β. Nonetheless, it remains less clear whether our argument applies to an interpretation that backs up the failure of property composition with a relational story about value assignments, such as Kochen appears to endorse.

readers with the stamina to carry on, we have included two appendices. In Appendix A we discuss the status of fundamental Lorentz-invariance in Bub's modal interpretation. In Appendix B we show why Hardy's argument against Lorentz-invariant hidden-variables theories is too blunt an instrument to establish the conclusions that we reach in this paper.

1 The EPR-Bohm experiment in a fully relativistic setting

In order to be as careful (and convincing) as we can, we will avoid most of the shortcuts common in discussions of Lorentz-invariance and the EPR-Bohm experiment. (We assume that the reader is generally familiar with the EPR-Bohm experiment in a non-relativistic setting.) Readers who are uninterested in the minutiae of relativistic quantum mechanics can start reading at the paragraph before Table 1.

We will consider three families of hyperplanes in space-time: an 'S-family,' in which the measurement-events are simultaneous (i.e., which is such that some member of the S-family intersects both measurement-events), an 'L-family,' in which the left-hand measurement occurs first, and an 'R-family', in which the right-hand measurement occurs first. Each of these three families may be associated with an inertial observer in a state of motion such that the hyperplanes of the family are its planes of simultaneity. The S-observer is stationary with respect to the measuring devices and the (center of mass of the) particles; the L-observer moves to the left at constant speed; and the R-observer to the right at the same constant speed. All three start their travels from the same point in space-time, and all three take themselves to be the spatial origin of their coordinates. We are primarily interested in just three members of each family, labeled $t^S = 0, 1, 2$, $t^L = 0, 1, 2$, and $t^R = 0, 1, 2$, as depicted in Figure 1. (The R-family is omitted — it is the obvious analogue of the L-family, obtained by reflecting our diagram through its central axis.)

The $1 + \varepsilon$ hyperplanes will be used only in this section. Note that

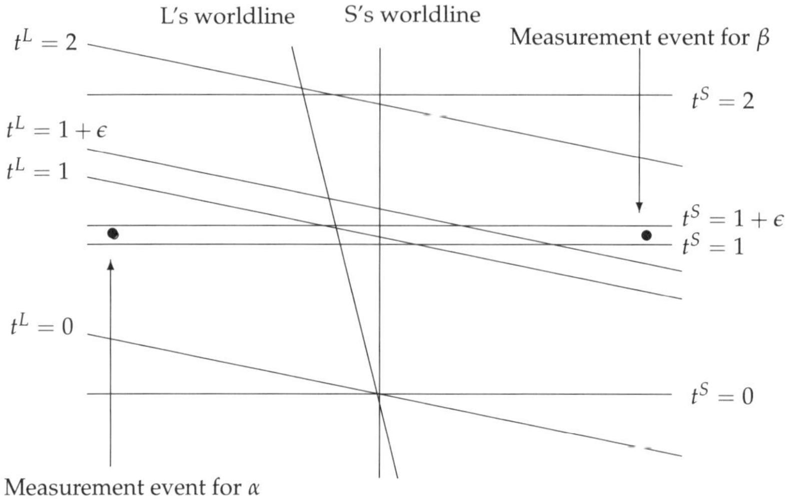

Figure 4.1: Two families of hyperplanes in the EPR-Bohm experiment

we have rigged L's clock so that the intersection of L's worldline with the $t^S = t$ hyperplane is at the time $t^L = t$. We could have instead allowed that these points of intersection happen at different times in the two frames — which would be the more natural thing to do, from the point of view of the theory of relativity — but the notation would then be somewhat messier.

It is usually easiest to think of the transformation from one family's hyperplanes to another's as a passive transformation. For example, the transformation from the $t^S = 0$ hyperplane to the $t^L = 0$ hyperplane may be considered to represent S's re-coordinatizing the space-time manifold. In general, the move from the S-frame to the L-frame can be read as S's re-coordinatizing space-time *as though* S were moving to the left. In that case, the transformation from the $t^S = 0$ hyperplane to the $t^L = 0$ hyperplane, for example, can be represented by S's asking: 'What would I say if I were using L-coordinates rather than S-coordinates?' Of course, there is also an active reading of these transformations, according to which the transformation from S to L represents a change from S's being at rest relative to the apparatuses and the (center of mass of the) particles, to S's being in motion relative to the apparatuses and particles. The equivalence between active and passive transformations allows us to switch back and forth between these ways of speaking as convenience dictates.

The reader will have noticed that (except for the time $t = 0$) L's hyperplanes do *not* intersect S's hyperplanes at S's spatial origin (i.e., at the point $x^S = 0$). Nonetheless, it will be convenient to represent the change from S's hyperplanes to L's hyperplanes (whether interpreted actively or passively) by a single transformation, which includes *both* the spatial translation (by S, out to where L is) and the pure Lorentz-boost. Although this transformation is not a pure Lorentz-boost about $x^S = 0$, we will use notation that is reminiscent of the Lorentz-transformation. We let Λ_t^F denote the transformation from S's $t^S = t$ hyperplane to F's $t^F = t$ hyperplane, where $F \in \{L, R\}$ and $t \in \{0, 1, 1 + \varepsilon, 2\}$.

This shortcut is justified because Λ_t^F is a pure boost about the point of intersection of F's and S's worldlines. For example, consider S and L at time t, and suppose that S wants to re-coordinatize space-time to use L's coordinates. First, S performs a spatial translation:

$$(t^S, x^S) \longrightarrow (t^S, x^S + vt), \tag{3}$$

and then a Lorentz-boost, the combination of the two giving:

$$(t^S, x^S) \longrightarrow (\gamma[t^S + v(x^S + vt)], \gamma[(x^S + vt) + vt^S]), \tag{4}$$

where $\gamma = \sqrt{1 - v^2}$ and the speed of light is 1. The transformation in (4) is just the expression for a Lorentz-boost about the point $-vt$. Hence our talk of Λ_t^F as a Lorentz-transformation is acceptable, as are statements such as 'the $t^S = 1$ hyperplane and the $t^L = 1$ hyperplane are related by a Lorentz-transformation'. In what follows, we will use this language freely.

Quantum-mechanically, these transformations are represented by unitary operators. (We will use the same notation to represent a transformation and its representation as a unitary operator on a Hilbert space.) The argument is familiar, but is worth reviewing quickly. We imagine a system at rest in S, in the quantum-mechanical state $|\varphi^S\rangle$. The probability, in S, that it yields the result $|n^S\rangle$ for a measurement of some observable having $|n^S\rangle$ as an eigenstate is $|\langle n^S|\varphi^S\rangle|^2$. In a new frame, F, these states are denoted '$|\varphi^F\rangle$' and '$|n^F\rangle$', and we assume that there is some transformation, T^F, such that $|\varphi^F\rangle = T^F|\varphi^S\rangle$ and $|n^F\rangle = T^F|n^S\rangle$. ($T^F$ is of course some operator on the Hilbert space, though there is no *a priori* demand that it be linear.) The probabilities for outcomes of a measurement should not depend on the coordinates we use to describe the system (or, in active terms, on the state of motion of the system and measuring device), and we therefore require that $|\langle n^S|\varphi^S\rangle|^2 = |\langle n^S|(T^F)^\dagger T^F|\varphi^S\rangle|^2$. Wigner (1931) showed that one can always choose T^F to be either linear and unitary or anti-linear and anti-unitary so as to make this equality hold for all $|\varphi^S\rangle$ and $|n^S\rangle$. The latter possibility is ruled out by the requirement that the quantum-mechanical representations of the Lorentz transformations mirror the group properties of the Lorentz group, and that they be continuously connected to the identity (hence ruling out consideration of reflections). (See Weinberg (1995, Chap. 2) for a recent discussion of these points and a proof of Wigner's theorem.) We will not need to make any assumptions about the specific form of these transformations, apart from their being unitary (and, of course, linear).

The discussion above applies to systems that evolve *freely*. If the *total* system undergoes some interaction between the $t^S = t$ hyperplane and the $t^L = t$ hyperplane, for example, then there is no reason to expect invariance. Hence, thus far we have been considering transformations of a total system (the measured systems plus the apparatuses) that is taken to be closed. However, we will also need some information about how components of the total system transform. In particular, we will need that Lorentz-transformations for compound systems whose components are not interacting are represented by tensor-products of the Lorentz-transformations of the constituent systems.

This assumption is standard in relativistic quantum mechanics, but we give a short argument for it here anyway. Imagine a pair of systems, α and β, and a pair of hyperplanes, η and ζ. Suppose that in the region between η and ζ, α and β evolve independently (i.e., there is no interaction Hamiltonian between them). Finally, suppose that the quantum state of the pair, $\alpha\&\beta$, on η is a product state, $|a^{\eta}\rangle|b^{\eta}\rangle$. It follows, of course, that α has its own statevector, $|a^{\eta}\rangle$, and β its own statevector, $|b^{\eta}\rangle$, and because α and β evolve independently of one another, we can consider how their states transform in isolation from one another. In particular, as we have already discussed, Wigner's theorem implies that there is some unitary operator, $\Lambda(\alpha)$, that is independent of $|b^{\eta}\rangle$ and that transforms α's state on η to its state on $\zeta : \Lambda(\alpha)|a^{\eta}\rangle = |a^{\zeta}\rangle$. Similarly, there exists a unitary operator $\Lambda(\beta)$ that is independent of $|a^{\eta}\rangle$ and such that $\Lambda(\beta)|b^{\eta}\rangle = |b^{\zeta}\rangle$. Therefore, independently of $|b^{\eta}\rangle$, α's state on ζ is $\Lambda(\alpha)|a^{\eta}\rangle$ and, independently of $|a^{\eta}\rangle$, β's state on ζ is $\Lambda(\beta)|b^{\eta}\rangle$.

To finish the argument, note that again by Wigner's theorem, there is some unitary operator, $\Lambda(\alpha\&\beta)$, that takes states of the composite system $\alpha\&\beta$ on η to states on ζ. By the argument of the previous paragraph, $\Lambda(\alpha\&\beta)$ acting on the product state $|a^{\eta}\rangle|b^{\eta}\rangle$ must therefore produce the product state $\Lambda(\alpha)|a^{\eta}\rangle \otimes \Lambda(\beta)|b^{\eta}\rangle$, no matter what $|a^{\eta}\rangle$ and $|b^{\eta}\rangle$ happen to be. But *every* state of the composite on η can be written as a (weighted) sum of product states, and therefore, by the linearity of $\Lambda(\alpha\&\beta)$, we must have $\Lambda(\alpha\&\beta) = \Lambda(\alpha) \otimes \Lambda(\beta)$, which is what we wanted to show.

We will also need to denote the unitary operators corresponding to time-evolution in a given frame, F, from time s to time t. We denote these operators by $U^F_{s,t}$ ($F \in \{S, L, R\}$). Finally, in the EPR-Bohm experiment, there are four systems of interest, the two measured systems, and the two measuring apparatuses. The former we call 'α' and 'β', and the latter we call 'A' and 'B'. It will happen sometimes that we must consider the Lorentz-boost, or the evolution operator, for just one or two of these systems. In that case, we write them as $\Lambda^F_t(\alpha)$, $U^F_{s,t}(\beta\&B)$, and so on. Without an indication of a system, these operators are taken to act on the (tensor-product) Hilbert space for the entire compound system. (Do not fear! Most of this ugly notation will disappear once we have established a few important results.)

We describe the experiment first in the S-family. The quantum state at $t^S = 0$ is:

$$|\Psi^S(0)\rangle = \sum_i c_i |a_i^S(0)\rangle |A_0^S(0)\rangle |b_i^S(0)\rangle |B_0^S(0)\rangle \tag{5}$$

where the summation is over $i = 1, 2$, and we have left the tensor-products implicit. (In this chapter, all summations from now on are over the values 1, 2.) The ket $|A_0^S(0)\rangle$ represents the 'ready' state of A, which has also two indicator-states, $|A_1^S(0)\rangle$, $|A_2^S(0)\rangle$. (Actually, the *time-evolutes* of these kets are the indicator-states of the apparatus. The kets $|A_1^S(0)\rangle$ and $|A_2^S(0)\rangle$ might be nothing like 'indicator-states', but their time-evolutes, $|A_1^S(1)\rangle$ and $|A_2^S(1)\rangle$, at the start of the measurements *are* indicator-states of A. The kets $|A_1^S(0)\rangle$ and $|A_2^S(0)\rangle$ will make no appearance here.) These three kets are the eigenkets of an 'apparatus-observable', $\mathbf{A}^S(0)$. The kets $|a_i^S(0)\rangle$ are similarly eigenkets of the observable $\mathbf{a}^S(0)$. Exactly analogous statements hold for B and β.

It might help the reader to have in mind the following picture. We can think of the observables $\mathbf{a}^F(t)$ as spin-observables for α in frame F, and the observables $\mathbf{A}^F(t)$ as certain discretized position-observables for α in frame F, so that the 'system' A is really just the spatial degrees of freedom of α. The observable $\mathbf{A}^F(t)$ measures whether α has been deflected 'up', 'down', or not at all, along a direction perpendicular to its line of flight, by a Stern-Gerlach field coupling α's spin to its position in frame F at time $t^F = t$. With this picture in mind, it should be fairly straightforward to understand how the $\mathbf{A}^F(t)$ behave in different frames.

To get the quantum state at $t^S = 1$, we apply the unitary time-evolution operator, $U_{0,1}^S$, to $|\Psi^S(0)\rangle$. To do so, note that there are no interactions between $t^S = 0$ and $t^S = 1$, so that $U_{0,1}^S$ factors as $U_{0,1}^S = U_{0,1}^S(\alpha) \otimes U_{0,1}^S(A) \otimes U_{0,1}^S(\beta) \otimes U_{0,1}^S(B)$. We may then write

$$|\Psi^S(1)\rangle = \sum_i c_i |a_i^S(1)\rangle |A_0^S(1)\rangle |b_i^S(1)\rangle |B_0^S(1)\rangle \tag{6}$$

where $|a_i^S(1)\rangle = U_{0,1}^S(\alpha)|a_i^S(0)\rangle$, and similarly for the other kets appearing in (6).

By $t^S = 2$, the measurements have occurred. We make the usual approximation that the measurement-interaction is impulsive, i.e., effectively instantaneous, and that it establishes a perfect correlation between the measured systems and the apparatuses. First consider the time interval $(1, 1+\varepsilon)$ in which the measurement occurs. We suppose that the measurement on α is of an observable, $\mathbf{a}'^S(1)$, whose eigenkets $|a_i'^S(1)\rangle$ are related to the kets $|a_i^S(1)\rangle$ by

$$|a_i^S(1)\rangle = e_{i1}|a_1'^S(1)\rangle + e_{i2}|a_2'^S(1)\rangle. \tag{7}$$

Similarly, the measurement on β is of an observable, $\mathbf{b}'^S(1)$, whose eigenkets are related to the kets $|b_j^S(1)\rangle$ by

$$|b_j^S(1)\rangle = d_{j1}|b_1'^S(1)\rangle + d_{j2}|b_2'^S(1)\rangle. \tag{8}$$

The quantum state at $t^S = 1+\varepsilon$ is then

$$\begin{aligned}|\Psi^S(1+\varepsilon)\rangle &= \sum_{i,j,k} c_i e_{ij} d_{ik} |a_j'^S(1)\rangle |A_j^S(1)\rangle |b_k'^S(1)\rangle |B_k^S(1)\rangle \\ &= \sum_{i,j,k} c_i e_{ij} d_{ik} |a_j'^S(1+\varepsilon)\rangle |A_j^S(1+\varepsilon)\rangle |b_k'^S(1+\varepsilon)\rangle |B_k^S(1+\varepsilon)\rangle.\end{aligned} \tag{9}$$

The second equality rests on the fact that the free evolutions of the measured and measuring systems are negligible because the interaction Hamiltonians dominate during impulsive measurements. The state at $t^S = 2$ is now readily obtained by unitarily evolving the state $|\Psi^S(1+\varepsilon)\rangle$

$$|\Psi^S(2)\rangle = \sum_{i,j,k} c_i e_{ij} d_{ik} |a_j'^S(2)\rangle |A_j^S(2)\rangle |b_k'^S(2)\rangle |B_k^S(2)\rangle, \tag{10}$$

where $|a_j'^S(2)\rangle = U^S_{1+\epsilon,2}(\alpha)|a_j'^S(1+\epsilon)\rangle$, and so on (because, after the measurement, every system freely evolves again, so that $U^S_{1+\varepsilon,2} = U^S_{1+\varepsilon,2}(\alpha) \otimes U^S_{1+\varepsilon,2}(A) \otimes U^S_{1+\varepsilon,2}(\beta) \otimes U^S_{1+\varepsilon,2}(B)$).

We move now to frame L. The state on the $t^L = 0$ hyperplane is easy to calculate. The Lorentz-transformation Λ_0^L factors as $\Lambda_0^L = \Lambda_0^L(\alpha) \otimes \Lambda_0^L(A) \otimes \Lambda_0^L(\beta) \otimes \Lambda_0^L(B)$, because it passes over only regions where the systems evolve freely. Hence

$$|\Psi^L(0)\rangle = \sum_i c_i |a_i^L(0)\rangle |A_0^L(0)\rangle |b_i^L(0)\rangle |B_0^L(0)\rangle, \tag{11}$$

where $|a_i^L(0)\rangle = \Lambda_0^L(\alpha)|a_i^S(0)\rangle$, and so on for the other systems.

Finding the state on the $t^L = 1$ hyperplane takes a bit more work. We have by definition that $|\Psi^L(1)\rangle = \Lambda_1^L|\Psi^S(1)\rangle$, but because Λ_1^L passes over the left-hand measurement-interaction, it only factors as $\Lambda_1^L = \Lambda_1^L(\alpha\&A) \otimes \Lambda_1^L(\beta) \otimes \Lambda_1^L(B)$, giving

$$|\Psi^L(1)\rangle = \sum_i c_i \left[\Lambda_1^L(\alpha\&A)|a_i^S(1)\rangle|A_0^S(1)\rangle\right] |b_i^L(1)\rangle|B_0^L(1)\rangle. \quad (12)$$

But we can *also* get $|\Psi^L(1)\rangle$ by transforming $|\Psi^S(1+\varepsilon)\rangle$ to $|\Psi^L(1+\varepsilon)\rangle$, then time evolving backwards to $|\Psi^L(1)\rangle$. I.e., $|\Psi^L(1)\rangle = U^L_{1+\varepsilon,1}\Lambda^L_{1+\varepsilon}|\Psi^S(1+\varepsilon)\rangle$, where $U^L_{1+\varepsilon,1} = (U^L_{1,1+\varepsilon})^{-1}$. The Lorentz-boost $\Lambda^L_{1+\varepsilon}$ passes through the right-hand measurement-interaction, so that it factors only as $\Lambda^L_{1+\varepsilon} = \Lambda^L_{1+\varepsilon}(\alpha) \otimes \Lambda^L_{1+\varepsilon}(A) \otimes \Lambda^L_{1+\varepsilon}(\beta\&B)$, while the evolution operator factors completely, as $U^L_{1+\varepsilon,1} = U^L_{1+\varepsilon,1}(\alpha) \otimes U^L_{1+\varepsilon,1}(A) \otimes U^L_{1+\varepsilon,1}(\beta) \otimes U^L_{1+\varepsilon,1}(B)$. (Figure 1 should make these points clear.) Hence we have

$$\begin{aligned}|\Psi^L(1)\rangle = \sum_{i,j,k} & c_i e_{ij} d_{ik} |a_j'^L(1)\rangle|A_j^L(1)\rangle \\ & \left[U^L_{1+\varepsilon,1}(\beta) \otimes U^L_{1+\varepsilon,1}(B)\Lambda^L_{1+\varepsilon}(\beta\&B)\right] |b_k'^S(1+\varepsilon)\rangle|B_k^S(1+\varepsilon)\rangle.\end{aligned} \quad (13)$$

Comparing (12) and (13) (in particular, the spectral projections of the reduced density operator for $\beta\&B$, which must come out the same in both cases) reveals that

$$\begin{aligned}\sum_k d_{ik} \left[U^L_{1+\varepsilon,1}(\beta) \otimes U^L_{1+\varepsilon,1}(B)\Lambda^L_{1+\varepsilon}(\beta\&B)\right] |b_k'^S(1+\varepsilon)\rangle|B_k^S(1+\varepsilon)\rangle \\ = |b_i^L(1)\rangle|B_0^L(1)\rangle,\end{aligned} \quad (14)$$

which yields exactly the result one would expect intuitively, namely,

$$|\Psi^L(1)\rangle = \sum_{i,j} c_i e_{ij} |a_j'^L(1)\rangle|A_j^L(1)\rangle|b_i^L(1)\rangle|B_0^L(1)\rangle \quad (15)$$

(up to phase factors, which are irrelevant for our argument). (In deriving (14), we have implicitly used the fact that $\sum_j e_{ij}e^*_{kj} = \delta_{ik}$, which follows from (7). Later we will also use $\sum_j d_{ij}d^*_{kj} = \delta_{ik}$.)

The state at $t^L = 2$ is readily obtained by Lorentz-transforming the state $|\Psi^S(2)\rangle$:

$$|\Psi^L(2)\rangle = \sum_{i,j,k} c_i e_{ij} d_{ik} |a_j'^L(2)\rangle |A_j^L(2)\rangle |b_k'^L(2)\rangle |B_k^L(2)\rangle \qquad (16)$$

where $|a_j'^L(2)\rangle = \Lambda_2^L(\alpha)|a_j'^S(2)\rangle$, and so on, because $\Lambda_2^L = \Lambda_2^L(\alpha) \otimes \Lambda_2^L(A) \otimes \Lambda_2^L(\beta) \otimes \Lambda_2^L(B)$.

By exactly analogous arguments, the states in the R-family are:

$$|\Psi^R(0)\rangle = \sum_i c_i |a_i^R(0)\rangle |A_0^R(0)\rangle |b_i^R(0)\rangle |B_0^R(0)\rangle \qquad (17)$$

$$|\Psi^R(1)\rangle = \sum_{i,k} c_i d_{ik} |a_i^R(1)\rangle |A_0^R(1)\rangle |b_k'^R(1)\rangle |B_k^R(1)\rangle \qquad (18)$$

$$|\Psi^R(2)\rangle = \sum_{i,j,k} c_i e_{ij} d_{ik} |a_j'^R(2)\rangle |A_j^R(2)\rangle |b_k'^R(2)\rangle |B_k^R(2)\rangle, \qquad (19)$$

with notational conventions as before.

Having carefully gone through this discussion of the EPR-Bohm experiment in the three families of hyperplanes, we now let our hair down with a few notational simplifications. First, we will often drop the superscripts referring to frames on the kets $|a_j^F(t)\rangle$, $|A_j^F(t)\rangle$, $|b_k^F(t)\rangle$, $|B_k^F(t)\rangle$, and so on. (No confusion should arise because whenever we mention a ket, it will be in the context of a clearly specified family of hyperplanes.) Second, we will often drop the times in the kets $|a_j^F(t)\rangle$, $|A_j^F(t)\rangle$, $|b_k^F(t)\rangle$, $|B_k^F(t)\rangle$ and so on. (Again, no confusion should arise when we do so.) Given these simplifications, we summarize the important results of this section in Table 4.1.

Finally, a terminological simplification: we will refer to a given family of hyperplanes as a 'frame'.

Given these quantum states, it is easy to calculate the definite properties of the systems α, A, β, and B. They are just the spectral projections of the reduced density operators for the subsystems in question. Assuming that the coefficients c_i, d_{ij}, e_{ij} do not lead to any degeneracies in the density operators of these subsystems, the definite properties in the three frames as well as their probabilities

State in Frame	at time 0	at time 1	at time 2
S	$\sum_i c_i\|a_i\rangle\|A_0\rangle\|b_i\rangle\|B_0\rangle$	$\sum_i c_i\|a_i\rangle\|A_0\rangle\|b_i\rangle\|B_0\rangle$	$\sum_{i,j,k} c_i e_{ij} d_{ik}\|a'_j\rangle\|A_j\rangle\|b'_k\rangle\|B_k\rangle$
L	$\sum_i c_i\|a_i\rangle\|A_0\rangle\|b_i\rangle\|B_0\rangle$	$\sum_{i,j} c_i e_{ij}\|a'_j\rangle\|A_j\rangle\|b_i\rangle\|B_0\rangle$	$\sum_{i,j,k} c_i e_{ij} d_{ik}\|a'_j\rangle\|A_j\rangle\|b'_k\rangle\|B_k\rangle$
R	$\sum_i c_i\|a_i\rangle\|A_0\rangle\|b_i\rangle\|B_0\rangle$	$\sum_{i,k} c_i d_{ik}\|a_i\rangle\|A_0\rangle\|b'_k\rangle\|B_k\rangle$	$\sum_{i,j,k} c_i e_{ij} d_{ik}\|a'_j\rangle\|A_j\rangle\|b'_k\rangle\|B_k\rangle$

Table 4.1: Quantum states in the EPR-Bohm experiment

Frame	$t=0$	$t=1$	$t=2$
S	$\|a_i\rangle, \|A_0\rangle, \|b_i\rangle, \|B_0\rangle$	$\|a_i\rangle, \|A_0\rangle, \|b_i\rangle, \|B_0\rangle$	$\|a'_j\rangle, \|A_j\rangle, \|b'_k\rangle, \|B_k\rangle$
prob	$\|c_i\|^2$	$\|c_i\|^2$	$\|\sum_i c_i e_{ij} d_{ik}\|^2$
L	$\|a_i\rangle, \|A_0\rangle, \|b_i\rangle, \|B_0\rangle$	$\|a'_j\rangle, \|A_j\rangle, \|b_i\rangle, \|B_0\rangle$	$\|a'_j\rangle, \|A_j\rangle, \|b'_k\rangle, \|B_k\rangle$
prob	$\|c_i\|^2$	$\|c_i e_{ij}\|^2$	$\|\sum_i c_i e_{ij} d_{ik}\|^2$
R	$\|a_i\rangle, \|A_0\rangle, \|b_i\rangle, \|B_0\rangle$	$\|a_i\rangle, \|A_0\rangle, \|b'_k\rangle, \|B_k\rangle$	$\|a'_j\rangle, \|A_j\rangle, \|b'_k\rangle, \|B_k\rangle$
prob	$\|c_i\|^2$	$\|c_i d_{ik}\|^2$	$\|\sum_i c_i e_{ij} d_{ik}\|^2$

Table 4.2: Vermaas-Dieks properties and their probabilities in the EPR-Bohm experiment

are as indicated in Table 4.2. (We have represented properties there with vectors, rather than projections. Strictly speaking, the properties should be represented by the projections onto the subspace spanned by the vectors that we have given.) In what follows, we will always assume that the coefficients avoid degeneracy in every frame. When it comes time to make a specific choice, we will choose coefficients that do avoid degeneracy in every frame.

Finally, the modal interpretation must provide joint transition probabilities from earlier possessed properties to later possessed properties. These joint transition probabilities will be frame-specific, because different frames disagree about the definite properties at a given time. (Compare the definite properties in R and L at time $t^R = 1$ and $t^L = 1$.) Note, however, that there is no hyperplane-dependence of properties in Fleming's (1989) sense — a component system's density operator (and therefore its set of definite properties) is frame-independent. The frame-dependence of joint properties, and therefore of joint transition probabilities, is a consequence of the fact that while the measurements occur simultaneously in frame S, they occur non-simultaneously in the other frames. We may write these frame-dependent joint transition probabilities as

$$\Pr^R\left(|a_i\rangle, |A_0\rangle, |b'_k\rangle, |B_k\rangle \text{ at } t^R = 1 \middle| \, |a_i\rangle, |A_0\rangle, |b_i\rangle, |B_0\rangle \text{ at } t^R = 0\right)$$
$$=_{df} p^R\left(a_i, b'_k \middle| \, a_i, b_i\right) \quad (20)$$

and so on. The right-hand side of (20) simplifies the notation considerably, and remains completely unambiguous, given the information in Table 4.2; in particular, the apparatus properties that are now left out of the notation can always be uniquely inferred from properties for the measured systems that have been left in.

2 Fundamental Lorentz-invariance

It is well known that, as a consequence of the quantum-mechanical no-signaling theorem, the orthodox quantum-mechanical account of the EPR-Bohm experiment is Lorentz-invariant at the statistical level.[3] The same holds in the modal interpretation. We will discuss this point further in Section 6, though even now it is clear that, because the modal interpretation lets the quantum-mechanical probabilities fix the distribution of properties at any time, its single-time probabilities are Lorentz-invariant.

[3] Actually, the no-signaling theorem is usually written down and proven in the framework of non-relativistic quantum mechanics. Its application in the context of relativistic quantum mechanics requires Wigner's theorem also. The no-signaling theorem tells us that a component system's statistics do not change when *other* components undergo interaction, and Wigner's theorem tells how to make the statistics for a single system Lorentz-invariant.

But orthodox quantum mechanics is *not* Lorentz-invariant at the level of the individual process. This point has been made often, and is a consequence of the apparent impossibility of making the projection postulate (collapse of the wavefunction) Lorentz-invariant in any empirically adequate way. (See, for example, Aharonov and Albert (1981), and Maudlin (1994).) One might hope that the modal interpretation, which denies the projection postulate, could avoid this problem.

What we want is a 'fundamentally Lorentz-invariant' theory, which we take to require the following condition:

> **Fundamental Lorentz-invariance:** For every pair of frames, $\Gamma, \Gamma' \in \{S, L, R\}$, and every system σ (compound or not):
> *if*
>
> (1) $|\varphi\rangle$ is a definite property for σ in F at $t^F = t$, and
>
> (2) $\Lambda_t^{F'}$ passes over only regions in which σ evolves freely, and
>
> (3) $\Lambda_t^{F'}(\sigma)|\varphi\rangle$ is a definite property for σ in F' at $t^{F'} = t$,
>
> *then* σ possesses $|\varphi\rangle$ in F at $t^F = t$ if and only if it possesses $\Lambda_t^{F'}(\sigma)|\varphi\rangle$ in F' at $t^{F'} = t$,

(Condition (2) is what allows us to write '$\Lambda_t^{F'}(\sigma)|\varphi\rangle$' in condition (3) because without it, $\Lambda_t^{F'}$ would not factor into a part for σ and a part for the rest.)

Note that we do not identify fundamental Lorentz-invariance with the impossibility of superluminal signaling. If no superluminal signaling were all we wanted, then the modal interpretation would already be Lorentz-invariant, as is standard quantum mechanics. (The no-signaling theorem holds in the modal interpretation just as it does in standard quantum mechanics, as will become clear in Section 6.) We have already alluded to this weaker form of Lorentz-invariance as 'empirical Lorentz-invariance', and we noted that it is compatible with the existence of a preferred frame.

But why should we adopt fundamental Lorentz-invariance? In particular, what is its relation to our earlier statement of the 'orthodox' interpretation of relativity, which says that there is no preferred frame? The idea is that, if there is no preferred frame, then there must be *some* sense in which what one frame says about the properties of a system is 'the same as' what any other frame says. Of course, we do not mean that two frames must attribute *identically* the same properties to a system — relativity theory makes no such requirement. Rather, frames must 'agree' up to some well-defined transformation between them.

But although the transformation between *quantum states* in different frames is given by the (unitary representation of the) corresponding Lorentz-transformation, why should that transformation also relate the *properties* ascribed in the different frames? Let the set of definite properties for a system, σ, in frame F at $t^F = t$ be DEF(F), and let the set of properties in frame F' at $t^{F'} = t$ be DEF(F'). Assume also that σ evolves freely in the region between the $t^F = t$ hyperplane and the $t^{F'} = t$ hyperplane. Then, if there is no preferred frame, there must be some map, μ : DEF(F) $\rightarrow$ DEF(F'), such that if frame F attributes $P \in$ DEF(F) to σ at $t^F = t$, then frame F''s attributing $\mu(P)$ to σ at $t^{F'} = t$ counts as 'the same state attribution', or, 'agreement' between F and F'.

An example might make the idea clearer: in standard relativity theory, if F attributes some length to a system, then we say that F' 'agrees' when it attributes a Lorentz-contracted length. In this case, the map between F-states and F'-states is given by the Lorentz-contraction. What we are after is the map, μ : DEF(F) $\rightarrow$ DEF(F'), that defines 'agreement' between F and F' on the possessed properties of σ. (Henceforth, single quotation marks around 'agreement' and its cognates indicates 'agreement up to the transformation, μ'.)

To see why μ must be the Lorentz-transformation, suppose it is not. Then, assuming that every element of DEF(F) has a different probability in F, the frames F and F' *must* sometimes 'disagree', in the sense that F attributes $P \in$ DEF(F) to the system, while the other fails to attribute $\mu(P)$. The reason is that if μ is not the Lorentz-transformation, then P and $\mu(P)$ have different probabilities in their respective frames. (This claim is justified by noting that to each projection in DEF(F), there corresponds a *unique* projection in DEF(F') with the same probability, because the quantum states in the two frames are unitarily equivalent.) Hence, as long as we consider cases in which the elements of DEF(F) have all different probabilities, fundamental Lorentz-invariance must hold. If it does not, then there are times when F and F' cannot both be 'right' about the property that they attribute to a system, meaning that one of them must be 'preferred'. (And note that in the Vermaas-Dieks interpretation, the elements of DEF(F) *must* all have different probabilities, because if two of them had the same probability, then they could not really be in the unique spectral resolution of σ's density operator.)

Finally, we will need the following condition, which is a consequence of fundamental Lorentz-invariance:

> **Invariant transition probabilities:** For every pair of frames, $F, F' \in \{S, L, R\}$, every pair of times, s, t with $s < t$, and every system σ (compound or not): *if*
>
> 1. $|\varphi\rangle$ and $|\chi\rangle$ are definite properties for σ in F at $t^F = s$ and $t^F = t$ respectively, and
> 2. both $\Lambda_s^{F'}$ and $\Lambda_t^{F'}$ pass over only regions in which σ evolves freely, and
> 3. $\Lambda_s^{F'}(\sigma)|\varphi\rangle$ and $\Lambda_t^{F'}(\sigma)|\chi\rangle$ are definite properties for σ in F' at $t^{F'} = s$ and $t^{F'} = t$ respectively,
>
> *then*

$$p^F\left(|\chi\rangle \text{ at } t^F = t \middle| \, |\varphi\rangle \text{ at } t^F = s\right)$$
$$= p^{F'}\left(\Lambda_t^{F'}(\sigma)|\chi\rangle \text{ at } t^{F'} = t \middle| \, \Lambda_t^{F'}(\sigma)|\varphi\rangle \text{ at } t^{F'} = s\right).$$

This condition follows immediately from fundamental Lorentz-invariance. Clause (3) plus fundamental Lorentz-invariance guarantees that F and F' will 'agree' (up to Lorentz-transformation) about the state of the system at times s and t. But they cannot *always* 'agree' about those states if they do not also 'agree' about transition probabilities. Later, we will use both fundamental Lorentz-invariance, and its consequence, invariant transition probabilities.

Note that by suppressing all of the Lorentz-transformations, our Table 4.2 reflects in a perspicuous way the sense of 'agreement' that we have been discussing.

To reiterate a point made earlier: all of the transformations that we are considering may be thought of as *passive* transformations (except, of course, the temporal evolutions in frame S). That is, we may imagine a single observer. This observer might begin by coordinatizing space-time using the hyperplanes of simultaneity of the S-family. The observer might then decide to change coordinate systems, using instead the coordinates given by the hyperplanes in the L-family. The condition that there be no preferred frame is just the condition that the observer can choose any such coordinate system, with no gain or loss in *physical content*. That is, the observer can choose any coordinate system to describe what is really the same physical situation and, in particular, what are really the same transition probabilities.

3 Stability

Prima facie, our condition of fundamental Lorentz-invariance may be satisfiable in the modal interpretation. After all, we already know from the quantum-mechanical no-signaling theorem that the reduced density operator of one particle (in the EPR-Bohm experiment) is unaffected by a measurement on the other. Hence, in particular, the definite properties for one system, and their *single time* probabilities, are frame-invariant. In fact, the no-signaling theorem motivates one easy — but ultimately unsatisfactory — attempt at a Lorentz-invariant dynamics: suppose that the transition probabilities from an earlier to a later time are given by the single-time probabilities at the later time. For example:

$$p^R(a_i, b'_j | a_k, b_k) = |c_i|^2 |d_{ij}|^2. \tag{21}$$

The right-hand side of (21) is just the quantum-mechanical probability at the later time ($t^R = 1$) that the system is in the state (a_i, b'_j). Defining all transition probabilities in this way, we would find that the invariance of transition probabilities automatically holds.

However, apart from being a rather unsatisfactory dynamics — earlier states have absolutely *no* influence on later states (because the right-hand side of (21) is independent of k) — the 'uncorrelated' dynamics exemplified in (21) violates stability, a condition for which we argue below.

Stability is the condition that freely evolving systems do not undergo transitions. (Healey (1989, p. 80) calls this condition 'free evolution', reserving the term 'stability' for a stronger condition concerning interactive evolution.) More precisely, they follow the evolution of the definite properties. When a system evolves freely, it evolves according to its own Hamiltonian, H (which we take to be time-independent), so that its density operator evolves according to

$$W(t) = e^{-iHt}W(0)e^{iHt}. \tag{22}$$

Because e^{-iHt} is a linear operator, equation (22) implies that the elements in the spectral decomposition of $W(t)$ evolve according to $P_j(t) = e^{-iHt}P_j(0)e^{iHt}$. In this case, stability requires that the system's possessed property 'follow' this evolution. (This requirement can be met because the probabilities of the P_j do not change under free evolution of the system.) For our purposes, stability means that a system cannot undergo a transition except as a consequence of its being measured or performing a measurement. (We are not counting a change from $|\sigma_i^F(s)\rangle$ to $|\sigma_i^F(t)\rangle$ for $s < t$ as a 'transition'.) In terms of transition probabilities:

> **Stability:** For any frame F, any system σ, initially definite property, s_k, and any later definite property s_i,
>
> $$p^F(s_i|s_k) = \delta_{ik} \tag{23}$$
>
> assuming that no measurement has been made on or by σ between the initial and later times.

Stability is clearly a substantive assumption, and one that we shall need later for our proof. Hence it requires some justification.

But first, we note that almost all existing density operator interpretations do adopt stability. Healey does so explicitly. Likewise, Kochen argues for stability in his interpretation (apparently on the basis of symmetry principles).[4] Finally, as we discuss below, Vermaas (1996) has in fact derived stability in the Vermaas-Dieks version. Only van Fraassen's interpretation is agnostic about stability, because it is in fact agnostic about dynamics in general — van Fraassen does not *deny* stability, but rather he leaves open all questions about dynamics. (See note 7.) Hence our assumption of stability is consistent with all existing density-operator interpretations, and is endorsed by every version that makes any commitments about dynamics.

But we find stability to be a plausible assumption in its own right, for several reasons. First, as we just noted, Vermaas has derived it from two quite plausible conditions. The first is that the modal interpretation reproduces the statistical predictions for measurement delivered by standard quantum mechanics. The second is that the transition probabilities for a given freely evolving system be functions only of the density operators for the system (at the two times), the sets of possible possessed properties (at the two times), and the evolution operator. This condition has the flavor of locality. From these two conditions, Vermaas derives stability. We admit, however, that in the present context Vermaas' derivation has limited interest. We are concerned here with Lorentz-invariance, not with locality. (We discuss this point in detail at the end of Section 5.) One of the lessons of our paper is that results concerning locality ought not necessarily be translated into results concerning Lorentz-invariance. Hence, in the context of our work, one can accept Vermaas' derivation as at best a plausibility argument.

But stability is plausible for other reasons as well. As Bacciagaluppi & Dickson (1999) point out, stability *must* hold for any freely evolving system in a pure state. (In that case, there is just one possible possessed property at each time, and it obviously follows the evolution given by (22).) Hence those who wish to deny stability (plausibly) must explain why stability fails for systems in mixed

---

[4]The reference is Kochen (1979, p. 35). Kochen does not phrase stability as we have done here. Instead, he says that the unitary operator for a freely evolving system should also be taken to generate the evolution of the ultrafilter (representing the possessed properties of the system) in the Boolean algebra of properties for the system. This condition is equivalent to our stability condition.

states. (The fact that it *can* fail is surely not sufficient explanation of why it does fail.) Indeed, consider two freely evolving systems, one whose density operator has just a single spectral projection, P (which may be multi-dimensional), the other whose density operator has several spectral projections, one of which is P. Suppose that both systems are governed by the same Hamiltonian. Now consider a situation where both systems possess P. The first system *must* obey stability, and there does not seem to be any relevant difference between it and the second system to account for a difference in their evolutions — they have the same possessed properties, and the same Hamiltonian.

Indeed, normally the Hamiltonian of a system is taken to be solely responsible for evolution, in such a way that transitions (whether deterministic or stochastic) are taken to be *induced* by changes in energy. But then consider a system in an eigenstate of the Hamiltonian. In that case, both classical mechanics and quantum mechanics say the sensible thing: the system makes no transitions. Stability also says the sensible thing. But a denial of stability in this case is committed to saying that a system's properties can change *without* any change in energy. And note that we can easily choose to consider an experiment in which the Hamiltonian is the identity. After all, Bohm's version of the EPR-Bohm experiment involves just the spin degrees of freedom of the particles, and the assumption is that (apart from the period during the measurement-interaction) the magnetic field is zero, so that the Hamiltonians for the two particles act like the identity on the spin part of the particles' statevectors.

Finally, we note that stability is quite plausible for macroscopic devices, such as measuring apparatuses. (Of course, we may allow that not all measuring apparatuses are macroscopic — indeed, our example earlier, of letting the spatial degrees of freedom of the measured system be the 'apparatus', uses a non-macroscopic apparatus — but equally well the apparatus in question *could* be macroscopic, and that possibility is all that we need for our argument.) Indeed, there are at least two reasons (apart from those already given) that stability for apparatuses — 'apparatus-stability' — should hold.

First, if apparatus-stability fails, then a very weak form of 'faithful measurement' must also fail. It is reasonable to hope that a well-functioning apparatus will provide a reliable (i.e., stable) *record* of the result of a measurement (regardless of whether that result is 'faithful' in a stronger sense, namely, in the sense of accurately indicating the 'true' state of the measured system). For example, if the apparatus prints 'the result is 5' at one time, then we expect that at a later time, the paper still reads 'the result is 5'. If apparatus-stability fails, then apparatuses do not in fact obey this very weak form of faithful measurement.[5]

But a more compelling reason for adopting stability is that to deny stability for (macroscopic) measuring apparatuses requires one to tell a rather convoluted story about why the evolution of apparatuses *appears* stable to us, and that story is necessarily going to involve substantive assumptions about how human memory is to be modeled in the framework of quantum mechanics, for it must explain why one's memory of the apparatus' early state is in fact radically mistaken.[6] Probably such a story can be told, but at the least, any interpretation forced to these lengths should reconsider whether fundamental Lorentz-invariance is worth that price. In any case, our argument suggests that fundamental Lorentz-invariance *does* come (if it comes at all) at that price.

But if stability is plausible for measuring apparatuses, then why is it not plausible as well for microscopic systems? After all, if we

[5]The real point here is not that without stability apparatuses would be unreliable, but that without determinism they would be unreliable. If an apparatus could randomly change its indication, then it would fail to provide a reliable record of the past. We focus on stability because in the present context it is determinism. Note also that we are not making a general argument for determinism. Our argument is just that the reliability of apparatuses requires determinism in the special case of free evolution of an apparatus after a measurement.

[6]Some authors — notably, van Fraassen (1997) — have been willing at least to remain agnostic about stability for apparatuses. We agree that the phenomena do not obviously *require* stability for macroscopic objects, but stability does seem to be the easiest explanation of certain phenomena. As we say, without stability, one will have to search for an explanation employing substantive assumptions about the quantum-mechanical representation of human memory. And even then, one would have to be willing to accept that the world — including the macroscopic world — is (or more precisely, was) nothing like what we think.

are willing to accept the distinction between 'measured system' and 'measuring apparatus' (or more generally, 'microscopic system' and 'macroscopic system') as a *fundamental* one — on which one could base a fundamental difference in laws of evolution — then why go to the length of modal interpretations? Accepting such a distinction, one can merely adopt some version of the projection postulate and be done with the measurement problem. If, on the other hand, we seek a theory in which such distinctions are *not* taken as fundamental, then it is difficult at best to see how to explain why stability should hold for measuring apparatuses, but not for measured systems. Doing so would amount to explaining why the fundamental laws of evolution governing 'large' objects is different from those governing 'small' objects.

The term 'fundamental' is crucial here. We are not claiming that properties of 'large' objects must always be shared by 'small' objects — we admit the possibility of genuinely emergent properties for compound systems. Our claim, instead, is that the fundamental laws governing the evolution of 'large' and 'small' objects be the same.

We admit that these arguments are only based on plausibility. Nonetheless, existing density-operator interpretations (van Fraassen's excepted) have done the right thing by adopting stability. As our arguments show, its denial brings consequences that seem difficult to swallow.

4 Contradiction

We turn now to the meat of our argument, which proceeds by assuming fundamental Lorentz-invariance and stability, then deriving a contradiction. (We have also, already in Table 4.2, assumed implicitly that the rule for determining definite properties takes the same form in every frame.)

To prove the result of this section, we begin by deriving transition probabilities between times 0 and 1 in frame R. After the measurement on β, the probability that the system possesses (a_i, b'_j) is $|c_i d_{ij}|^2$. Now consider how a system could come to possess (a_i, b'_j).

By stability, there is only one way, namely, by initially possessing (a_i, b_i), and the right-hand side making a transition from b_i to b'_j. We can therefore derive the probability of a transition from (a_i, b_i) to (a_i, b'_j) after the measurement on the right-hand side. We have that $|c_i|^2$ of the time, the system initially possesses (a_i, b_i) and that after the measurement on β, it possesses (a_i, b'_j) $|c_i d_{ij}|^2$ of the time, so that given it initially possesses (a_i, b_i), it must make the transition to (a_i, b'_j) exactly $|c_i d_{ij}|^2 / |c_i|^2$ of the time. (We are of course assuming that neither c_1 nor c_2 is zero.)

Carrying through this idea formally, we can use Bayes' rule to write

$$\begin{aligned} &p^R(a_i, b'_j \text{ at } t^R = 1 | a_k, b_k \text{ at } t^R = 0) \\ &= \frac{p^R(a_k, b_k \text{ at } t^R = 0 | a_i, b'_j \text{ at } t^R = 1) \times p^R(a_i, b'_j \text{ at } t^R = 1)}{p^R(a_k, b_k \text{ at } t^R = 0)}. \end{aligned} \tag{24}$$

By stability,

$$p^R(a_k, b_k \text{ at } t^R = 0 | a_i, b'_j \text{ at } t^R = 1) = \delta_{ki} \tag{25}$$

and using Table 4.2 to fill in the other probabilities in (24), we get

$$p^R(a_i, b'_j | a_k, b_k) = \delta_{ki} \frac{|c_i d_{ij}|^2}{|c_k|^2}. \tag{26}$$

An analogous argument applied in frame L gives

$$p^L(a'_i, b_j | a_k, b_k) = \delta_{kj} \frac{|c_j e_{ji}|^2}{|c_k|^2}. \tag{27}$$

Return now to frame R, and consider the probabilities for transitions between times $t^R = 1$ and $t^R = 2$. We need not derive expressions for these transition probabilities, except to note that by stability, β can make no transitions. Hence the marginal probability for a transition from (a_i, b_i) at time $t^R = 0$ to b'_j at time $t^R = 2$ is the same as the marginal probability for a transition from (a_i, b_i) at time $t^R = 0$ to b'_j at time $t^R = 1$. Therefore

$$p^R(b'_j \text{ at } t^R = 2 | a_k, b_k \text{ at } t^R = 0) = \sum_i \delta_{ki} \frac{|c_i d_{ij}|^2}{|c_k|^2} = |d_{kj}|^2. \tag{28}$$

Similarly, in frame L, using (27) we get:

$$p^L(a'_j \text{ at } t^L = 2|a_k, b_k \text{ at } t^L = 0) = |e_{kj}|^2. \tag{29}$$

The marginal transition probabilities (28) and (29) must, by the invariance of transition probabilities, be frame-independent. (Actually, these 'marginal' transition probabilities are not explicitly covered by the condition of invariance of transition probabilities, because they have behind the conditionalization stroke a state for the total system. However, because of the strict correlation between a_k and b_k in all frames at time 0, the transition probabilities in (28) and (29) are equivalent to truly marginal transition probabilities, with just b_k (in (28)) or a_k (in (29)) behind the conditionalization stroke.) In particular, these marginals must be derivable from the joint transition probabilities in frame S. Hence the transition probabilities from time $t^S = 0$ to time $t^S = 2$ must obey the following equations:

$$\begin{aligned} p^S(a'_1, b'_1|a_2, b_2) + p^S(a'_1, b'_2|a_2, b_2) &= |e_{21}|^2 & (30)\\ p^S(a'_2, b'_1|a_2, b_2) + p^S(a'_2, b'_2|a_2, b_2) &= |e_{22}|^2 & (31)\\ p^S(a'_1, b'_1|a_2, b_2) + p^S(a'_2, b'_1|a_2, b_2) &= |d_{21}|^2 & (32)\\ p^S(a'_1, b'_2|a_2, b_2) + p^S(a'_2, b'_2|a_2, b_2) &= |d_{22}|^2. & (33) \end{aligned}$$

(There are four additional equations, involving conditionals on (a_1, b_1), but they are derivable from (30)–(33). Indeed, of (30)–(33), only two are independent equations.) We can solve all of these equations in terms of $p^S(a'_1, b'_1|a_2, b_2)$. For simplicity of notation, we will from now on write $p^S(a'_i, b'_j|a_k, b_k) = p^k_{ij}$. The solution is:

$$\begin{aligned} p^2_{12} &= |e_{21}|^2 - p^2_{11} & (34)\\ p^2_{21} &= |d_{21}|^2 - p^2_{11} & (35)\\ p^2_{22} &= |d_{22}|^2 - |e_{21}|^2 + p^2_{11}. & (36) \end{aligned}$$

But (34)–(36) do not take account of the fact that the p^2_{ij} must be non-negative. (They do imply that the p^2_{ij} sum to 1, however.) For that, we need to restrict p^2_{11} to just those values that make each of the p^2_{ij} lie between 0 and 1. The restrictions are:

$$\max[|d_{21}|^2 + |e_{21}|^2 - 1, 0] \leq p^2_{11} \leq \min[|d_{21}|^2, |e_{21}|^2]. \tag{37}$$

The first inequality follows from (36), because if $|d_{22}|^2 - |e_{21}|^2$ is negative, then p^2_{11} must be greater than or equal to $-(|d_{22}|^2 - |e_{21}|^2)$, which is equal to $|e_{21}|^2 + |d_{21}|^2 - 1$. The second inequality obviously follows from (34) and (35).

The inequalities (37) are satisfiable for every value of the c_i, d_{ij}, and e_{ij}, but there are further constraints on the p^k_{ij}. After averaging over the possible properties at $t^S = 0$, empirical adequacy requires that these transitions yield the single-time distributions at time $t^S = 2$ prescribed by the quantum state. From Table 4.2 we get that the probability that (a'_i, b'_j) is possessed at time $t^S = 2$ is $|c_1 e_{1i} d_{1j} + c_2 e_{2i} d_{2j}|^2$. Empirical adequacy therefore requires

$$|c_1|^2 p^1_{11} + |c_2|^2 p^2_{11} = |c_1 e_{11} d_{11} + c_2 e_{21} d_{21}|^2. \tag{38}$$

(There are of course other empirical adequacy conditions, but we will need just (38).)

The inequalities (37) impose, via (38), inequalities on p^1_{11}. It remains to be seen whether the inequalities imposed on p^1_{11} are themselves satisfiable. As it turns out, they are not.

It is sufficient to consider a specific case. Choose $c_1 = 1/2$ and $c_2 = -\sqrt{3}/2$. Now we choose a specific measurement on each side by choosing the d_{ij} and the e_{ij}:

$$\begin{aligned} d_{11} &= d_{22} = 1/2 \qquad & d_{21} &= -d_{12} = \sqrt{3}/2 \\ e_{11} &= e_{22} = 1/2 \qquad & e_{21} &= -e_{12} = \sqrt{3}/2. \end{aligned} \tag{39}$$

(Note that these choices are consistent with the definition of the e_{ij} and the d_{ij} in (7) and (8), and in particular, with the orthonormality of the eigenstates of the measured observables.) We pause to note that for these values, the modal interpretation really does attribute nontrivial states to the systems. There are no degeneracies in any frame. For the initial state, $|\Psi(0)\rangle$, this point is obvious. Hence, because the coefficients are the same in $|\Psi^L(0)\rangle$ and $|\Psi^R(0)\rangle$, there is no problem at time 0. At $t^S = 1$, $|\Psi^S(1)\rangle$ has the same coefficients, so that again there is no degeneracy. In frame R at $t^R = 1$, the reduced states for the four systems are, taking partial traces of

$|\Psi^R(1)\rangle\langle\Psi^R(1)|$,

$$\begin{aligned} W_\alpha^R(1) &= \tfrac{1}{4}|a_1\rangle\langle a_1| + \tfrac{3}{4}|a_2\rangle\langle a_2| \\ W_\beta^R(1) &= \tfrac{5}{8}|b_1'\rangle\langle b_1'| + \tfrac{3}{8}|b_2'\rangle\langle b_2'| \\ W_A^R(1) &= |A_0\rangle\langle A_0| \\ W_B^R(1) &= \tfrac{5}{8}|B_1\rangle\langle B_1| + \tfrac{3}{8}|B_2\rangle\langle B_2|. \end{aligned} \tag{40}$$

In frame L at time $t^L = 1$, the situation is the same, because the values for the e_{ij} are the same as those for the d_{ij}. Finally, we check the states at time 2. Because all coefficients in the three frames are the same, it suffices to check the reduced states derived from $|\Psi^S(2)\rangle$:

$$\begin{aligned} W_\alpha^S(2) &= \tfrac{5}{8}|a_1'\rangle\langle a_1'| + \tfrac{3}{8}|a_2'\rangle\langle a_2'| \\ W_\beta^S(2) &= \tfrac{5}{8}|b_1'\rangle\langle b_1'| + \tfrac{3}{8}|b_2'\rangle\langle b_2'| \\ W_A^S(2) &= \tfrac{5}{8}|A_1\rangle\langle A_1| + \tfrac{3}{8}|A_2\rangle\langle A_2| \\ W_B^S(2) &= \tfrac{5}{8}|B_1\rangle\langle B_1| + \tfrac{3}{8}|B_2\rangle\langle B_2|. \end{aligned} \tag{41}$$

Therefore, there are no degeneracies in any frame at any time. (Note that the expressions in (41) actually follow from those in (40) and the analogous ones for frame L, from the no-signaling theorem.)

We now show that the inequality imposed on p_{11}^1 by (37) via (38) requires p_{11}^1 to be negative. In the example we are considering, the inequality (37) is

$$1/2 \le p_{11}^2 \le 3/4. \tag{42}$$

We now use this inequality to impose an inequality on p_{11}^1 via (38). Rewrite (38) as

$$|c_1|^2 p_{11}^1 = -|c_2|^2 p_{11}^2 + |c_1 e_{11} d_{11} + c_2 e_{21} d_{21}|^2. \tag{43}$$

Using (42), the right-hand side of (43) is at its *largest* when $p_{11}^2 = 1/2$ (because then p_{11}^2 is at its smallest, so that $-p_{11}^2$ is at its largest). Hence we get the inequality:

$$|c_1|^2 p_{11}^1 \le -|c_2|^2(1/2) + |c_1 e_{11} d_{11} + c_2 e_{21} d_{21}|^2. \tag{44}$$

Plugging in the values from (39), (44) becomes

$$(1/4)p_{11}^1 \leq -(3/4)(1/2) + \left[(1/2)(1/2)(1/2) + (-\sqrt{3}/2)(\sqrt{3}/2)(\sqrt{3}/2)\right]^2 \tag{45}$$

which gives

$$\begin{aligned} p_{11}^1 \leq & -(3/2) + 4[(1/8) - (3\sqrt{3}/8)]^2 \\ = & -(24/16) + [1/16 + 27/16 - 6\sqrt{3}/16] \\ = & 4/16 - 6\sqrt{3}/16 < 0 \end{aligned} \tag{46}$$

so that we end up having to say that p_{11}^1 is negative.[7] Modal interpretations therefore cannot maintain fundamental Lorentz-invariance (unless they are willing to give up on stability).

Note that we have nowhere assumed Jarrett's (1984) 'completeness' condition, which, in terms of the modal interpretation, says:

$$p(a_i', b_j' | a_k, b_k) = p(a_i' | a_k, b_k) p(b_j' | a_k, b_k). \tag{47}$$

We have *permitted* the failure of (47) because we derived expressions in the S-frame for marginal probabilities of the type that occur on the right-hand side of (47) but fell short of demanding (47)'s equality. The fact is that we just did not need to. The demand that the expressions we got for the probabilities on the right (from stability and fundamental Lorentz-invariance) be recovered as marginals from joint probabilities of the type on the left, coupled with the demand that such joints recover the quantum-mechanically predicted joint probability after averaging over 'hidden-variables' (cf. (38)), was enough to get our contradiction.

This point is important, because many authors have suggested that Lorentz-invariance is won by denying completeness and maintaining Jarrett's locality condition, which says, roughly, that transitions between time 0 and time 2 on one side do not depend on what

[7] Guido Bacciagaluppi has noticed a slight simplification in the argument leading to this contradiction. His simplification relies on Pitowsky's (1989) work on correlation polytopes. In order to avoid explaining Pitowsky's work, we have not included the simplification here.

observable was measured on the other side. (As Jarrett showed (cf. Shimony (1986)), Bell's locality condition is equivalent to the conjunction of Jarrett's completeness and locality conditions,[8] so that giving up one or the other is sufficient to avoid Bell's theorem.) For example, Healey (1989) seems to have expressed a version of this view, specifically in the context of his modal interpretation, when he remarks (pp. 58–9): 'The key question concerning locality is therefore whether or not this failure of completeness *itself* implies any physically or metaphysically worrying action at a distance for the present approach to quantum mechanics [italics his].' But violations (or not) of completeness cannot be the end of the story about Lorentz-invariance: one must consider transition probabilities explicitly. It is unclear how Healey's interpretation can do so and still avoid our argument.

Nonetheless, while Healey maintains (as orthodox relativity does) that the existence of a preferred frame 'would certainly violate the spirit of relativity' (p. 149), he claims that his version of the modal interpretation (which satisfies stability) does not do so, and argues for that claim in the case $|c_1| = |c_2|$. In that case, Healey's version does not pick out the definite properties given in Table 4.2, and thereby avoids our argument.[9] However, we have considered a nontrivial case (where $c_1 \neq c_2$) and see no reason why Healey's interpretation is not committed to Table 4.2 (given suitable assumptions about the constitution of the four systems A, B, α, and β, to eliminate the effects of the provisos he puts on his general rules for property ascription).

Whatever else is true, the result of this section makes clear that for existing density operator modal interpretations (i.e., those sufficiently similar to Vermaas' and Dieks'), denying completeness can

---

[8] Actually, Jarrett only applied his analysis under the condition that the hidden state does not evolve between the emission-event and the measurement-event. Stability effectively secures this condition for us.

[9] Indeed, in that case, Healey's interpretation says that α and β have *only* the trivial property, corresponding to the identity operators on the subspaces spanned by $\{|a_1\rangle, |a_2\rangle\}$ and $\{|b_1\rangle, |b_2\rangle\}$. Many other density-operator interpretations agree with Healey's in this case. (The exception is the interpretation of Bacciagaluppi and Dickson (1997); cf. Bacciagaluppi, Donald, and Vermaas (1995).)

achieve Lorentz-invariance in the 'empirical' sense at best, *not* in the 'fundamental' sense. It also shows that issues about how violations of completeness (or indeed locality) affect Lorentz-invariance should be decided in the context of a specific dynamical model of the EPR-Bohm experiment, rather than on the basis of general arguments. This point was anticipated by Jones and Clifton (1993), who argued that violations of completeness may not be as benign as they are often taken to be. They show that in some theories, a violation of completeness *entails* nonlocality, making it clear once again that the consequences of violating completeness are best not assessed outside the context of a given theory.

5 Empirical Lorentz-invariance

What should we make of this result? We have already hinted that it is reminiscent of Bohm's (1952) theory. There too, there is a preferred frame of reference. However, Bohm's theory remains empirically Lorentz-invariant, in the sense that it is a theorem of the theory that nobody could ever find the preferred frame. We will show that the same result holds in the modal interpretation. (We focus again on the Vermaas-Dieks version, but similar arguments can be made for other versions.)

As we already noted, it is obvious that the single-time probabilities of the modal interpretation are Lorentz-invariant, because they are just the quantum-mechanical ones. But our discussion at the end of Section 5 makes it clear that this form of Lorentz-invariance is not enough — Lorentz-invariance can (and in this case does) fail for the transition probabilities, which entails a failure of (fundamental) Lorentz-invariance for at least some runs of the experiment. The question here is whether this failure can ever be detected.

The same question arises in the Bohm theory. The proof that the failure of Lorentz-invariance for the Bohmian trajectories is experimentally undetectable relies essentially on the fact that any *measurement* of the initial position of a system will destroy the very correlations that create the problem for Lorentz-invariance in the first place. Therefore, any attempt to *discover* (via measurement) the trajectory of a particle will ruin the condition that required the failure of Lorentz-invariance. These points were already anticipated by Bohm (1952) and are discussed as well by Valentini (1991*a*,*b*), Bohm and Hiley (1993), and Cushing (1994), among others.[10]

The same argument can be made in the modal interpretation. The main point is that in order to find a preferred frame, one must first know the initial hidden state of the system (otherwise the predictions of the modal interpretation are identical to those of standard quantum mechanics), but to do so, apparently one must make a measurement on the system at the initial time, thereby disrupting the correlations that were a threat to Lorentz invariance in the first place.

Let us see what happens when we make such a measurement. Our account will proceed in frame S. Consider again the initial state $|\Psi^S(0)\rangle$, and the measurements made by A and B. In order to discover empirically the transition probabilities $p(a_i', b_j'|a_k, b_k)$, for example, we must know and remember the initially possessed property, and therefore, we must somehow record it, so that we may later compare it with the final possessed property. To do so, introduce another pair of apparatuses, M and N, to make a measurement of the observables, $\mathbf{a}$ and $\mathbf{b}$ (with eigenvectors $\{|a_i\rangle\}$ and $\{|b_i\rangle\}$, respectively), to record the possessed property at the earlier time. After this measurement, the state of the entire system is

$$|\Psi^S(\text{earlier})\rangle = \sum_i c_i|a_i\rangle|M_i\rangle|A_0\rangle|b_i\rangle|N_i\rangle|B_0\rangle. \tag{48}$$

After the measurement on the system at the later time, the state is

$$|\Psi^S(\text{later})\rangle = \sum_{i,j,k} c_i e_{ij} d_{ik}|a_j'\rangle|M_i\rangle|A_j\rangle|b_k'\rangle|N_i\rangle|B_k\rangle. \tag{49}$$

Equation (49) reveals that the first measurement has ruined the correlation between α and β. In particular, we have that the joint prob-

[10] Actually, the problem of finding a (fundamentally) Lorentz-invariant Bohm-like theory (i.e., one that has the Bohm theory as its non-relativistic limit) is much more complex than our brief comments suggest. See Berndl *et al.* (1996) for a discussion.

ability of the property $(a'_i, M_k, A_i, b'_j, N_k, B_j)$ at the later time is

$$p^S(a'_i, M_k, A_i, b'_j, N_k, B_j) = |c_k|^2|e_{ki}|^2|d_{kj}|^2. \tag{50}$$

If we assume stability (so that the properties possessed by M and N at the later time are also possessed at the earlier time), then (50) gives us an easy expression for transition probabilities, and one that is, moreover, entirely factorizable:

$$p^S(a'_i, b'_j|a_k, b_k) = |e_{ki}|^2|d_{kj}|^2 = p^S(a'_i|a_k, b_k)p^S(b'_j|a_k, b_k). \tag{51}$$

The initial measurements by M and N allow the initially possessed property, (a_k, b_k), to screen each of α and β off from each other. In other words, it destroys the peculiarly quantum-mechanical nature of the correlations between α and β, thus making possible a Lorentz-invariant dynamics. (It is a trivial exercise to extend the expression in (51) to all frames, and to show that the resulting dynamics is empirically adequate in all frames.)

But how, one might ask, could this result be true? After all, from a mathematical point of view, the initial measurement just 'mapped' the vectors $\{|a_i\rangle\}$ and $\{|b_j\rangle\}$ to the vectors $\{|a_i\rangle|M_i\rangle\}$ and $\{|b_j\rangle|N_j\rangle\}$, respectively. How could that change make any *ultimate* difference?

To answer this question, note that to reinstate the argument of Section 5, we would have to consider a measurement not of α and β at the later time, but of the compound systems $\alpha\&M$ and $\beta\&N$ at the later time. Moreover, the measurements would have to be of observables with eigenvectors of the form

$$(e_{11}|a_1\rangle + e_{12}|a_2\rangle) \otimes (m_{11}|M_1\rangle + m_{12}|M_2\rangle), \tag{52}$$

for example (where none of the coefficients are zero). But then, to measure transition probabilities, we would once again have to know properties initially possessed by the compound systems $\alpha\&M$ and $\beta\&N$. To do so, we would, again, have to introduce some measuring, or recording, device. Clearly the process would repeat itself. To sum up: if we use our initial 'probe' into the hidden properties of the system *as* a recording device, then it ruins the correlations that we need, whereas if we consider it to be a part of the system, then we will need some other 'probe', and we will need to use *it* as a recording device, so that it will ruin the correlations, and so on, and so on.

We are therefore left in the following situation: granted stability, there must be a preferred frame, but there is apparently no way for any experiment to determine what this frame is; without stability, modal interpretations must either again violate fundamental Lorentz-invariance, or accept some fairly unpalatable consequences, noted in Section 4. The modal interpretation therefore requires stability to maintain plausibility, but cannot have fundamental Lorentz-invariance. Just as the Bohm theory does, the modal interpretation needs a preferred frame that, nevertheless, cannot be found. As Bell (1987, p. 180) said of the Bohm theory: 'It seems an eccentric way to make a world.'

Appendix A: Lorentz-invariance in Bub's modal interpretation

The question immediately arises whether our result can be extended to other modal interpretations, and in particular, to Bub's (1992, 1995) interpretation (cf. Bub and Clifton (1996)). In his interpretation, the definite properties for the compound system $\alpha\&A\&\beta\&B$ are given by the nonzero projections of the quantum-mechanical state onto the eigenspaces of some preferred observable, **R**. Call these mutually orthogonal projections the 'Bub-projections'. From the Bub-projections, $\{P_i\}$, we generate an ortholattice, L, of properties for $\alpha\&A\&\beta\&B$ by setting

$$L = \{P|\forall i : P_i \leq P \text{ or } P_i \perp P\} \tag{53}$$

where, as usual, '$P_i \leq P$' indicates that the subspace corresponding to P_i is a subspace of the one corresponding to P and '$P_i \perp P$' indicates orthogonality of the corresponding subspaces. L is the set of definite properties for the compound system, $\alpha\&A\&\beta\&B$.[11]

[11] Such ortholattices, or ones like them, have been called variously 'faux-Boolean algebras' (Dickson 1995*b*,*a*), '(maximal) quasiBoolean algebras' (Bell and Clifton

For a subsystem, for example α, the definite properties are derived by restriction

$$L^\alpha = \{P^\alpha | P^\alpha \otimes \mathbf{1} \in L\} \tag{54}$$

where P^α is a projection on H^α (the Hilbert space for α), and $\mathbf{1}$ is the identity on $H^A \otimes H^\beta \otimes H^B$. We denote this set by L^α because it is in fact an ortholattice as well, and indeed takes the same form as L in (53). That is, there is a set of mutually orthogonal projections, $\{P_j^\alpha\}$, such that

$$L^\alpha = \{P^\alpha | \forall j : P_j^\alpha \leq P^\alpha \text{ or } P_j^\alpha \perp P^\alpha\}. \tag{55}$$

The set $\{P_j^\alpha\}$ is obtained as follows. First, from the set $\{P_i\}$ (the Bub-projections for the compound system), form the set $\{\widetilde{P}_i^\alpha\}$, where each $\widetilde{P}_i^\alpha$ satisfies

$$P_i \leq \widetilde{P}_i^\alpha \otimes \mathbf{1} \text{ and there is no } P^\alpha \text{ such that } P_i \leq P^\alpha \otimes \mathbf{1} < \widetilde{P}_i^\alpha \otimes \mathbf{1}. \tag{56}$$

In other words, $\widetilde{P}_i^\alpha$ is the 'smallest' projection on H^α such that $P_i \leq \widetilde{P}_i^\alpha \otimes \mathbf{1}$. Now partition the set $\{\widetilde{P}_i^\alpha\}$ into equivalence classes, indexed by j, taking two projections in $\{\widetilde{P}_i^\alpha\}$ to be equivalent whenever there is a sequence of projections in $\{\widetilde{P}_i^\alpha\}$, beginning with one of the projections and ending with the other, such that adjacent projections of the sequence are nonorthogonal. (One can check that this definition indeed yields an equivalence relation on $\{\widetilde{P}_i^\alpha\}$.) Finally, to get the set $\{P_j^\alpha\}$ from the set $\{\widetilde{P}_i^\alpha\}$, take the join (i.e., the span) of the elements in the jth equivalence class of $\{\widetilde{P}_i^\alpha\}$ to get the jth projection in the set $\{P_j^\alpha\}$. By construction, the projections in $\{P_j^\alpha\}$ are mutually orthogonal, and in fact are exactly the projections that

1995), and 'X-form sets' (Zimba and Clifton 1998). Some density-operator modal interpretations also choose such ortholattices as the set of definite properties, letting the spectral projections of a system's density operator take the place of the Bub-projections. However, we did not need to worry about the *full* set of definite properties for our previous argument, which focused just on the spectral projections themselves.

define the Bub subsystem property lattice in (55).[12] Because this al-

[12]We include the proof that the algorithm does in fact construct L^α as defined in (54) by restriction, because to our knowledge the algorithm and its proof have not appeared elsewhere. From (54) and (53), we have

$$L^\alpha = \{P^\alpha | \forall i : P_i \leq P^\alpha \otimes \mathbf{1} \text{ or } P_i \perp P^\alpha \otimes \mathbf{1}\}.$$

We must demonstrate that this lattice is precisely the lattice given in (55) — i.e., we must demonstrate the equivalence

$$\forall i : P_i \leq P^\alpha \otimes \mathbf{1} \text{ or } P_i \perp P^\alpha \otimes \mathbf{1} \iff \forall j : P_j^\alpha \leq P^\alpha \text{ or } P_j^\alpha \perp P^\alpha$$

for arbitrary P^α — under the assumption that the P_j^αs are as defined in the algorithm. The proof is tedious but elementary, in that it consists entirely in keeping the logic of the argument straight and the definitions of the various projections in clear view.

To prove $\Leftarrow$ above, fix arbitrary i and consider the equivalence class in which $\widetilde{P}_i^\alpha$ lies, along with the projection P_j^α spanned by that class. By definition, $P_i \leq \widetilde{P}_i^\alpha \otimes \mathbf{1}$ and $\widetilde{P}_i^\alpha \leq P_j^\alpha$, hence $P_i \leq P_j^\alpha \otimes \mathbf{1}$. But given the right-hand side above, either $P_j^\alpha \otimes \mathbf{1} \leq P^\alpha \otimes \mathbf{1}$ or $P_j^\alpha \otimes \mathbf{1} \perp P^\alpha \otimes \mathbf{1}$. Therefore, either $P_i \leq P^\alpha \otimes \mathbf{1}$ or $P_i \perp P^\alpha \otimes \mathbf{1}$, which is just the left-hand side (given that i was arbitrary).

For the proof of $\Rightarrow$, we shall assume the negation of the right-hand side above in order to deduce a contradiction with the left. Suppose, then, that for some j, $P_j^\alpha \not\leq P^\alpha$ and $P_j^\alpha \not\perp P^\alpha$. By definition of P_j^α, there must exist $\widetilde{P}_i^\alpha$ and $\widetilde{P}_{i'}^\alpha$ in the jth equivalence class such that

$$\widetilde{P}_i^\alpha \not\leq P^\alpha \text{ and } \widetilde{P}_{i'}^\alpha \not\perp P^\alpha, \tag{$*$}$$

where recall that equivalence of $\widetilde{P}_i^\alpha$ and $\widetilde{P}_{i'}^\alpha$ means that there is a sequence

$$\{\widetilde{P}_i^\alpha, \widetilde{P}_{i_1}^\alpha, \ldots, \widetilde{P}_{i_n}^\alpha, \widetilde{P}_{i'}^\alpha\}$$

with adjacent members of the sequence nonorthogonal. On the other hand, the left-hand side above is equivalent to $\forall i : P_i \leq P^\alpha \otimes \mathbf{1}$ or $P_i \leq (\mathbf{1} - P^\alpha) \otimes \mathbf{1}$. Because $\widetilde{P}_i^\alpha \otimes \mathbf{1}$ is the 'smallest' projection of the form $P^\alpha \otimes \mathbf{1}$ containing P_i, we therefore have $\forall i : \widetilde{P}_i^\alpha \leq P^\alpha$ or $\widetilde{P}_i^\alpha \leq (\mathbf{1} - P^\alpha)$, or equivalently

$$\forall i : \widetilde{P}_i^\alpha \leq P^\alpha \text{ or } \widetilde{P}_i^\alpha \perp P^\alpha.$$

In particular, this condition holds for all the indices in our sequence, so we have

$$\begin{aligned}
&\widetilde{P}_i^\alpha \leq P^\alpha \text{ or } \widetilde{P}_i^\alpha \perp P^\alpha, \text{ and} \\
&\widetilde{P}_{i_1}^\alpha \leq P^\alpha \text{ or } \widetilde{P}_{i_1}^\alpha \perp P^\alpha, \text{ and} \\
&\vdots \\
&\widetilde{P}_{i_n}^\alpha \leq P^\alpha \text{ or } \widetilde{P}_{i_n}^\alpha \perp P^\alpha, \text{ and} \\
&\widetilde{P}_{i'}^\alpha \leq P^\alpha \text{ or } \widetilde{P}_{i'}^\alpha \perp P^\alpha.
\end{aligned} \tag{$**$}$$

gorithm gives a quick method for constructing the lattice of definite properties for a given subsystem (particularly in the cases we shall deal with), we use it below.

It is immediately evident that, in general, Bub will not agree with Table 4.2, and it is therefore an open question whether our argument applies to his interpretation. Indeed, there is a version of Bub's interpretation that does escape our argument: take the preferred observable to be $\mathbf{1} \otimes \mathbf{A} \otimes \mathbf{1} \otimes \mathbf{B}$. (Recall that the observables **A** and **B** have the apparatus pointer-states as eigenstates.) However, given stability, any preferred observable of the form $\mathbf{u} \otimes \mathbf{A} \otimes \mathbf{v} \otimes \mathbf{B}$, where **u** and **v** are *not* the identity (but can be anything else) is subject to our argument. The case where **R** cannot be written as a tensor-product of observables on H^{α}, H^{A}, H^{β} and H^{B} respectively is up in the air. (But note that if one does not at least choose an observable that has factor-observables on each of H^{A} and H^{B}, then it could easily happen that for some initial quantum-mechanical states, the apparatuses end up with no nontrivial properties, or anyway, with properties other than eigenstates of **A** and **B**. Hence there might be good reason to choose only factorizable preferred observables.)

In other words, at least for observables that can be written as a tensor-product on $H^{\alpha} \otimes H^{A} \otimes H^{\beta} \otimes H^{B}$, Bub's interpretation must make a fundamental distinction between the measuring apparatuses and the measured systems to avoid our argument. It is not clear that having to make such a distinction is much better than simply violating fundamental Lorentz-invariance (but, of course, keeping empirical Lorentz-invariance) — but the choice of alternatives here may ultimately be a matter of taste.

We begin with the case where $\mathbf{R} = \mathbf{1} \otimes \mathbf{A} \otimes \mathbf{1} \otimes \mathbf{B}$. However, we are going to allow (indeed, insist) that the preferred observable, **R**,

---

We now obtain our sought-after contradiction, using $(*)$ and $(**)$. Suppose, as the first conjunct in $(*)$ asserts, that $\widetilde{P}_i^{\alpha} \not\leq P^{\alpha}$. Then the first line (i.e. conjunct) of $(**)$ requires $\widetilde{P}_i^{\alpha} \perp P^{\alpha}$. But since (by definition of the sequence) $\widetilde{P}_i^{\alpha} \not\perp \widetilde{P}_{i_1}^{\alpha}$, the second line of $(**)$ and $\widetilde{P}_i^{\alpha} \perp \widetilde{P}^{\alpha}$ force $\widetilde{P}_i^{\alpha} \perp P^{\alpha}$. Repeating the same argument with the third line (suppressed) forces $\widetilde{P}_{i_2}^{\alpha} \perp P^{\alpha}$, and so on all the way down the list in $(**)$. At the end, we reach the conclusion that $\widetilde{P}_{i'}^{\alpha} \perp P^{\alpha}$, in conflict with the second conjunct of $(*)$.

is time-dependent. The reason is that the apparatus-states are themselves time-dependent — recall that they are given by (in frame S) $|A_i^S(t)\rangle$ and $|B_j^S(t)\rangle$. Hence if Bub's interpretation is going to meet the minimal requirement that it make the apparatus-states definite at all times, it will, minimally, have to allow that **A** and **B** evolve in time so that they have the $|A_i^S(t)\rangle$ and the $|B_j^S(t)\rangle$ as eigenvectors at all times.

Another way to secure this result — and doing so this way would make no difference to our argument — is to allow that $\mathbf{A}^S$ and $\mathbf{B}^S$ are time-independent, but have *all* of the $|A_i^S(t)\rangle$ and $|B_j^S(t)\rangle$, respectively, as eigenvectors. That is, for all t, i, and j, $|A_i^S(t)\rangle$ is an eigenvector of (the time-independent) $\mathbf{A}^S$ and $|B_j^S(t)\rangle$ is an eigenvector of $\mathbf{B}^S$. Such observables would be degenerate (unless $|A_i^S(t)\rangle$ and $|B_j^S(t)\rangle$ are time-independent). For example, we could let $\mathbf{A}^S$ and $\mathbf{B}^S$ be the discretized position-observables described in Section 2. In that case, supposing that, for example, $\mathbf{A}^S$ measures deflection (positively or negatively) along the axis perpendicular to the line of flight of α, the eigenvectors of $\mathbf{A}^S$ will in fact be time-independent. (Note that the state of α is not, however, time-independent. After all, α moves along its line of flight; or more precisely, α's state is a wavepacket moving along the line of flight. But the degree of freedom along the line of flight is irrelevant to the value α has for $\mathbf{A}^S$, which depends only on α's deflection from the line of flight.)

In the case where **u** and **v** are the identity, then, we will write the definite-valued observable in frame S as $\mathbf{R}^S(t) = \mathbf{1} \otimes \mathbf{A}^S(t) \otimes \mathbf{1} \otimes \mathbf{B}^S(t)$. Of course, we must secure definiteness of the apparatus-observables in other frames too. The obvious way to do so is to choose the definite-valued observable in another frame, F, to be $\mathbf{R}^F(t) = \mathbf{1} \otimes \mathbf{A}^F(t) \otimes \mathbf{1} \otimes \mathbf{B}^F(t)$, where $\mathbf{A}^F(t) = (\Lambda_t^F)^{-1}\mathbf{A}^S(t)\Lambda_t^F$. As before, we will usually not include frames and times in our expressions, writing simply $\mathbf{R} = \mathbf{1} \otimes \mathbf{A} \otimes \mathbf{1} \otimes \mathbf{B}$ for the preferred observable in all frames and at all times. (Again, in the case where $\mathbf{A}^S$ and $\mathbf{B}^S$ are the discretized position-observables described in Section 2, it is especially perspicuous what the observables $\mathbf{A}^F$ and $\mathbf{B}^F$ are for $F \neq S$.)

Frame	$t = 0$	$t = 1$	$t = 2$
S	$\sum_i c_i \lvert a_i\rangle \lvert A_0\rangle \lvert b_i\rangle \lvert B_0\rangle$	$\sum_i c_i \lvert a_i\rangle \lvert A_0\rangle \lvert b_i\rangle \lvert B_0\rangle$	$\sum_i c_i e_{ij} d_{ik} \lvert a'_j\rangle \lvert A_j\rangle \lvert b'_k\rangle \lvert B_k\rangle$
L	$\sum_i c_i \lvert a_i\rangle \lvert A_0\rangle \lvert b_i\rangle \lvert B_0\rangle$	$\sum_i c_i e_{ij} \lvert a'_j\rangle \lvert A_j\rangle \lvert b_i\rangle \lvert B_0\rangle$	$\sum_i c_i e_{ij} d_{ik} \lvert a'_j\rangle \lvert A_j\rangle \lvert b'_k\rangle \lvert B_k\rangle$
R	$\sum_i c_i \lvert a_i\rangle \lvert A_0\rangle \lvert b_i\rangle \lvert B_0\rangle$	$\sum_i c_i d_{ik} \lvert a_i\rangle \lvert A_0\rangle \lvert b'_k\rangle \lvert B_k\rangle$	$\sum_i c_i e_{ij} d_{ik} \lvert a'_j\rangle \lvert A_j\rangle \lvert b'_k\rangle \lvert B_k\rangle$

Table 4.3: Bub-projections for $\mathbf{R} = \mathbf{1} \otimes \mathbf{A} \otimes \mathbf{1} \otimes \mathbf{B}$

For the case $\mathbf{R} = \mathbf{1} \otimes \mathbf{A} \otimes \mathbf{1} \otimes \mathbf{B}$, the Bub-projections (or rather, the vectors that generate the rays onto which the Bub-projections project) are, for $j, k = 1, 2$, as shown in Table 4.3.

Actually, whether the vectors in Table 4.3 in fact correspond to the Bub-projections depends on whether they are nonzero, i.e., on whether the coefficients attached to the vectors in Table 4.3 are nonzero. But no harm is done by assuming that all of the vectors in Table 4.3 correspond to Bub-projections, because any that are in fact the zero vector have probability zero anyway. (The transition probability from any state with nonzero probability to a state with probability zero must be zero in every frame, and transition probabilities from states with zero probability to states with zero probability can be defined however we like. Therefore, including states of probability zero will not get in the way of Lorentz-invariance.)

We must now use the algorithm described earlier to find the definite properties for subsystems. Consider first the case $t = 0$ in frame S. The (one and only) Bub-projection in this case is not factorizable as $P^\alpha \otimes \mathbf{1}$. Hence the smallest projection of the form $P^\alpha \otimes \mathbf{1}$ that contains the Bub-projection must be the identity, $\mathbf{1} \otimes \mathbf{1}$. The reasoning for other frames at $t = 0$ is the same. (If the dimension of H^α is greater than two, then the smallest projection of the form $P^\alpha \otimes \mathbf{1}$ that contains the Bub-projection is $P \otimes \mathbf{1}$, where P is the projection onto the two-dimensional subspace of H^α spanned by the $|a_i\rangle$. We will consider only the case where the dimension of H^α and of H^β is two, but this assumption could be dropped without affecting our argument.)

The case of frame L at time $t = 1$ is slightly different. For a given

Frame	$t=0$	$t=1$	$t=2$
S	$\mathbf{1}(\alpha), \lvert A_0\rangle, \mathbf{1}(\beta), \lvert B_0\rangle$	$\mathbf{1}(\alpha), \lvert A_0\rangle, \mathbf{1}(\beta), \lvert B_0\rangle$	$\lvert a'_j\rangle, \lvert A_j\rangle, \lvert b'_k\rangle, \lvert B_k\rangle$
L	$\mathbf{1}(\alpha), \lvert A_0\rangle, \mathbf{1}(\beta), \lvert B_0\rangle$	$\lvert a'_j\rangle, \lvert A_j\rangle, \mathbf{1}(\beta), \lvert B_0\rangle$	$\lvert a'_j\rangle, \lvert A_j\rangle, \lvert b'_k\rangle, \lvert B_k\rangle$
R	$\mathbf{1}(\alpha), \lvert A_0\rangle, \mathbf{1}(\beta), \lvert B_0\rangle$	$\mathbf{1}(\alpha), \lvert A_0\rangle, \lvert b'_k\rangle, \lvert B_k\rangle$	$\lvert a'_j\rangle, \lvert A_j\rangle, \lvert b'_k\rangle, \lvert B_k\rangle$

Table 4.4: Bub's definite properties for $\mathbf{R} = \mathbf{1} \otimes \mathbf{A} \otimes \mathbf{1} \otimes \mathbf{B}$.

j, the projection

$$\mathbf{1}(\alpha) \otimes \mathbf{1}(A) \otimes \Big(\sum_{i,k} c_i e_{ij} c_k^* e_{kj}^* |b_i\rangle\langle b_k| \Big) \otimes \mathbf{1}(B) \tag{57}$$

is the minimal projection containing the corresponding Bub-projection (for the same value of j), where $\mathbf{1}(\sigma)$ is the identity on H^σ. There are now two cases to consider. Letting $\sum_{i,k} c_i e_{ij} c_k^* e_{kj}^* |b_i\rangle\langle b_k| = \widetilde{P}_j^\beta$, we have:

Case 1 — non trivial properties: $\widetilde{P}_1^\beta \perp \widetilde{P}_2^\beta$ or $\widetilde{P}_1^\beta = \widetilde{P}_2^\beta$
Case 2 — trivial property only: $\widetilde{P}_1^\beta \not\perp \widetilde{P}_2^\beta$ and $\widetilde{P}_1^\beta \neq \widetilde{P}_2^\beta$

In case 1, the equivalence classes mentioned in the algorithm are just the projections $\widetilde{P}_1^\beta$ and $\widetilde{P}_2^\beta$. There are then two sub-cases. Either $\sum_{i,k} c_i e_{i1} c_k^* e_{k1}^* = \sum_{i,k} c_i e_{i2} c_k^* e_{k2}^*$, in which case $\widetilde{P}_1^\beta = \widetilde{P}_2^\beta$ or $\sum_i |c_i|^2 e_{i1} e_{i2}^* = 0$, in which case $\widetilde{P}_1^\beta \perp \widetilde{P}_2^\beta$. Either way, β has $\widetilde{P}_1^\beta$ and $\widetilde{P}_2^\beta$ (which might be the same) as possible properties at $t=1$. In case 2, the algorithm dictates that we take the span of $\widetilde{P}_1^\beta$ and $\widetilde{P}_2^\beta$ to get a possible property, which, in this case, is $\mathbf{1}(\beta)$. Hence it is possible for β to come to have nontrivial possible properties in L as a result of the measurement on α. We will return to this point below, but for now we consider just the generic case, where β can have only the trivial property $\mathbf{1}(\beta)$ in L at $t=1$. In this case, following the algorithm in the same way for the other frames and times, we get the results in Table 4.4 (again for $j, k = 1, 2$).

We can derive all of the transition probabilities from the information in the table, plus an assumption of apparatus-stability. In frame S, the $0 \to 1$ transition probabilities are obviously trivial, and the $1 \to 2$ transition probabilities must be just the quantum-mechanical probabilities at time $t^S = 2$, because there is just one possible modal state at $t^S = 1$. Similarly, in both L and R, the $0 \to 1$ transition probabilities must just be the quantum-mechanical probabilities at time 1. To get the $1 \to 2$ transition probabilities in these frames, we assume apparatus-stability. In frame L, for example, we assume that if A possessed $|A_j\rangle$ at $t^L = 1$, then it also must possess $|A_j\rangle$ at $t^L = 2$. It follows that the probability of a transition from $|a'_j\rangle, |A_j\rangle, \mathbf{1}(\beta), |B_0\rangle$ to $|a'_j\rangle, |A_j\rangle, |b'_k\rangle, |B_k\rangle$ is just the *conditional* probability of $|b'_k\rangle$ given $|a'_j\rangle$, where this conditional probability is calculated in the usual way using the quantum state at $t^L = 2$. (The reasoning here is the same sort of reasoning that we used when we applied Bayes' rule to get the $0 \to 1$ transitions in the Vermaas-Dieks version.)

The same account applies even when β or α has nontrivial properties in L and R, respectively. For β in L, this point is obvious in the case where the projections of (57) are the same, because the transitions for β must remain trivial. When the projections in (57) are orthogonal, β is perfectly correlated with α, and then again the account given above for the transition probabilities must hold. An analogous argument applies in the case where α has nontrivial properties in R.

There is no need to write down the explicit expressions for these transition probabilities. We know already that they must turn out to be Lorentz-invariant, because they are all just quantum-mechanical probabilities in some form or other, and we know that quantum-mechanical probabilities are Lorentz-invariant.

But as we mentioned, it can happen, for example, that the possible states of β *change* as a result of the measurement on α. This feature of Bub's interpretation might be unsettling to some. Indeed, it looks very much like a form of Fleming's (1989) view that properties of systems are hyperplane-dependent. (And recall that the Vermaas-Dieks version avoids hyperplane dependence of this sort.)

But even if the hyperplane-dependence of definite properties is not worrisome, one might still be worried about the choice $\mathbf{R} = \mathbf{1} \otimes \mathbf{A} \otimes \mathbf{1} \otimes \mathbf{B}$, for it seems to require a fundamental distinction between the measuring apparatuses and the measured systems. (How else would we know where to put the identity operators?) The question naturally arises whether Bub can get Lorentz-invariance while allowing $\mathbf{u}$ and $\mathbf{v}$ to be nontrivial.

Of course, *one* choice for the $\mathbf{u}$ and $\mathbf{v}$ we know will not work, namely, the one that commits Bub to Table 4.2. Bub and Clifton (1996) show how to choose $\mathbf{R}$ for a two-component compound system, $\alpha\&\beta$, that recovers the Vermaas-Dieks interpretation for *one* of the components. For example, if W^{α} is α's density operator, then choose $\mathbf{R} = W^{\alpha} \otimes \mathbf{1}(\beta)$ to recover the Vermaas-Dieks interpretation for α. Their result is easily extended to our case, where we have four component systems, each of whose spectral projections is one-dimensional.[13] We will not bother with the details here — obviously if we choose an $\mathbf{R}$ that reproduces Table 4.2, then Bub's interpretation is not fundamentally Lorentz-invariant.

But what about other choices for $\mathbf{R}$, still of the form $\mathbf{R} = \mathbf{u} \otimes \mathbf{A} \otimes \mathbf{v} \otimes \mathbf{B}$? The calculations in this case would, we fear, try the patience of any reader. Suffice it to say that the procedure is exactly as it was in the case of the Vermaas-Dieks interpretation. Defining $\mathbf{u}$ and $\mathbf{v}$ so that their eigenvectors obey

$$\langle u^S_m(t)|a^S_i(t)\rangle =_{df} \mu^S_{mi}(t) \qquad \langle v^S_n(t)|b^S_j(t)\rangle =_{df} \nu^S_{nj}(t) \tag{58}$$

we get a table of definite properties for the subsystems, along with their probabilities, as in Table 5.

Now just follow the steps of Section 5. The calculations in this case tend to be an order of magnitude more complex, but nothing conceptually new enters the picture. The result is that for some quantum states, and some choices of $\mathbf{a}'$ and $\mathbf{b}'$, Bub's interpretation (with apparatus-stability) cannot be fundamentally Lorentz-invariant.

[13] To recover the Vermaas-Dieks properties for *all four* subsystems at once, it suffices to take $\mathbf{R}$ to be the (factorizable) observable whose eigenspaces are given by the various 4-fold tensor products one can form from the eigenspaces of each of the four reduced states of the subsystems.

Frame	$t = 0$	$t = 1$	$t = 2$
S	$\|u_m\rangle, \|A_0\rangle, \|v_n\rangle, \|B_0\rangle$	$\|u_m\rangle, \|A_0\rangle, \|v_n\rangle, \|B_0\rangle$	$\|u_m\rangle, \|A_f\rangle, \|v_n\rangle, \|B_g\rangle$
prob	$\|\sum_i c_i \mu_{mi} \nu_{ni}\|^2$	$\|\sum_i c_i \mu_{mi} \nu_{ni}\|^2$	$\|\sum_i c_i e_{if} d_{ig} \mu_{mf} \nu_{ng}\|^2$
L	$\|u_m\rangle, \|A_0\rangle, \|v_n\rangle, \|B_0\rangle$	$\|u_m\rangle, \|A_f\rangle, \|v_n\rangle, \|B_0\rangle$	$\|u_m\rangle, \|A_f\rangle, \|v_n\rangle, \|B_g\rangle$
prob	$\|\sum_i c_i \mu_{mi} \nu_{ni}\|^2$	$\|\sum_i c_i e_{if} \mu_{mf} \nu_{ni}\|^2$	$\|\sum_i c_i e_{if} d_{ig} \mu_{mf} \nu_{ng}\|^2$
R	$\|u_m\rangle, \|A_0\rangle, \|v_n\rangle, \|B_0\rangle$	$\|u_m\rangle, \|A_0\rangle, \|v_n\rangle, \|B_g\rangle$	$\|u_m\rangle, \|A_f\rangle, \|v_n\rangle, \|B_g\rangle$
prob	$\|\sum_i c_i \mu_{mi} \nu_{ni}\|^2$	$\|\sum_i c_i d_{ig} \mu_{mi} \nu_{ng}\|^2$	$\|\sum_i c_i e_{if} d_{ig} \mu_{mf} \nu_{ng}\|^2$

Table 4.5: Bub's definite properties for $\mathbf{R} = \mathbf{u} \otimes \mathbf{A} \otimes \mathbf{v} \otimes \mathbf{B}$ and their probabilities

A yet more general case is the one where $\mathbf{R}$ is not even factorizable as $\mathbf{R} = \mathbf{R}^{\alpha} \otimes \mathbf{R}^{A} \otimes \mathbf{R}^{\beta} \otimes \mathbf{R}^{B}$ for *any* $\mathbf{R}^{\alpha}, \ldots \mathbf{R}^{B}$. We have not dared to try the calculation for this case, which is order of magnitude more complex than the case $\mathbf{R} = \mathbf{u} \otimes \mathbf{A} \otimes \mathbf{v} \otimes \mathbf{B}$. However, as we noted earlier, it is not obvious how this general case is going to recover the definiteness of apparatus-states.

Appendix B: Hardy's argument against Lorentz-invariance

We have suggested that discussions of locality and Lorentz-invariance sometimes suffer from over-generality, to the extent that a given dynamical model of the EPR-Bohm experiment might slip through the cracks of a very general argument. A nice example to illustrate our suggestion is Hardy's argument that no hidden-variables theory can be Lorentz-invariant. In this Appendix, we show how the modal interpretation slips through the cracks in Hardy's argument and note that the Bohm theory does as well. (Of course, we do not mean to suggest that either is fundamentally Lorentz-invariant, but only that Hardy's argument does not establish that they are not.)

Hardy (1992) gave his argument in terms of a double-interferometry experiment on neutrons, but it can be conveniently restated in the terms that we have already used (cf. Clifton and Niemann (1992)). The state of the pair of particles is

$$|\Psi|\rangle = \tfrac{1}{\sqrt{3}}(|a_1\rangle|b_2\rangle + |a_2\rangle|b_1\rangle + |a_2\rangle|b_2\rangle), \tag{59}$$

and Hardy assumes that this state does not change in time. (Hardy does not include the apparatuses explicitly, and we will not either.) We consider the measurement of observables $\mathbf{a}'$ and $\mathbf{b}'$, related to the observables $\mathbf{a}$ and $\mathbf{b}$ (whose eigenvectors are the $|a_i\rangle$ and the $|b_j\rangle$) as

$$|a_1\rangle = \tfrac{1}{\sqrt{2}}(|a'_1\rangle + |a'_2\rangle) \qquad |a_2\rangle = \tfrac{1}{\sqrt{2}}(|a'_1\rangle - |a'_2\rangle) \tag{60}$$

$$|b_1\rangle = \tfrac{1}{\sqrt{2}}(|b'_1\rangle + |b'_2\rangle) \qquad |b_2\rangle = \tfrac{1}{\sqrt{2}}(|b'_1\rangle - |b'_2\rangle). \tag{61}$$

(For example, one may think of $\mathbf{a}$ and $\mathbf{b}$ as x-spin observables, and $\mathbf{a}'$ and $\mathbf{b}'$ as z-spin observables.) The frames S, L, and R, and the various times in those frames, are defined as before. Given these definitions, we have the following quantum-mechanical probabilities:

$$p^S(a'_2, b'_2 \text{ at } t^S = 2) > 0 \tag{62}$$

$$p^S(a_1, b_1 \text{ at } t^S = 1) = 0 \tag{63}$$

$$p^L(b_1 \text{ at } t^L = 1 | a'_2 \text{ at } t^L = 1) = 1 \tag{64}$$

$$p^R(a_1 \text{ at } t^R = 1 | b'_2 \text{ at } t^R = 1) = 1. \tag{65}$$

Now, Hardy argues, (62) entails that sometimes, in frame S, the result (a'_2, b'_2) occurs at $t^S = 2$. Implicitly invoking stability, Hardy then claims that, by Lorentz-invariance (for us, fundamental Lorentz-invariance), when a'_2 occurs at $t^S = 2$, frame L must attribute a'_2 to α at $t^L = 1$. Then (64) entails that L must assign b_1 to β at $t^L = 1$. The reasoning is similar to that of Einstein, Podolsky, and Rosen (1935). They say that when the value of a physical quantity can be predicted with certainty, then the system *possesses* that value, regardless of whether the presence of that value is actually confirmed by a measurement. Hardy says that if a given projection has probability 1, then it is possessed. Hence, because the occurrence of a'_2 confers probability 1 on (the projection corresponding to) b_1, β must possess b_1 whenever a'_2 occurs. By a similar argument, and using (65), frame R must attribute a_1 to α at $t^R = 1$. Again invoking fundamental Lorentz-invariance, Hardy concludes that S must assign (a_1, b_1) to the pair at $t^S = 1$. But by (63), it cannot do so.

Hardy's argument does not work against either the Bohm theory or modal interpretations. The case of the Bohm theory is discussed in a reply to Hardy's argument by Berndl and Goldstein (1994). Of course, we are not claiming (nor do Berndl and Goldstein) that the Bohm theory is fundamentally Lorentz-invariant. The point is that the non-invariance of the Bohm theory must be shown by other means. As Clifton *et al.* (1992) pointed out, what allows the Bohm theory to escape arguments such as Hardy's is its *contextuality* — the properties of a system depend on what observables are measured, both on it and on other systems. (This point and the following one were also made by Clifton and Niemann (1992).) Of course, the Bohm theory's contextualism may be considered a form of nonlocality, but again, nonlocality is not at issue here (and was not Hardy's main concern either). We are interested in Lorentz-invariance. Nonlocality does not in general imply non-Lorentz-invariance. Whether it does in a specific case depends on the details of the theory being considered.

As for modal interpretations, the main point is that using the quantum probabilities (64) and (65) to establish the possessed properties in L and R presupposes that b_1 is one of the definite properties in L at $t^L = 1$, and similarly for a_1 in R at $t^R = 1$. But they are *not*, in the Vermaas-Dieks interpretation, and they need not be in Bub's interpretation. In other words, modal interpretations in general *deny* Einstein, Podolsky, and Rosen's criterion for the possession of properties. For modal interpretations, to be able to predict with (conditional) probability 1 that the value of a physical quantity is x means only that *if the quantity were measured, we would get the result* x, nothing more.

For example, in the Vermaas-Dieks interpretation, the reduced state of β in L at $t^L = 1$ is

$$W_\beta^L = \tfrac{1}{3}|b_2\rangle\langle b_2| + \tfrac{2}{3}|b_2'\rangle\langle b_2'|. \tag{66}$$

The right-hand side of (66) is not the spectral resolution of W_β^L, but it is clear from (66) that $|b_1\rangle\langle b_1|$ is not in the spectral resolution of W_β^L. Indeed, a calculation shows that the spectral resolution of W_β^L is

$$(\tfrac{3+\sqrt{5}}{6})|\varphi_1\rangle\langle\varphi_1| + (\tfrac{3-\sqrt{5}}{6})|\varphi_2\rangle\langle\varphi_2| \tag{67}$$

where

$$\begin{aligned} |\varphi_1\rangle &= (\tfrac{1}{\sqrt{2}}\sqrt{1-\tfrac{1}{\sqrt{5}}})|b_1\rangle + (\tfrac{1}{\sqrt{2}}\sqrt{1+\tfrac{1}{\sqrt{5}}})|b_2\rangle \\ |\varphi_2\rangle &= -(\tfrac{1}{\sqrt{2}}\sqrt{1+\tfrac{1}{\sqrt{5}}})|b_1\rangle + (\tfrac{1}{\sqrt{2}}\sqrt{1-\tfrac{1}{\sqrt{5}}})|b_2\rangle. \end{aligned} \tag{68}$$

In other words, despite (64), we cannot conclude from the attribution of a_2' to α in L at $t^L = 1$, that β possesses b_1. For Vermaas-Dieks, β won't possess b_1 unless **b** is *actually* measured; but then, of course, the conditions necessary to set up Hardy's argument can no longer be satisfied.

A similar point can be made for Bub. In his case, of course, which properties are among the definite properties depends on how we choose the definite-valued observable. In fact, from Table 4.5, we see that we would have to choose **v=b** to get Hardy's argument off the ground. So again, Bub's interpretation can escape Hardy's argument with a judicious choice for **R**. Indeed, we have already exhibited a choice for **R** that yields a fundamentally Lorentz-invariant theory, modulo worries about hyperplane dependence.

Bibliography

Aharonov, Y. & Albert, A. (1981), 'Can we make sense of the measurement process in relativistic quantum mechanics?', *Physical Review Letters* **24**, 337–59.

Bacciagaluppi, G. & Dickson, M. (1999), 'Dynamics for modal interpretations', *Foundations of Physics* **29**(8), 1165–201.

Bacciagaluppi, G., Donald, M. & Vermaas, P. (1995), 'Continuity and discontinuity of definite properties in the modal interpretation', *Helvetica Physica Acta* **68**, 679–704.

Bell, J. L. & Clifton, R. (1995), 'QuasiBoolean algebras and simultaneously definite properties in quantum mechanics', *International Journal of Theoretical Physics* **34**, 2409–21.

Bell, J. S. (1987), Beables for quantum field theory, in 'Speakable and Unspeakable in Quantum Mechanics', Cambridge University Press, Cambridge, pp. 173–80.

Berndl, K. & Goldstein, S. (1994), 'Comment on "Quantum mechanics, local realistic theories, and Lorentz-invariance realistic theories"', *Physical Review Letters* **72**, 780.

Berndl, K., Dürr, D., Goldstein, S. & Zanghì, N. (1996), 'Nonlocality, Lorentz invariance, and Bohmian quantum theory', *Physical Review A* **53**, 2062–73.

Bohm, D. (1952), 'A suggested interpretation of the quantum theory in terms of 'hidden' variables, I and II', *Physical Review* **85**, 166–93.

Bohm, D. & Hiley, B. (1993), *The Undivided Universe: An Ontological Interpretation of Quantum Theory*, Routledge, London.

Bub, J. (1992), 'Quantum mechanics without the projection postulate', *Foundations of Physics* **22**, 737–54.

Bub, J. (1995), 'On the structure of quantal proposition systems', **24**, 1261–79.

Bub, J. & Clifton, R. (1996), 'A uniqueness theorem for "No collapse" interpretations of quantum mechanics', *Studies in History and Philosophy of Modern Physics* **27**, 181–219. Chapter 2 of this volume.

Clifton, R. (1995), 'Independently motivating the Kochen-Dieks modal interpretation of quantum mechanics', *British Journal for the Philosophy of Science* **46**, 33–57. Chapter 1 of this volume.

Clifton, R. (1996), 'The properties of modal interpretations of quantum mechanics', *British Journal for the Philosophy of Science* **47**(3), 371–98.

Clifton, R. & Niemann, P. (1992), 'Locality, Lorentz invariance, and linear algebra: Hardy's theorem for two entangled spin-*s* particles', *Physics Letters. A* **166**, 177–84.

Clifton, R., Pagonis, C. & Pitowksy, I. (1992), Relativity, quantum mechanics, and EPR, in D. Hull, M. Forbes & K. Okruhlik, eds, 'Proceedings of the Philosophy of Science Association 1992, Vol. I', Philosophy of Science Assocation, East Lansing, MI, pp. 114–28.

Cushing, J. (1994), Locality/separability: Is this necessarily a useful distinction?, in D. Hull, M. Forbes & R. Burian, eds, 'The Philosophy of Science Association 1994, Vol. I', Philosophy of Science Association, East Lansing, MI, pp. 107–16.

Dickson, W. M. (1995*a*), 'Faux-Boolean algebras and classical models', *Foundations of Physics Letters* **8**(5), 401–15.

Dickson, W. M. (1995*b*), 'Faux-Boolean algebras, classical probability, and determinism', *Foundations of Physics Letters* **8**(3), 231–42.

Dickson, W. M. (1997), On the plurality of dynamics: Transition probabilities and modal interpretations, in R. Healey & G. Hellman, eds, 'Quantum Measurement: Beyond Paradox', University of Minnesota Press, Minneapolis, pp. 160–82.

Dieks, D. (1988), 'The formalism of quantum theory: an objective description of reality?', *Annalen der Physik* **45**(3), 174–90.

Dieks, D. (1989), 'Quantum mechanics without the projection postulate and its realistic interpretation', *Foundations of Physics* **19**(11), 1397–423.

Einstein, A., Podolsky, B. & Rosen, N. (1935), 'Can quantum-mechanical description of physical reality be considered complete?', *Physical Review* **47**, 777–80.

Fine, A. (1970), 'Insolubility of the quantum measurement problem', *Physical Review D* **2**, 2783–7.

Fleming, G. (1989), Lorentz invariant state reduction, and localization, in A. Fine & J. Leplin, eds, 'The Philosophy of Science Association 1988, Vol. II', Philosophy of Science Association, East Lansing, MI, pp. 112–26.

Hardy, L. (1992), 'Quantum mechanics, local realistic theories, and Lorentz-invariant realistic theories', *Physical Review Letters* **68**(20), 2981–4.

Healey, R. (1989), *The philosophy of quantum mechanics*, Cambridge University Press, Cambridge.

Jarrett, J. P. (1984), 'On the physical significance of the locality conditions in the Bell arguments', *Noûs* **18**(4), 569–89.

Jones, M. & Clifton, R. (1993), Against experimental metaphysics, in P. French, T. Uehling & H. Wettstein, eds, 'Midwest Studies in Philosophy, Vol. 18', University of Notre Dame Press, Notre Dame, IN, pp. 295–316.

Kochen, S. (1979), The interpretation of quantum mechanics, unpublished notes, Princeton University.

Kochen, S. (1985), A new interpretation of quantum mechanics, in P. Lahti & P. Mittelstaedt, eds, 'Symposium on the Foundations of Modern Physics', World Scientific, Singapore, pp. 151–70.

Maudlin, T. (1994), *Quantum non-locality and relativity*, Blackwell, Oxford.

Pitowsky, I. (1989), *Quantum probability — quantum logic*, Springer-Verlag, Berlin.

Redhead, M. L. G. (1986), Relativity and quantum mechanics — conflict or peaceful coexistence?, in 'New techniques and ideas in quantum measurement theory', New York Acad. Sci., New York, pp. 14–20.

Shimony, A. (1986), Events and processes in the quantum world, in 'Quantum concepts in space and time', Oxford University Press, New York, pp. 182–203.

Valentini, A. (1991*a*), 'Signal-locality, uncertainty, and the subquantum *H*-theorem. I', *Physics Letters. A* **156**, 5–11.

Valentini, A. (1991*b*), 'Signal-locality, uncertainty, and the subquantum *H*-theorem. II', *Physics Letters. A* **158**, 1–8.

van Fraassen, B. C. (1979), 'Hidden variables and the modal interpretation of quantum theory', *Synthese* **42**, 155–65.

van Fraassen, B. C. (1991), *Quantum mechanics: An empiricist view*, Oxford University Press, New York.

van Fraassen, B. C. (1997), 'Modal interpretation of repeated measurement. A rejoinder to S. Leeds and R. Healey: "A note on van Fraassen's modal interpretation of quantum mechanics"', *Philosophy of Science* **64**, 669–76.

Vermaas, P. E. (1996), 'Unique transition probabilities in the modal interpretation', *Studies in History and Philosophy of Modern Physics* **27**, 133–59.

Vermaas, P. E. & Dieks, D. (1995), 'The modal interpretation of quantum mechanics and its generalization to density operators', *Foundations of Physics* **25**(1), 145–58.

Weinberg, S. (1996), *The quantum theory of fields. Vol. I*, Cambridge University Press, Cambridge.

Wigner, E. (1931), *Gruppentheorie und ihre Anwendung auf die Quantenmechanik der Atomspektren*, F. Vieweg und Sohn Akt.-Ges., Braunschweig.

Zimba, J. & Clifton, R. (1998), Valuations on functionally closed sets of quantum-mechanical observables and von Neumann's no-hidden-variables theorem, in D. Dieks & P. E. Vermaas, eds, 'The Modal Interpretation of Quantum Mechanics', Kluwer, Dordrecht, pp. 69–101.

Part II

Foundations of Algebraic Quantum Field Theory

Chapter 5

The modal interpretation of algebraic quantum field theory

1 The modal interpretation of nonrelativistic quantum theory

The modal interpretation is actually a family of interpretations sharing the feature that a system's density operator constrains the possibilities for assigning definite values to its observables (Healey 1989, Dieks & Vermaas 1998). We start with a brief review of the basic idea behind these interpretations from an algebraic point of view (Zimba & Clifton 1998, Halvorson & Clifton 1999).

Consider a 'universe' comprised of a quantum system U, with finitely many degrees of freedom, represented by a Hilbert space $\mathsf{H}_U = \bigotimes_i \mathsf{H}_i$. At any given time, U will occupy a pure vector state $x \in \mathsf{H}_U$ that determines a reduced density operator D_S on the Hilbert space $\mathsf{H}_S = \bigotimes_{i \in S} \mathsf{H}_i$ of any subsystem $S \subseteq U$. Let $\mathcal{B}(\mathsf{H}_S)$ de-

The author wishes to thank Hans Halvorson (for supplying the argument immediately following Eq. (2) and suggesting the use of the KMS condition in the proof of Proposition 1), and Reinhard Werner (for helpful correspondence about centralizers and inspiring Proposition 2).

note the algebra of all bounded operators on H_S, and for any single operator or family of operators $T \subseteq \mathcal{B}(\mathsf{H}_S)$, let T' denote its commutant (i.e., all operators on H_S that commute with those in T). Let P_S denote the projection onto the range of D_S, and consider the subalgebra of $\mathcal{B}(\mathsf{H}_S)$ given by the direct sum

$$\mathcal{M}_S \equiv P_S^{\perp}\mathcal{B}(\mathsf{H}_S)P_S^{\perp} + D_S''P_S. \tag{1}$$

When S is not entangled with its environment $\overline{S}$ (represented by $\mathsf{H}_{\overline{S}} = \bigotimes_{i \notin S}\mathsf{H}_i$), D_S will itself be a pure state, induced by a unit vector $y \in \mathsf{H}_S$. In this case, $\mathcal{M}_S$ consists of all operators with y as an eigenvector — the self-adjoint members of which are taken, in orthodox quantum theory, to be the observables of S with definite values. On the other hand, when there *is* entanglement and D_S is mixed — in particular, in the extreme case $P_S = I$ (where essentially every vector state $y \in \mathsf{H}_S$ is a component of the mixture) — then $\mathcal{M}_S = D_S''$, and $\mathcal{M}_S$ consists simply of all functions of D_S. In this case, orthodox quantum theory has nothing to say about the properties of S. Thus, when S is Schrödinger's cat entangled with some potentially cat killing device $\overline{S}$, we get the infamous measurement problem.

In nonatomic versions of the modal interpretation (van Fraassen 1991, Kochen 1985, Clifton 1995), there is no preferred partition of the universe into subsystems. Any particular subsystem $S \subseteq U$ is taken to have definite values for all the self-adjoint operators that lie in $\mathcal{M}_S$, and this applies *whether or not* D_S is pure. The values of these observables are taken to be distributed according to the usual Born rule. Thus, the expectation of any observable $A \in \mathcal{M}_S$ is $\mathrm{Tr}(D_S A)$, and the probability that A possesses some particular value a_j is $\mathrm{Tr}(D_S P^j)$, where $P^j \in \mathcal{M}_S$ is the corresponding eigenprojection of A. As in orthodox quantum theory, *which* precise value for A occurs on a given occasion (from amongst those with nonzero probability in state D_S) is not fixed by the interpretation. However, the occurrence of a value does not require that it be 'measured'. And, unlike orthodox quantum theory, no miraculous collapse is needed to solve the measurement problem. Instead, after a typical unitary 'measurement' interaction between two parts of the universe — O the 'measured' system, and A the apparatus — decoherence induced by A's coupling to the environment $\overline{O \cup A}$ will force the density operator D_A of the apparatus to diagonalize in a basis extremely close to one which diagonalizes the pointer observable of A (Bacciagaluppi & Hemmo 1996). Thus, after the measurement, the definite-valued observables in $\mathcal{M}_A$ will be such that the pointer points!

In atomic versions of the modal interpretation (Healey 1989, Bacciagaluppi & Dickson 1999, Spekkens & Sipe 2001), one does not tell a separate story about the definite-valued observables for each subsystem $S \subseteq U$. Rather, S is taken to inherit properties from those of its atomic components, represented by the individual Hilbert spaces in $\bigotimes_{i \in S} \mathsf{H}_i$. In the approach favored by Dieks (1998) (cf. Bacciagaluppi & Dickson 1999), each atomic system i possesses definite values for all the observables in $\mathcal{M}_i$, as determined by the corresponding atomic density operator D_i in accordance with (1). The definite-valued observables of S are then built up by embedding each $\mathcal{M}_i$ in $\mathcal{B}(\mathsf{H}_S)$ (via tensoring it with the identity on $\mathsf{H}_{\overline{S}}$), and taking the von Neumann subalgebra of $\mathcal{B}(\mathsf{H}_S)$ generated by all these embeddings. The definite properties of S will therefore include all projections $\bigotimes_{i \in S} P_i^{j(i)}$ that are tensor products of the spectral projections of the individual atomic density operators, and their joint probabilities are again taken to be given by the usual Born rule $\mathrm{Tr}[D_S(\bigotimes_{i \in S} P_i^{j(i)})]$. In the presence of decoherence, the expectation is that atomic versions of the modal interpretation can yield essentially the same resolution of the measurement problem as do nonatomic versions (Dieks 1998, Vermaas 1998).

In both versions of the modal interpretation, the observables with definite values must change over time as a function of the (generally, non-unitary) evolution of the reduced density operators of the systems involved. In principle, this evolution can be determined from the (unitary) Schrödinger evolution of the universal state vector $x \in \mathsf{H}_U$ (Bacciagaluppi *et al.* 1995). But the evolution of the precise values of the definite-valued observables themselves is not determined by the Schrödinger equation. Various more or less natural proposals have been made for 'completing' the modal interpretation with a dynamics for values (Bacciagaluppi & Dickson 1999, Vermaas 1996, Ax & Kochen 1999). Unfortunately, it has been shown that the most natural proposals for a dynamics, particularly in the case of atomic modal interpretations, must break Lorentz-invariance (Dickson & Clifton 1998). Most of Dieks' recent Letter (Dieks 2000) is concerned to address this dynamics problem by appropriating ideas from the decoherent histories approach to quantum theory (cf. Hemmo 1998). However, we shall focus here entirely on the viability of Dieks' new proposal for picking out definite-valued observables of a relativistic quantum field that are associated with approximately point-sized regions of Minkowski spacetime (Dieks 2000, Sec. 5).

2 Critique of Dieks' proposal

Dieks' stated aim is to see if the modal interpretation can achieve sensible results in the context of quantum field theory. For this purpose, he adopts the formalism of algebraic quantum field theory because of its generality (Haag 1992, Baumgärtel & Wollenberg 1992). In the concrete 'Haag-Araki' approach, one supposes that a quantum field on Minkowski spacetime M will associate to each bounded open region $O \subseteq M$ a von Neumann algebra $\mathcal{R}(O)$ of observables measurable in that region, where the collection $\{\mathcal{R}(O) : O \subseteq M\}$ acts irreducibly on some fixed Hilbert space H. It is then natural to treat each open region O and associated algebra $\mathcal{R}(O)$ as a quantum system in its own right. Given any (normal) state ρ of the field (where ρ is a state functional on $\mathcal{B}(\mathsf{H})$), we can then ask which observables in $\mathcal{R}(O)$ are picked out as definite-valued by the restriction, ρ_O, of the state ρ to $\mathcal{R}(O)$.

The difficulty for the modal interpretation (though Dieks himself does not put it this way) is that when O has nonempty spacelike complement O', $\mathcal{R}(O)$ will typically be a type III factor that contains no nonzero finite projections (Haag 1992, Sec. V.6; Baumgärtel & Wollenberg 1992, Sec. 17.2). Because of this, $\mathcal{R}(O)$ cannot contain compact operators, like density operators, all of whose (non-null) spectral projections are finite-dimensional. As a result, there is no density operator in $\mathcal{R}(O)$ that can represent ρ_O. Moreover, if we try to apply the standard modal prescription based on Eq. (1) to a density operator in $\mathcal{B}(\mathsf{H})$ that agrees with ρ_O, there is no guarantee that the resulting set of observables will pick out a subalgebra of $\mathcal{R}(O)$, and we will be left with nothing to say about which observables have definite values in O. The moral Dieks draws from this is that 'We can therefore not take the open spacetime regions and their algebras as fundamental, if we want an interpretation in terms of (more or less) localized systems whose properties would specify an event' (Dieks 2000, p. 322). We shall see in the next section that this conclusion is overly pessimistic.

In any case, Dieks' strategy for dealing with the problem is to exploit the fact that, in most models of the axioms of algebraic quantum field theory, the local algebras associated with diamond shaped spacetime regions (i.e., regions given by the interior of the intersection of the causal future and past of two spacetime points) have the split property. The property is that for any two concentric diamond shaped regions $\diamondsuit_r, \diamondsuit_{r+\epsilon} \subseteq M$, with radii r and $r+\epsilon$, there is a type I 'interpolating' factor $\mathcal{N}_{r+\epsilon}$ such that $\mathcal{R}(\diamondsuit_r) \subset \mathcal{N}_{r+\epsilon} \subset \mathcal{R}(\diamondsuit_{r+\epsilon})$. Now since $\mathcal{N}_{r+\epsilon} \approx \mathcal{B}(\overline{\mathsf{H}})$ for some Hilbert space $\overline{\mathsf{H}}$, we know there is always a unique density operator $D_{r+\epsilon} \in \mathcal{B}(\overline{\mathsf{H}})$ that agrees with ρ on $\mathcal{N}_{r+\epsilon}$, and therefore with $\rho_{\diamondsuit_r}$. The proposal is, then, to take both r and ϵ to be fixed small numbers and apply the prescription in Eq. (1) to this density operator $D_{r+\epsilon}$, yielding a definite-valued subalgebra $\mathcal{M}_{r+\epsilon} \subseteq \mathcal{N}_{r+\epsilon}$. This, according to Dieks, should give an approximate indication of which observables have definite values at the common origin of the two diamonds in 'the classical limiting situation in which classical field and particle concepts become approximately applicable' (Dieks 2000, p. 323). Thus, Dieks proposes to build up an atomic modal interpretation of the field as follows. (i) We subdivide (approximately) the whole of spacetime M into a collection of non-overlapping diamond regions $\diamondsuit_r$, with some fixed small radius r. (ii) We choose some fixed small ϵ and an interpolating factor $\mathcal{N}_{r+\epsilon}$ for each diamond, using its ρ-induced density operator $D_{r+\epsilon}$ and Eq. (1) to determine the definite-valued observables $\mathcal{M}_{r+\epsilon} \subset \mathcal{R}(\diamondsuit_{r+\epsilon})$ to be loosely associated with the origin.

(iii) Finally, we build up definite-valued observables associated with collections of diamonds, and define their joint probabilities in the usual way via Born's rule (defining transition probabilities between values of observables associated with timelike-separated diamonds using the familiar multi-time generalization of that rule employed in the decoherent histories approach).

As things stand, there is much arbitrariness in this proposal that enters into the stages (i) and (ii). Dieks himself recognizes the arbitrariness in the size of the partition of M chosen. He also acknowledges that this arbitrariness cannot be eliminated by passing to the limit $r, \epsilon \to 0$, because the intersection of the algebras associated with any collection of concentric diamonds is always the trivial algebra $\mathfrak{C}I$ (Wightman 1964). Indeed, one would have thought that this undermines any attempt to formulate an *atomic* modal interpretation in this context, because it forces the choice of 'atomic diamonds' in the partition of M to be essentially arbitrary. Dieks appears to suggest that this arbitrariness will become unimportant in some classical limit of relativistic quantum field theory in which we should recover 'the classical picture according to which field values are attached to spacetime points' (Dieks 2000, p. 325). But it is not sufficient for the success of a proposal for interpreting a relativistic *quantum* field theory that the interpretation give sensible results in the limit of *classical* relativistic (or nonrelativistic) field theory. Indeed, the only relevant limit would appear to be the nonrelativistic limit; i.e., Galilean quantum field theory. But, there, one still needs to spatially smear "operator-valued" fields at each point to obtain a well-defined algebra of observables in a spatial region (Requardt 1982), so there will again be no natural choice to make for atomic spatial regions or algebras.

There is also another, more troubling, degree of arbitrariness at step (ii) in the choice of the type I interpolating factor $\mathcal{N}_{r+\epsilon}$ about each origin point. For any fixed partition and fixed $r, \epsilon > 0$, we can always sub-divide the interval $(r, r+\epsilon)$ further, and then the split property implies the existence of a pair of interpolating type I factors satisfying

$$\mathcal{R}(\Diamond_r) \subset \mathcal{N}_{r+\epsilon/2} \subset \mathcal{R}(\Diamond_{r+\epsilon/2}) \subset \mathcal{N}_{r+\epsilon} \subset \mathcal{R}(\Diamond_{r+\epsilon}). \tag{2}$$

The problem is that we now face a nontrivial choice deciding which of these factors' ρ-induced density operators to use to pick out the definite-valued observables in the state ρ associated with the origin. If we pick $D_{r+\epsilon} \in \mathcal{N}_{r+\epsilon}$, then by (1) all observables $A \in \mathcal{N}_{r+\epsilon}$ that share the same spectral projections as $D_{r+\epsilon}$ will have definite values. However, *no* such A can lie in $\mathcal{M}_{r+\epsilon/2} \subseteq \mathcal{N}_{r+\epsilon/2}$, nor even in $\mathcal{R}(\Diamond_{r+\epsilon/2})$. The reason is that A's spectral projections are *finite in* $\mathcal{N}_{r+\epsilon}$. So if those projections were also in the type III algebra $\mathcal{R}(\Diamond_{r+\epsilon/2})$, they would have to be infinite in $\mathcal{R}(\Diamond_{r+\epsilon/2})$, and therefore also infinite projections in $\mathcal{N}_{r+\epsilon}$ — which is impossible.

Clearly we can sub-divide the interval $(r, r+\epsilon)$ arbitrarily many times in this way and obtain a monotonically decreasing sequence of type I factors satisfying $\mathcal{R}(\Diamond_{r+\epsilon/2^{n+1}}) \subset \mathcal{N}_{r+\epsilon/2^n} \subset \mathcal{R}(\Diamond_{r+\epsilon/2^n})$ that all interpolate between $\mathcal{R}(\Diamond_r)$ and $\mathcal{R}(\Diamond_{r+\epsilon})$. The sequence $\{\mathcal{N}_{r+\epsilon/2^n}\}_{n=0}^{\infty}$ has no least member, and its greatest member, $\mathcal{N}_{r+\epsilon}$, is arbitrary, because we could also further sub-divide the interval $(r+\epsilon/2, r+\epsilon)$ ad infinitum. Thus there is no natural choice of interpolating factor for picking out the observables definite at the origin, even if we restrict ourselves to a 'nice' decreasing sequence of interpolating factors of the form $\{\mathcal{N}_{r+\epsilon/2^n}\}_{n=0}^{\infty}$.

On the other hand, it can actually be shown that $\mathcal{R}(\Diamond_r) = \bigcap_{n=0}^{\infty} \mathcal{N}_{r+\epsilon/2^n}$ (Horuzhy 1988, pp. 12–13), (Baumgärtel & Wollenberg 1992, p. 426). Furthermore, suppose $\{\mathcal{N}_{r+\epsilon_n}\}_{n=0}^{\infty}$ is *any other* decreasing type I sequence satisfying

$$\mathcal{R}(\Diamond_{r+\epsilon_{n+1}}) \subset \mathcal{N}_{r+\epsilon_n} \subset \mathcal{R}(\Diamond_{r+\epsilon_n}),\ \epsilon_0 = \epsilon,\ \epsilon_n > \epsilon_{n+1},\ \lim \epsilon_n = 0. \tag{3}$$

Then, since for any n there will be a sufficiently large n' such that $\mathcal{N}_{r+\epsilon_{n'}} \subset \mathcal{N}_{r+\epsilon/2^n}$ (and vice versa), clearly $\mathcal{R}(\Diamond_r) = \bigcap_{n=0}^{\infty} \mathcal{N}_{r+\epsilon_n}$, and this intersection will also be independent of ϵ. It would seem, then, that the natural way to avoid choosing between the myriad type I factors that interpolate between $\mathcal{R}(\Diamond_r)$ and $\mathcal{R}(\Diamond_{r+\epsilon})$ is to take the observables definite-valued at the origin to be those in the intersection $\bigcap_{n=0}^{\infty} \mathcal{M}_{r+\epsilon_n} \subseteq \mathcal{R}(\Diamond_r)$ (where, as before, $\mathcal{M}_{r+\epsilon_n}$ is the modal subalgebra of $\mathcal{N}_{r+\epsilon_n}$ determined via (1) by the density operator in $\mathcal{N}_{r+\epsilon_n}$ that represents ρ). Indeed, Dieks' suggestion appears to be that when we take successively smaller values for ϵ (holding r fixed), and choose a type I interpolating factor at each stage, we should be getting progressively better approximations to the set of observables that are truly definite at the origin. What better candidate for that set can there be than an intersection like $\bigcap_{n=0}^{\infty} \mathcal{M}_{r+\epsilon_n}$?

Unfortunately, we have no guarantee that *this* intersection, unlike $\bigcap_{n=0}^{\infty} \mathcal{N}_{r+\epsilon_n}$ itself, is independent of the particular sequence $\{\epsilon_n\}$ or its starting value $\epsilon_0 = \epsilon$. The reason any intersection of form $\bigcap_{n=0}^{\infty} \mathcal{N}_{r+\epsilon_n}$ is so independent is because $\mathcal{N}_{r+\epsilon_n} \supset \mathcal{N}_{r+\epsilon_{n+1}}$ for all n. But this does *not* imply $\mathcal{M}_{r+\epsilon_n} \supset \mathcal{M}_{r+\epsilon_{n+1}}$. To take just a trivial example: when ρ is a pure state of $\mathcal{N}_{r+\epsilon_n}$ that induces a mixed state on the proper subalgebra $\mathcal{N}_{r+\epsilon_{n+1}}$ of $\mathcal{N}_{r+\epsilon_n}$, $\mathcal{M}_{r+\epsilon_{n+1}}$ will contain observables with dispersion in the state ρ, but $\mathcal{M}_{r+\epsilon_n}$ will not.

We conclude that there is little prospect of eliminating the arbitrariness in Dieks' proposal and making it well-defined. One ought to look for another, *intrinsic* way to pick out the definite-valued observables in $\mathcal{R}(\diamondsuit_r)$ that does not depend on special assumptions such as the split property.

3 The modal interpretation for arbitrary von Neumann algebras

There are two salient features of the algebra $\mathcal{M}_S$ in Eq. (1) that make it an attractive set of definite-valued observables to modal interpreters. First, $\mathcal{M}_S$ is locally determined by the quantum state D_S of system S together with the structure of its algebra of observables. In particular, there is no need to add any additional structure to the standard formalism of quantum theory to pick out S's properties. Second, the restriction of the state D_S to the subalgebra $\mathcal{M}_S$ is a mixture of dispersion-free states (given by the density operators one obtains by renormalizing the (non-null) spectral projections of D_S). This second feature is what makes it possible to think of the observables in $\mathcal{M}_S$ as possessing definite values distributed in accordance with standard Born rule statistics (Clifton 1999). Let us see, then, whether we can generalize these two features to come up with a proposal for the definite-valued observables of a system described by an arbitrary von Neumann algebra $\mathcal{R}$ (acting on some Hilbert space H) in an arbitrary state ρ of $\mathcal{R}$.

Generally, a state ρ of $\mathcal{R}$ will be a mixture of dispersion-free states on a subalgebra $\mathcal{S} \subseteq \mathcal{R}$ just in case there is a probability measure μ_ρ on the space Λ of dispersion-free states of $\mathcal{S}$ such that

$$\rho(A) = \int_\Lambda \omega_\lambda(A) d\mu_\rho(\lambda), \text{ for all } A \in \mathcal{S}, \tag{4}$$

where $\omega_\lambda(A^2) = \omega_\lambda(A)^2$ for all self-adjoint elements $A \in \mathcal{S}$. This somewhat cumbersome condition turns out to be equivalent (Halvorson & Clifton 1999, Prop. 2.2(ii)) to simply requiring that

$$\rho([A,B]^*[A,B]) = 0 \text{ for all } A,B \in \mathcal{S}. \tag{5}$$

In particular, ρ can always be represented as a mixture of dispersion-free states on any abelian subalgebra $\mathcal{S} \subseteq \mathcal{R}$. Conversely, if ρ is a faithful state of $\mathcal{R}$, i.e., ρ maps no nonzero positive elements of $\mathcal{R}$ to zero, then the *only* subalgebras that allow ρ to be represented as a mixture of dispersion-free states are the abelian ones. There is now an easy way to pick out a subalgebra $\mathcal{S} \subseteq \mathcal{R}$ with this property, using only ρ and the algebraic operations available within $\mathcal{R}$.

Consider the following two mathematical objects explicitly defined in terms of $\mathcal{R}$ and ρ. First, the support projection of the state ρ in $\mathcal{R}$, defined by

$$P_{\rho,\mathcal{R}} \equiv \wedge\{P = P^2 = P^* \in \mathcal{R} : \rho(P) = 1\}, \tag{6}$$

which is simply the smallest projection in $\mathcal{R}$ that the state ρ 'makes true'. Second, there is the centralizer subalgebra of the state ρ in $\mathcal{R}$, defined by

$$\mathcal{C}_{\rho,\mathcal{R}} \equiv \{A \in \mathcal{R} : \rho([A,B]) = 0 \text{ for all } B \in \mathcal{R}\}. \tag{7}$$

For any von Neumann algebra $\mathcal{K}$, let $\mathcal{Z}(\mathcal{K}) \equiv \mathcal{K} \cap \mathcal{K}'$, the center algebra of $\mathcal{K}$. Then it is reasonable for the modal interpreter to take as definite-valued all the observables that lie in the direct sum

$$\mathcal{S} = \mathcal{M}_{\rho,\mathcal{R}} \equiv P^{\perp}_{\rho,\mathcal{R}}\mathcal{R}P^{\perp}_{\rho,\mathcal{R}} + \mathcal{Z}(\mathcal{C}_{\rho,\mathcal{R}})P_{\rho,\mathcal{R}} \subseteq \mathcal{R}, \tag{8}$$

where the algebra in the first summand acts on the subspace $P^{\perp}_{\rho,\mathcal{R}}\mathsf{H}$ and that of the second acts on $P_{\rho,\mathcal{R}}\mathsf{H}$. The state ρ is a mixture of dispersion-free states on $\mathcal{M}_{\rho,\mathcal{R}}$, by (5), because ρ maps all elements of the form $P^{\perp}_{\rho,\mathcal{R}}\mathcal{R}P^{\perp}_{\rho,\mathcal{R}}$ to zero, and the product of the commutators of any two elements of $\mathcal{Z}(\mathcal{C}_{\rho,\mathcal{R}})P_{\rho,\mathcal{R}}$ also gets mapped to zero, for the trivial reason that $\mathcal{Z}(\mathcal{C}_{\rho,\mathcal{R}})$ is abelian.

The set $\mathcal{M}_{\rho,\mathcal{R}}$ directly generalizes the algebra of Eq. (1) to the non-type I case where the algebra of observables of the system does not contain a density operator representative of the state ρ. Assuming the type I case, $\mathcal{R} \approx \mathcal{B}(\overline{\mathsf{H}})$ for some Hilbert space $\overline{\mathsf{H}}$, ρ is given by a density operator D on $\overline{\mathsf{H}}$, $P_{\rho,\mathcal{R}}$ is equivalent to the range projection of D, and $\mathcal{Z}(\mathcal{C}_{\rho,\mathcal{R}}) \approx \mathcal{Z}(\mathcal{C}_{D,\mathcal{B}(\overline{\mathsf{H}})})$. So to show that $\mathcal{M}_{\rho,\mathcal{R}}$ is isomorphic to the algebra of Eq. (1), it suffices to establish that $\mathcal{Z}(\mathcal{C}_{D,\mathcal{B}(\overline{\mathsf{H}})}) = D''$. It is easy to see that $\mathcal{C}_{D,\mathcal{B}(\overline{\mathsf{H}})} = D'$ (invoking cyclicity and positive-definiteness of the trace), thus $\mathcal{Z}(\mathcal{C}_{D,\mathcal{B}(\overline{\mathsf{H}})}) = D' \cap D''$. However, since D' always contains a maximal abelian subalgebra of $\mathcal{B}(\overline{\mathsf{H}})$ (viz., that generated by the projections onto any complete orthonormal basis of eigenvectors for D), we always have $D'' \subseteq D'$.

Choosing $\mathcal{M}_{\rho,\mathcal{R}}$ is certainly not the only way to pick a subalgebra $\mathcal{S} \subseteq \mathcal{R}$ that is definable in terms of ρ and $\mathcal{R}$ and allows ρ to be represented as a mixture of dispersion-free states. There is the obvious orthodox alternative one can always consider, viz., the definite algebra of ρ in $\mathcal{R}$,

$$\mathcal{S} = \mathcal{O}_{\rho,\mathcal{R}} \equiv \{A \in \mathcal{R} : \rho(AB) = \rho(A)\rho(B) \text{ for all } B \in \mathcal{R}\}, \tag{9}$$

which coincides with the complex span of all self-adjoint members of $\mathcal{R}$ on which ρ is dispersion-free (Halvorson & Clifton 1999, p. 2445). Note, however, that we always have $\mathcal{O}_{\rho,\mathcal{R}} \subseteq \mathcal{M}_{\rho,\mathcal{R}}$. Indeed the problem is that the orthodox choice $\mathcal{O}_{\rho,\mathcal{R}}$ generally will contain far too few definite-valued observables to solve the measurement problem. For example, when ρ is faithful — and there will always be a norm dense set of states of $\mathcal{R}$ that are — we get just $\mathcal{O}_{\rho,\mathcal{R}} = \mathfrak{C}I$. Thus it is natural for a modal interpreter to require that the choice of $\mathcal{S} \subseteq \mathcal{R}$ be maximal. In the case where ρ is faithful, we now show that this singles out the choice $\mathcal{S} = \mathcal{M}_{\rho,\mathcal{R}} = \mathcal{Z}(\mathcal{C}_{\rho,\mathcal{R}})$ uniquely (and we conjecture that a similar uniqueness result holds for the more general expression for $\mathcal{M}_{\rho,\mathcal{R}}$ in Eq. (8), using the fact that an arbitrary state ρ always renormalizes to a faithful state on $P_{\rho,\mathcal{R}}\mathcal{R}P_{\rho,\mathcal{R}}$).

Proposition 1. *Let $\mathcal{R}$ be a von Neumann algebra and ρ a faithful normal state of $\mathcal{R}$ with centralizer $\mathcal{C}_{\rho,\mathcal{R}} \subseteq \mathcal{R}$. Then $\mathcal{Z}(\mathcal{C}_{\rho,\mathcal{R}})$, the center of $\mathcal{C}_{\rho,\mathcal{R}}$, is the unique subalgebra $\mathcal{S} \subseteq \mathcal{R}$ such that:*

1. *The restriction of ρ to $\mathcal{S}$ is a mixture of dispersion-free states.*
2. *$\mathcal{S}$ is definable solely in terms of ρ and the algebraic structure of $\mathcal{R}$.*
3. *$\mathcal{S}$ is maximal with respect to properties 1. and 2.*

Proof. By 3., it suffices to show than any $\mathcal{S} \subseteq \mathcal{R}$ satisfying 1. and 2. is contained in $\mathcal{Z}(\mathcal{C}_{\rho,\mathcal{R}})$. And for this, it suffices (because von Neumann algebras are generated by their projections) to show that $\underline{\mathcal{S}}$, the subset of projections in $\mathcal{S}$, is contained in $\mathcal{Z}(\mathcal{C}_{\rho,\mathcal{R}})$. Recall also that, as a consequence of 1. and the faithfulness of ρ, $\mathcal{S}$ must be abelian. And in virtue of 2., any automorphism $\sigma : \mathcal{R} \longrightarrow \mathcal{R}$ that preserves the state ρ in the sense that $\rho \circ \sigma = \rho$, must leave the set $\mathcal{S}$ (not necessarily pointwise) invariant, i.e., $\sigma(\mathcal{S}) = \mathcal{S}$.

$\underline{\mathcal{S}} \subseteq \mathcal{C}'_{\rho,\mathcal{R}}$. Any unitary operator $U \in \mathcal{C}_{\rho,\mathcal{R}}$ defines an inner automorphism on $\mathcal{R}$ that leaves ρ invariant, therefore $U\mathcal{S}U^{-1} = \mathcal{S}$. Since $\mathcal{S}$ is abelian, $[UPU^{-1}, P] = 0$ for each $P \in \underline{\mathcal{S}}$ and all unitary $U \in \mathcal{C}_{\rho,\mathcal{R}}$. By Lemma 4.2 of Halvorson & Clifton (1999) (with $\mathfrak{V} = \mathcal{C}''_{\rho,\mathcal{R}} = \mathcal{C}_{\rho,\mathcal{R}}$), this implies that $P \in \mathcal{C}'_{\rho,\mathcal{R}}$.

$\underline{\mathcal{S}} \subseteq \mathcal{C}_{\rho,\mathcal{R}}$. Since ρ is faithful, there is a one-parameter group $\{\sigma_t : t \in \mathfrak{R}\}$ of automorphisms of $\mathcal{R}$ — the modular automorphism group of $\mathcal{R}$ determined by ρ (Kadison & Ringrose 1997, Sec. 9.2) — leaving ρ invariant. Since $\mathcal{C}_{\rho,\mathcal{R}}$ consists precisely of the fixed points of the modular group (Kadison & Ringrose 1997, Prop. 9.2.14), it suffices to show that it leaves the individual elements of $\underline{\mathcal{S}}$ fixed. For this, we use the fact that the modular group satisfies the KMS condition with respect to ρ: for each $A, B \in \mathcal{R}$, there is a complex-valued function f, bounded and continuous on the strip $\{z \in \mathfrak{C} : 0 \leq \text{Im}z \leq 1\}$ in the complex plane, and analytic on the interior of that strip, such that

$$f(t) = \rho(\sigma_t(A)B),\ f(t+i) = \rho(B\sigma_t(A)),\ t \in \mathfrak{R}. \tag{10}$$

In fact, we shall need only one simple consequence of the KMS condition, viz., if $f(t) = f(t+i)$ for all $t \in \mathfrak{R}$, then f is constant (Kadison & Ringrose 1997, p. 611).

Fix an arbitrary projection $P \in \underline{\mathcal{S}}$. Since the modular automorphism group must leave $\mathcal{S}$ as a whole invariant, and $\mathcal{S}$ is abelian, $[\sigma_t(P), P^{\perp}] = 0$ for all $t \in \mathfrak{R}$. However, there exists a function f with the above properties such that

$$f(t) = \rho(\sigma_t(P)P^{\perp}) = \rho(P^{\perp}\sigma_t(P)) = f(t+i),\ t \in \mathfrak{R},$$

so it follows that f is constant. In particular, since $\sigma_0(P)P^{\perp} = PP^{\perp} = 0$, f is identically zero, and $\rho(\sigma_t(P)P^{\perp}) = 0$ for all $t \in \mathfrak{R}$. And since $\sigma_t(P)$ and $P^{\perp}$ are commuting projections, their product is a (positive) projection, so that the faithfulness of ρ requires that $\sigma_t(P)P^{\perp} = 0$, or equivalently $\sigma_t(P) = \sigma_t(P)P$, for all $t \in \mathfrak{R}$. Running through the exact same argument, starting with $P^{\perp} \in \underline{\mathcal{S}}$ in place of P, yields $\sigma_t(P^{\perp}) = \sigma_t(P^{\perp})P^{\perp}$, or equivalently, $\sigma_t(P)P = P$, for all $t \in \mathfrak{R}$. Together with $\sigma_t(P) = \sigma_t(P)P$, this implies that $\sigma_t(P) = P$ for all $t \in \mathfrak{R}$. □

The choice $\mathcal{M}_{\rho,\mathcal{R}} = \mathcal{Z}(\mathcal{C}_{\rho,\mathcal{R}})$ has another feature that generalizes a natural consequence of the modal interpretation of nonrelativistic quantum theory. Suppose the universal state $x \in \mathsf{H}_U$ defines a faithful state ρ_x on both $\mathcal{B}(\mathsf{H}_S)$ and $\mathcal{B}(\mathsf{H}_{\overline{S}})$. This requires that $\dim \mathsf{H}_S = \dim \mathsf{H}_S = n$ (possibly ∞), and, furthermore, that any Schmidt decomposition of the state vector x relative to the factorization $\mathsf{H}_U = \mathsf{H}_S \otimes \mathsf{H}_{\overline{S}}$ takes the form

$$x = \sum_{i=1}^{n} c_i v_i \otimes w_i,\ c_i \neq 0 \text{ for all } i = 1 \text{ to } n, \tag{11}$$

where the vectors v_i and w_i are complete orthonormal bases in their respective spaces. As is well known, for each distinct eigenvalue $\widetilde{\lambda}_j$ for D_S, the span of the vectors v_i for which $|c_i|^2 = \widetilde{\lambda}_j$ coincides with the range of the $\widetilde{\lambda}_j$-eigenprojection of D_S, and similarly for $D_{\overline{S}}$. Consequently, there is a natural bijective correspondence between the properties represented by the projections in the two sets $\mathcal{M}_S = D_S''$ and $\mathcal{M}_{\overline{S}} = D_{\overline{S}}''$: any definite property S happens to possess is strictly correlated to a unique property of its environment $\overline{S}$ that occurs with the same frequency. More formally, for any $P \in \mathcal{M}_S$, there is a unique $\overline{P} \in \mathcal{M}_{\overline{S}}$ satisfying

$$(x, P\overline{P}x) = (x, Px) = (x, \overline{P}x). \tag{12}$$

To see this, note that any $P \in \mathcal{M}_S$ (in this case, the set of all functions of D_S) is a sum of spectral projections of D_S. Let $\overline{P} \in \mathcal{M}_{\overline{S}}$ be the sum of the corresponding spectral projections of $D_{\overline{S}}$ for the same eigenvalues. Then it is evident from the form of the state expansion in (11) that $\overline{P}$ has the property in (12), and no other projection in $\mathcal{M}_{\overline{S}}$ does. This has led some nonatomic modal interpreters, such as Kochen (1985), to interpret each property P of S, not as a property that S possesses absolutely, but only in relation to its environment $\overline{S}$ possessing the corresponding property $\overline{P}$.

For a general von Neumann factor $\mathcal{R}$, $(\mathcal{R} \cup \mathcal{R}')'' = \mathcal{B}(\mathsf{H})$ need not be isomorphic to the tensor product $\mathcal{R} \otimes \mathcal{R}'$ (particularly when $\mathcal{R}$ is type III, for then $\mathcal{R} \otimes \mathcal{R}'$ must be type III as well). Therefore, there is no direct analogue of a Schmidt decomposition for a pure state $x \in \mathsf{H}$ relative to the factorization $(\mathcal{R} \cup \mathcal{R}')''$ of $\mathcal{B}(\mathsf{H})$. Nevertheless, we show next that there is still the same strict correlation between definite properties in $\mathcal{M}_{\rho_x,\mathcal{R}} = \mathcal{Z}(\mathcal{C}_{\rho_x,\mathcal{R}})$ and $\mathcal{M}_{\rho_x,\mathcal{R}'} = \mathcal{Z}(\mathcal{C}_{\rho_x,\mathcal{R}'})$.

Proposition 2. *Let $\mathcal{R}$ be a von Neumann algebra acting on a Hilbert space H, and suppose $x \in \mathsf{H}$ induces a state ρ_x that is faithful on both $\mathcal{R}$ and $\mathcal{R}'$. Then for any projection $P \in \mathcal{Z}(\mathcal{C}_{\rho_x,\mathcal{R}})$, there is a unique projection $\overline{P} \in \mathcal{Z}(\mathcal{C}_{\rho_x,\mathcal{R}'})$ such that $(x, P\overline{P}x) = (x, Px) = (x, \overline{P}x)$.*

Proof. For any fixed $A \in \mathcal{R}$, call an element $B \in \mathcal{R}'$ a double for A (in state x) just in case $Ax = Bx$ and $A^*x = B^*x$. By an elementary application of modular theory, Werner (1999, Sec. II) has shown that $\mathcal{C}_{\rho_x,\mathcal{R}}$ consists precisely of those elements of $\mathcal{R}$ with doubles in $\mathcal{R}'$ (with respect to x). Moreover, the double of any element of $\mathcal{R}$ clearly has to be unique, by the faithfulness of ρ_x on $\mathcal{R}'$. Now it is easy to see (again using the faithfulness of ρ_x) that the double of any projection $P \in \mathcal{C}_{\rho_x,\mathcal{R}}$ is a projection $\overline{P} \in \mathcal{C}_{\rho_x,\mathcal{R}'}$ satisfying (12). We claim that whenever $P \in \mathcal{C}'_{x,\mathcal{R}}$, we have $\overline{P} \in \mathcal{C}'_{x,\mathcal{R}'}$. For this, it suffices to show $P \in \mathcal{C}'_{x,\mathcal{R}}$ implies that for arbitrary $B \in \mathcal{C}_{x,\mathcal{R}'}$, $[\overline{P}, B]x = 0$ (and then $[\overline{P}, B]$ itself is zero, since ρ_x is faithful). Letting $A \in \mathcal{C}_{x,\mathcal{R}}$ be the double of B in $\mathcal{R}$, we get

$$\overline{P}Bx - B\overline{P}x = \overline{P}Ax - BPx = A\overline{P}x - PBx = APx - PAx = 0, \quad (13)$$

as required. Finally, were there another projection $\widetilde{P} \in \mathcal{Z}(\mathcal{C}_{\rho_x,\mathcal{R}'})$ satisfying (12), then by exploiting the fact that $\overline{P}$ is P's double in $\mathcal{R}'$, we get $(x, \overline{P}\widetilde{P}x) = (x, \overline{P}x) = (x, \widetilde{P}x)$; or, equivalently,

$$(x, \overline{P}\widetilde{P}^{\perp}x) = (x, \overline{P}^{\perp}\widetilde{P}x) = 0. \quad (14)$$

Since $\mathcal{Z}(\mathcal{C}_{\rho_x,\mathcal{R}'})$ is abelian, both $\overline{P}\widetilde{P}^{\perp}$ and $\overline{P}^{\perp}\widetilde{P}$ are (positive) projections in $\mathcal{R}$. But as ρ_x is faithful on $\mathcal{R}$, Eqs. (14) entail that $\overline{P}\widetilde{P}^{\perp} = \overline{P}^{\perp}\widetilde{P} = 0$, which in turn implies that $\overline{P} = \widetilde{P}$, as required for uniqueness. □

Let us return now to the problem of picking out a set of definite-valued observables localized in a diamond region with associated algebra $\mathcal{R}(\diamondsuit_r)$. Let ρ be any pure state of the field that induces a faithful state on $\mathcal{R}(\diamondsuit_r)$; for example, ρ could be the vacuum or any one of the dense set of states of a field with bounded energy (by the Reeh-Schlieder theorem — see (Horuzhy 1988, Thm. 1.3.1)). By Proposition 1, the definite-valued observables in $\mathcal{R}(\diamondsuit_r)$ are simply those in the subalgebra $\mathcal{Z}(\mathcal{C}_{\rho,\mathcal{R}(\diamondsuit_r)})$. Note that this proposal yields observables all of which have an exact spacetime localization within the open set $\diamondsuit_r$ and are picked out intrinsically by the local algebra $\mathcal{R}(\diamondsuit_r)$ and the field state ρ. Contrary to Dieks' pessimistic conclusion, we *can* take open spacetime regions as fundamental for determining the definite-valued observables. In fact, this proposal works independent of the size of r, and so could also be embraced by *non*atomic modal interpreters not wishing to commit themselves to a particular partition of the field into subsystems (or to thinking from the outset in terms of approximately point-localized field observables). Finally, note that since the algebra of a diamond region $\diamondsuit_r$ satisfies duality with respect to the algebra of its spacelike complement $\diamondsuit_r'$, i.e., $\mathcal{R}(\diamondsuit_r)' = \mathcal{R}(\diamondsuit_r')$ (Haag 1992, p. 145), Proposition 2 tells us that there is a natural bijective correspondence between the properties in $\mathcal{Z}(\mathcal{C}_{\rho,\mathcal{R}(\diamondsuit_r)})$ and strictly correlated properties in $\mathcal{Z}(\mathcal{C}_{\rho,\mathcal{R}(\diamondsuit_r')})$ associated with the complement region.

4 A potential difficulty with ergodic states

We have seen that there *is*, after all, a well-motivated and unambiguous prescription extending the standard modal interpretation of nonrelativistic quantum theory to the local algebras of quantum field theory. We also, now, have a natural standard of comparison with Galilean quantum field theory. At least in the case of free fields, it is possible to build up local algebras in M from spatially smeared 'field algebras' defined on spacelike hyperplanes in M. A diamond region corresponds to the domain of dependence of a spatial region in a hyperplane, and it can be shown that the algebra of that spatial region will also be type III and, indeed, *coincide* with its domain of dependence algebra (Horuzhy 1988, Prop. 3.3.2, Thm. 3.3.4). These type III spatial algebras in M, and the definite-valued observables therein, are what should be compared, in the nonrelativistic limit, to the corresponding equal time spatial algebras defined on simultaneity slices of Galilean spacetime. Unfortunately, since the algebras in the Galilean case are invariably type I (Horuzhy 1988, p. 35), this limit is bound to be mathematically singular, and its *physical* characterization needs to be dealt with carefully. But this is a problem for *any* would-be interpreter of relativistic quantum field theory, not just modal interpreters. All we should require of them, at this stage, is that they be able to say something sensible in the relativistic case about the local observables with definite values (which was, indeed, Dieks' original goal). However, as we now explain, it is not clear whether even this goal can be attained.

If $\mathcal{R} \approx \mathcal{B}(\overline{\mathsf{H}})$ is type I, it possesses at most one faithful state ρ such that $\mathcal{Z}(\mathcal{C}_{\rho,\mathcal{R}}) = \mathfrak{C}I$. This is easy to see, because if $D \in \mathcal{B}(\overline{\mathsf{H}})$ represents ρ, $\mathcal{Z}(\mathcal{C}_{\rho,\mathcal{R}}) \approx D''$, and $D'' = \mathfrak{C}I$ implies that D itself must be a multiple of the identity. So when $\overline{\mathsf{H}}$ is finite-dimensional, we must have $D = I/\dim \overline{\mathsf{H}}$, the unique maximally mixed state, and in the infinite-dimensional case, no such density operator even exists. Elsewhere Dieks (1994) has argued convincingly that there is no problem when a system, occupying a maximally mixed state, possesses only trivial properties, because such states are rare and highly unstable under environmental decoherence (cf. Healey 1989, pp. 99–100). However, the situation is quite different for the local algebras of algebraic quantum field theory.

In all physically reasonable models of the axioms of the theory, every local algebra $\mathcal{R}(O)$ is isomorphic to the unique (up to isomorphism) hyperfinite type III_1 factor (Haag 1992, Sec. V.6; Baumgärtel & Wollenberg 1992, Sec. 17.2). In that case, there is a novel way to obtain $\mathcal{Z}(\mathcal{C}_{\rho,\mathcal{R}(O)}) = \mathfrak{C}I$, namely, when the state ρ of $\mathcal{R}(O)$ is an ergodic state (Herman & Takesaki 1970, Barnett 1995), i.e., ρ possesses a trivial centralizer in $\mathcal{R}(O)$. (Were $\mathcal{R}$ a nonabelian type I factor, this would be impossible, since $D' = \mathfrak{C}I$ implies $D'' = \mathcal{B}(\overline{\mathsf{H}}) \approx \mathcal{R}$, which is patently false.) In fact, we have the following result.

Proposition 3. *If $\mathcal{R}$ is the hyperfinite type III_1 factor, there is a norm dense set of unit vectors in the Hilbert space* H *on which $\mathcal{R}$ acts that induce faithful states on $\mathcal{R}$ with trivial centralizers (i.e., ergodic states).*

Proof. First recall the following facts provable from the axioms of algebraic quantum field theory: (i) the vacuum state of a field on M has a trivial centralizer in the algebra of any Rindler wedge (Baumgärtel & Wollenberg (1992), Sec. 16.1.1); (ii) the vacuum is faithful for any wedge algebra (by the Reeh-Schlieder theorem); and (iii) wedge algebras are hyperfinite type III_1 factors (Baumgärtel & Wollenberg 1992, Ex. 16.2.14, pp. 426–7). Since being faithful and having a trivial centralizer are isomorphic invariants, it follows that *any* instantiation $\mathcal{R}$ of the hyperfinite type III_1 factor possesses at least one faithful normal state ρ with trivial centralizer (even when

$\mathcal{R}$ is the algebra of a *bounded* open region, like a diamond). Now since $\mathcal{R}$ is type III, all its states are vector states (combine Sakai (1971), Cor. 2.9.28 with Kadison & Ringrose (1997), Thm. 7.2.3); in particular, $\rho = \rho_x$ for some unit vector $x \in \mathsf{H}$. Furthermore, by the homogeneity of the state space of type III_1 factors (Connes & Størmer 1978, Cor. 6), the set of all unit vectors of the form $UU'x$, with $U \in \mathcal{R}$ and $U' \in \mathcal{R}'$ unitary operators, lies dense in H. But clearly any such vector must again induce a faithful state on $\mathcal{R}$ with trivial centralizer. □

Combining Propositions 1 and 3, there will be a whole host of states of any relativistic quantum field in which the modal interpreter is forced to assert that *no* nontrivial local observables have definite values! Note, however, that while the set of field states ergodic for any given type III_1 local algebra is always dense, this does not automatically imply that such states are typical or generic. Indeed, results of Summers & Werner (1988, particularly Cor. 2.4) imply that for any local diamond algebra $\mathcal{R}(\Diamond_r)$, there will *also* always be a dense set of field states whose centralizers in $\mathcal{R}(\Diamond_r)$ contain the hyperfinite type II_1 factor, and so will *not* be trivial. Still, the modal interpreter needs to provide some physical reason for neglecting the densely many field states that *do* yield trivial definite-valued observables locally. Obviously instability under decoherence is no longer relevant.

Perhaps one could try to bypass Proposition 1 by exploiting extra structure not contained in the particular field state and local algebra to pick out the definite-valued observables in a region. For example, one might try to exploit the field's total energy-momentum operator, and, in particular, its generator of time evolution. In the context of the nonrelativistic modal interpretation, Bacciagaluppi *et al.* (1995) (cf. Bacciagaluppi & Dickson 1999, p. 1181) have successfully invoked the analytic properties of the time evolution of the spectral projections of a system's reduced density operator D_S to avoid discontinuities that occur in the definite-valued set $\mathcal{M}_S$ at moments of time where the multiplicity of the eigenvalues of D_S changes. Their methods yield a natural dynamical way, independent of instability considerations, to avoid the trivial definite-valued sets determined by maximally mixed density operators. So one might hope that these same dynamical methods could be extended to type III_1 algebras so as to yield a richer set of properties in a local region than Proposition 1 allows for ergodic states. In any case, modal interpreters need to do more work to show that their interpretation yields sensible local properties in quantum field theory (even before one considers, with Dieks, how to define Lorentz invariant decoherent histories of properties).

Bibliography

Ax, J. & Kochen, S. (1999), 'Extension of quantum mechanics to individual systems'. quant-ph/9905077.

Bacciagaluppi, G. & Dickson, M. (1999), 'Dynamics for modal interpretations', *Foundations of Physics* **29**, 1165–201.

Bacciagaluppi, G. & Hemmo, M. (1996), 'Modal interpretations, decoherence and measurements', *Studies in History and Philosophy of Modern Physics* **27B**, 239–77.

Bacciagaluppi, G., Donald, M. & Vermaas, P. (1995), 'Continuity and discontinuity of definite properties in the modal interpretation', *Helvetica Physica Acta* **68**, 679–704.

Barnett, L. (1995), 'Free product von Neumann algebras of type III', *Proceedings of the American Mathematical Society* **123**, 543–53.

Baumgärtel, H. & Wollenberg, M. (1992), *Causal nets of operator algebras*, Akademie Verlag, Berlin.

Clifton, R. (1995), 'Making sense of the Kochen-Dieks "No-collapse" interpretation of quantum mechanics independent of the measurement problem', *Annals of the New York Academic of Science* **755**, 570–8.

Clifton, R. (1999), Beables in algebraic quantum theory, in J. Butterfield & C. Pagonis, eds, 'From Physics to Philosophy', Cambridge Univ. Press, Cambridge, pp. 12–44.

Connes, A. & Størmer, E. (1978), 'Homogeneity of the state space of factors of type III_1', *Journal of Functional Analysis* **28**, 187–96.

Dickson, M. & Clifton, R. (1998), Lorentz-invariance in modal interpretations, in D. Dieks & P. Vermaas, eds, 'The Modal Interpretation of Quantum Mechanics', Kluwer, Dordrecht, pp. 9–47.

Dieks, D. (1994), 'Modal interpretation of quantum mechanics, measurements, and macroscopic behavior', *Physical Review A* **49**, 2290–300.

Dieks, D. (1998), Preferred factorizations and consistent property attribution, in 'Quantum measurement: Beyond paradox', University of Minnesota Press, Minneapolis, pp. 144–60.

Dieks, D. (2000), 'Consistent histories and relativistic invariance in the modal interpretation of quantum mechanics', *Physics Letters A* **265**, 317–25.

Dieks, D. & Vermaas, P. (1998), *The Modal Interpretation of Quantum Mechanics*, Kluwer, Dordrecht.

Haag, R. (1992), *Local Quantum Physics*, 2nd ed., Springer, New York.

Halvorson, H. & Clifton, R. (1999), 'Maximal beable subalgebras of quantum-mechanical observables', *International Journal of Theoretical Physics* **38**, 2441–84.

Healey, R. (1989), *The Philosophy of Quantum Mechanics: An Interactive Interpretation*, Cambridge Univ. Press, New York.

Hemmo, M. (1998), Quantum histories in the modal interpretation, in D. Dieks & P. Vermaas, eds, 'The Modal Interpretation of Quantum Mechanics', Kluwer, Dordrecht, pp. 253–77.

Herman, R. H. & Takesaki, M. (1970), 'States and automorphism groups of operator algebras', *Communications in Mathematical Physics* **19**, 142–60.

Horuzhy, S. S. (1988), *Introduction to Algebraic Quantum Field Theory*, Kluwer, Dordrecht.

Kadison, R. & Ringrose, J. (1997), *Fundamentals of the Theory of Operator Algebras*, American Mathematical Society, Providence, RI.

Kochen, S. (1985), A new interpretation of quantum mechanics, in 'Symposium on the foundations of modern physics', World Scientific Publishing, Singapore, pp. 151–69.

Requardt, M. (1982), 'Spectrum condition, analyticity, Reeh-Schlieder and cluster properties in non-relativistic Galilei-invariant quantum theory', *Journal of Physics A* **15**, 3715–23.

Sakai, S. (1971), *C^*-algebras and W^*-algebras*, Springer, Berlin.

Spekkens, R. W. & Sipe, J. E. (2001), 'Non-orthogonal core projectors for modal interpretations of quantum mechanics', *Foundations of Physics* **31**, 1403–30.

Summers, S. J. & Werner, R. (1988), 'Maximal violation of Bell's inequalities for algebras of observables in tangent spacetime regions', *Annales de l'Institut Henri Poincaré. Physique Théorique* **49**, 215–43.

van Fraassen, B. C. (1991), *Quantum Mechanics: An Empiricist View*, Clarendon Press, Oxford.

Vermaas, P. (1998), The pros and cons of the Kochen-Dieks and the atomic modal interpretation, in D. Dieks & P. Vermaas, eds, 'The Modal Interpretation of Quantum Mechanics', Kluwer, Dordrecht, pp. 103–48.

Vermaas, P. E. (1996), 'Unique transition probabilities in the modal interpretation', *Studies in History and Philosophy of Science. B. Studies in History and Philosophy of Modern Physics* **27**, 133–59.

Werner, R. F. (1999), EPR states for von Neumann algebras, quant-ph/9910077.

Wightman, A. S. (1964), 'La théorie quantique locale et la théorie quantique des champs', *Annales de l'Institut Henri Poincaré Sect. A (N.S.)* **1**, 403–20.

Zimba, J. & Clifton, R. (1998), Valuations on functionally closed sets of quantum-mechanical observables and von Neumann's no-hidden-variables theorem, in 'The Modal Interpretation of Quantum Mechanics', Kluwer, Dordrecht, pp. 69–101.

Chapter 6

Generic Bell correlation between arbitrary local algebras in quantum field theory

with Hans Halvorson

1 Introduction

There are many senses in which the phenomenon of Bell correlation, originally discovered and investigated in the context of elementary nonrelativistic quantum mechanics (Bell 1987, Clauser *et al.* 1969), is 'generic' in quantum field theory models. For example, it has been shown that every pair of commuting nonabelian von Neumann algebras possesses *some* normal state with maximal Bell correlation (Summers 1990; see also Landau 1987*b*). Moreover, in most standard quantum field models, *all* normal states are maximally Bell correlated across spacelike separated tangent wedges

We are extremely grateful to David Malament for encouraging our work on this project. We wish also to thank the anonymous referee for helping us to understand the scope of our results (which led to significant improvements in Sec. 4), and Rainer Verch for providing helpful correspondence.

or double cones (Summers 1990). Finally, every bounded energy state in quantum field theory sustains maximal Einstein-Podolsky-Rosen (EPR) correlations across arbitrary spacelike separated regions (Redhead 1995), and has a form of nonlocality that may be evinced by means of the state's violation of a conditional Bell inequality (Landau 1987*a*). (We also note that the study of Bell correlation in quantum field theory has recently borne fruit in the introduction of a new algebraic invariant for an inclusion of von Neumann algebras (Summers & Werner 1995, Summers 1997).)

Despite these numerous results, it remains an open question whether 'most' states will have some or other Bell correlation relative to *arbitrary* spacelike separated regions. Our main purpose in this paper is to verify that this is so: for any two spacelike separated regions, there is an open dense set of states which have Bell correlations across those two regions.

In Section 2 we prove the general result that for any pair of mutually commuting von Neumann algebras of infinite type, a dense set of vectors will induce states which are Bell correlated across these two algebras. In Section 3 we introduce, following Werner (1989), a notion of 'nonseparability' of states that generalizes, to mixed states, the idea of an entangled pure state vector. We then show that for a pair of nonabelian von Neumann algebras, a vector cyclic for either algebra induces a nonseparable state. Finally, in Section 4 we apply these results to algebraic quantum field theory.

2 Bell correlation between infinite von Neumann algebras

Let $\mathcal{H}$ be a Hilbert space, let $\mathcal{S}$ denote the set of unit vectors in $\mathcal{H}$, and let $\mathfrak{B}(\mathcal{H})$ denote the set of bounded linear operators on $\mathcal{H}$. We will use the same notation for a projection in $\mathfrak{B}(\mathcal{H})$ and for the subspace in $\mathcal{H}$ onto which it projects. If $x \in \mathcal{S}$, we let ω_x denote the state of $\mathfrak{B}(\mathcal{H})$ induced by x. Let $\mathcal{R}_1, \mathcal{R}_2$ be von Neumann algebras acting on $\mathcal{H}$ such that $\mathcal{R}_1 \subseteq \mathcal{R}_2'$, and let $\mathcal{R}_{12}$ denote the von Neumann algebra $\{\mathcal{R}_1 \cup \mathcal{R}_2\}''$ generated by $\mathcal{R}_1$ and $\mathcal{R}_2$. Following Summers &

Werner (1995), we set

$$\mathcal{T}_{12} \equiv \Big\{(1/2)[A_1(B_1+B_2)+A_2(B_1-B_2)] : \\ A_i = A_i^* \in \mathcal{R}_1, B_i = B_i^* \in \mathcal{R}_2, -I \leq A_i, B_i \leq I\Big\}. \quad (1)$$

Elements of $\mathcal{T}_{12}$ are called *Bell operators* for $\mathcal{R}_{12}$. For a given state ω of $\mathcal{R}_{12}$, let

$$\beta(\omega) \equiv \sup\{\,|\omega(R)| : R \in \mathcal{T}_{12}\}. \quad (2)$$

If $\omega = \omega_x|_{\mathcal{R}_{12}}$ for some $x \in \mathcal{S}$, we write $\beta(x)$ to abbreviate $\beta(\omega_x|_{\mathcal{R}_{12}})$. From (2), it follows that the map $\omega \to \beta(\omega)$ is norm continuous from the state space of $\mathcal{R}_{12}$ into $[1,\sqrt{2}]$ (Summers & Werner 1995, Lemma 2.1). Since the map $x \to \omega_x|_{\mathcal{R}_{12}}$ is continuous from $\mathcal{S}$, in the vector norm topology, into the (normal) state space of $\mathcal{R}_{12}$, in the norm topology, it also follows that $x \to \beta(x)$ is continuous from $\mathcal{S}$ into $[1,\sqrt{2}]$. If $\beta(\omega) > 1$, we say that ω violates a Bell inequality, or is *Bell correlated*. In this context, Bell's theorem (Bell 1964) is the statement that a local hidden variable model of the correlations that ω dictates between $\mathcal{R}_1$ and $\mathcal{R}_2$ is only possible if $\beta(\omega) = 1$. Note that the set of states ω on $\mathcal{R}_{12}$ that violate a Bell inequality is open (in the norm topology) and, similarly, the set of vectors $x \in \mathcal{S}$ that induce Bell correlated states on $\mathcal{R}_{12}$ is open (in the vector norm topology).

We assume now that the pair $\mathcal{R}_1, \mathcal{R}_2$ satisfies the *Schlieder property*. That is, if $A \in \mathcal{R}_1$ and $B \in \mathcal{R}_2$ such that $AB = 0$, then either $A = 0$ or $B = 0$. Let $V \in \mathcal{R}_1$ and $W \in \mathcal{R}_2$ be nonzero partial isometries.[1] Suppose that the initial space V^*V of V is orthogonal to the final space VV^* of V; or, equivalently, that $V^2 = 0$. Similarly, suppose $W^2 = 0$. Consider the projections

$$E = V^*V + VV^*, \qquad F = W^*W + WW^*. \quad (3)$$

We show that there is a Bell operator $\widetilde{R}$ for $\mathcal{R}_{12}$ such that $\widetilde{R}y = \sqrt{2}y$ for some unit vector $y \in EF$, and $\widetilde{R}(I-E)(I-F) = (I-E)(I-F)$.

[1] A partial isometry V is an operator on a Hilbert space $\mathcal{H}$ that maps some particular closed subspace $C \subseteq \mathcal{H}$ isometrically onto another closed subspace $C' \subseteq \mathcal{H}$, and maps $C^{\perp}$ to zero.

Let

$$\begin{aligned} A_1 &= V + V^* & B_1 &= W + W^* \\ A_2 &= i(V^* - V) & B_2 &= i(W^* - W) \\ A_3 &= [V, V^*] & B_3 &= [W, W^*]. \end{aligned} \tag{4}$$

Note that $A_i^2 = E$, the A_i are self-adjoint contractions in $\mathcal{R}_1$, $A_iE = EA_i = A_i$, and $[A_1, A_2] = 2iA_3$. Similarly, $B_i^2 = F$, the B_i are self-adjoint contractions in $\mathcal{R}_2$, $B_iF = FB_i = B_i$, and $[B_1, B_2] = 2iB_3$. If we let R denote the Bell operator constructed from A_i, B_i, a straightforward calculation shows that (cf. Landau 1987*b*)

$$R^2 = EF - \frac{1}{4}[A_1, A_2][B_1, B_2] = EF + A_3B_3. \tag{5}$$

Note that $P \equiv VV^* \neq 0$ is the spectral projection for A_3 corresponding to eigenvalue 1, and $Q \equiv WW^* \neq 0$ is the spectral projection for B_3 corresponding to eigenvalue 1. Since $\mathcal{R}_1, \mathcal{R}_2$ satisfy the Schlieder property, there is a unit vector $y \in PQ$, and thus $A_3B_3y = y$. Since $PQ < EF$, it follows from (5) that $R^2y = 2y$. Thus, we may assume without loss of generality that $Ry = \sqrt{2}y$. (If $Ry \neq \sqrt{2}y$, then interchange B_1, B_2 and replace A_1 with $-A_1$. Note that the resulting Bell operator $R' = -R$ and $R'y_0 = \sqrt{2}y_0$, where $y_0 \equiv (\sqrt{2}y - Ry)/\|\sqrt{2}y - Ry\| \in EF$.)

Now for $i = 1, 2$, let $\widetilde{A}_i = (I - E) + A_i$ and $\widetilde{B}_i = (I - F) + B_i$. It is easy to see that $\widetilde{A}_i^2 = I$ and $\widetilde{B}_i^2 = I$, so the $\widetilde{A}_i$ and $\widetilde{B}_i$ are again self-adjoint contractions in $\mathcal{R}_1$ and $\mathcal{R}_2$ respectively. If we let $\widetilde{R}$ denote the corresponding Bell operator, a straightforward calculation shows that

$$\widetilde{R} = (I - E)(I - F) + (I - E)B_1 + A_1(I - F) + R. \tag{6}$$

Since the $\sqrt{2}$ eigenvector y for R lies in EF, we have $\widetilde{R}y = Ry = \sqrt{2}y$. Furthermore, since $A_i(I - E) = 0$ and $B_i(I - F) = 0$, we have $\widetilde{R}(I - E)(I - F) = (I - E)(I - F)$ as required.

A special case of the following result, where $\mathcal{R}_1$ and $\mathcal{R}_2$ are type I_∞ factors, was proved as Proposition 1 of Clifton *et al.* (2000). Recall that $\mathcal{R}$ is said to be *of infinite type* just in case the identity I is equivalent, in $\mathcal{R}$, to one of its proper subprojections.

Proposition 1. *Let $\mathcal{R}_1, \mathcal{R}_2$ be von Neumann algebras acting on $\mathcal{H}$ such that $\mathcal{R}_1 \subseteq \mathcal{R}_2'$, and $\mathcal{R}_1, \mathcal{R}_2$ satisfy the Schlieder property. If $\mathcal{R}_1, \mathcal{R}_2$ are of infinite type, then there is an open dense subset of vectors in $\mathcal{S}$ which induce Bell correlated states for $\mathcal{R}_{12}$.*

Note that the hypotheses of this proposition are invariant under isomorphisms of $\mathcal{R}_{12}$. Thus, by making use of the universal normal representation of $\mathcal{R}_{12}$ (Kadison & Ringrose 1997, p. 458), in which all normal states are vector states, it follows that the set of states Bell correlated for $\mathcal{R}_1, \mathcal{R}_2$ is norm dense in the normal state space of $\mathcal{R}_{12}$.

Proof of the proposition: Since $\mathcal{R}_1$ is infinite, there is a properly infinite projection $P \in \mathcal{R}_1$ (Kadison & Ringrose 1997, Prop. 6.3.7). Since P is properly infinite, we may apply the halving lemma (Kadison & Ringrose 1997, Lemma 6.3.3) repeatedly to obtain a countably infinite family $\{P_n\}$ of mutually orthogonal projections such that $P_n \sim P_{n+1}$ for all n and $\sum_{n=1}^{\infty} P_n = P$. (Halve P as $P_1 + F_1$; then halve F_1 as $P_2 + F_2$, and so on. Now replace P_1 by $P - \sum_{n=2}^{\infty} P_n$; cf. Kadison & Ringrose 1997, Lemma 6.3.4.) Let $P_0 \equiv I - P$. For each $n \in \mathbb{N}$, let V_n denote the partial isometry with initial space $V_n^* V_n = P_n$ and final space $V_n V_n^* = P_{n+1}$. By the same reasoning, there is a countable family $\{Q_n\}$ of mutually orthogonal projections in $\mathcal{R}_2$ and partial isometries W_n with $W_n^* W_n = Q_n$ and $W_n W_n^* = Q_{n+1}$. For each $n \in \mathbb{N}$, let

$$\begin{array}{ll} A_{1,n} = V_{n+1} + V_{n+1}^* & B_{1,n} = W_{n+1} + W_{n+1}^*, \\ A_{2,n} = i(V_{n+1}^* - V_{n+1}) & B_{2,n} = i(W_{n+1}^* - W_{n+1}), \end{array} \tag{7}$$

and let

$$E_n = V_{n+1}^* V_{n+1} + V_{n+1} V_{n+1}^* = P_{n+1} + P_{n+2}, \tag{8}$$

$$F_n = W_{n+1}^* W_{n+1} + W_{n+1} W_{n+1}^* = Q_{n+1} + Q_{n+2}. \tag{9}$$

Define $\widetilde{A}_{i,n}$ and $\widetilde{B}_{i,n}$ as in the discussion preceding this proposition, let $\widetilde{R}_n$ be the corresponding Bell operator, and let the unit vector $y_n \in E_n F_n$ be the $\sqrt{2}$ eigenvector for $\widetilde{R}_n$.

Now, let x be any unit vector in $\mathcal{H}$. Since $\sum_{i=0}^{n} P_i \leq I - E_n$ and since $\sum_{i=0}^{\infty} P_i = I$, we have $(I - E_n) \to I$ in the strong-operator topology. Similarly, $(I - F_n) \to I$ in the strong-operator topology. Therefore if we let

$$x_n \equiv \frac{(I - E_n)(I - F_n)x}{\|(I - E_n)(I - F_n)x\|}, \tag{10}$$

we have

$$x = \lim_n (I - E_n)(I - F_n)x = \lim_n x_n. \tag{11}$$

Note that the inner product $\langle x_n, y_n \rangle = 0$, and thus

$$z_n \equiv (1 - n^{-1})^{1/2} x_n + n^{-1/2} y_n \tag{12}$$

is a unit vector for all n. Since $\lim_n z_n = x$, it suffices to observe that each z_n is Bell correlated for $\mathcal{R}_{12}$. Recall that $\widetilde{R}_n(I - E_n)(I - F_n) = (I - E_n)(I - F_n)$, and thus $\widetilde{R}_n x_n = x_n$. A simple calculation then reveals that

$$\beta(z_n) \geq \langle \widetilde{R}_n z_n, z_n \rangle = (1 - n^{-1}) + n^{-1}\sqrt{2} > 1. \tag{13}$$

□

3 Cyclic vectors and entangled states

Proposition 1 establishes that Bell correlation is generic for commuting pairs of infinite von Neumann algebras. However, we are given no information about the character of the correlations of particular states. We provide a partial remedy for this in the next proposition, where we show that any vector cyclic for $\mathcal{R}_1$ (or for $\mathcal{R}_2$) induces a state that is not classically correlated; i.e., it is 'nonseparable'.

Again, let $\mathcal{R}_1, \mathcal{R}_2$ be von Neumann algebras on $\mathcal{H}$ such that $\mathcal{R}_1 \subseteq \mathcal{R}_2'$. Recall that a state ω of $\mathcal{R}_{12}$ is called a *normal product state* just in case ω is normal, and there are states ω_1 of $\mathcal{R}_1$ and ω_2 of $\mathcal{R}_2$ such that

$$\omega(AB) = \omega_1(A)\omega_2(B), \tag{14}$$

for all $A \in \mathcal{R}_1, B \in \mathcal{R}_2$. Werner (1989), in dealing with the case of $\mathfrak{B}(\mathbb{C}^n) \otimes \mathfrak{B}(\mathbb{C}^n)$, defined a density operator D to be *classically correlated* — the term *separable* is now more commonly used — just in case D can be approximated in trace norm by convex combinations of density operators of form $D_1 \otimes D_2$. Although Werner's definition of nonseparable states directly generalizes the traditional notion of pure entangled states, he showed that a nonseparable mixed state need not violate a Bell inequality; thus, Bell correlation is in general a sufficient, though not necessary condition for a state's being nonseparable. On the other hand, it has since been shown that nonseparable states often possess more subtle forms of nonlocality, which may be indicated by measurements more general than the single ideal measurements which can indicate Bell correlation (Popescu 1995). (See Clifton *et al.* (2000) and Clifton & Halvorson (2000) for further discussion.)

In terms of the linear functional representation of states, Werner's separable states are those in the norm closed convex hull of the product states of $\mathfrak{B}(\mathbb{C}^n) \otimes \mathfrak{B}(\mathbb{C}^n)$. However, in case of the more general setup — i.e., $\mathcal{R}_1 \subseteq \mathcal{R}_2'$, where $\mathcal{R}_1, \mathcal{R}_2$ are arbitrary von Neumann algebras on $\mathcal{H}$ — the choice of topology on the normal state space of $\mathcal{R}_{12}$ will yield in general different definitions of separability. Moreover, it has been argued that norm convergence of a sequence of states can never be verified in the laboratory, and as a result, the appropriate notion of physical approximation is given by the (weaker) weak* topology (Emch 1972, Haag 1992). And the weak* and norm topologies do not generally coincide *even* on the normal state space (Dell'Antonio 1967).

For the next proposition, then, we will define the *separable* states of $\mathcal{R}_{12}$ to be the normal states in the weak* closed convex hull of the normal product states. Note that $\beta(\omega) = 1$ if ω is a product state, and since β is a convex function on the state space, $\beta(\omega) = 1$ if ω is a convex combination of product states (Summers & Werner 1995, Lemma 2.1). Furthermore, since β is lower semicontinuous in the weak* topology (Summers & Werner 1995, Lemma 2.1), $\beta(\omega) = 1$ for any separable state. Contrapositively, any Bell correlated state must be nonseparable.

We now introduce some notation that will aid us in the proof of our result. For a state ω of the von Neumann algebra $\mathcal{R}$ and an operator $A \in \mathcal{R}$, define the state ω^A on $\mathcal{R}$ by

$$\omega^A(X) \equiv \frac{\omega(A^*XA)}{\omega(A^*A)}, \tag{15}$$

if $\omega(A^*A) \neq 0$, and let $\omega^A = \omega$ otherwise. Suppose now that $\omega(A^*A) \neq 0$ and ω is a convex combination of states:

$$\omega = \sum_{i=1}^{n} \lambda_i \omega_i. \tag{16}$$

Then, letting $\lambda_i^A \equiv \omega(A^*A)^{-1}\omega_i(A^*A)\lambda_i$, ω^A is again a convex combination

$$\omega^A = \sum_{i=1}^{n} \lambda_i^A \omega_i^A. \tag{17}$$

Moreover, it is not difficult to see that the map $\omega \to \omega^A$ is weak* continuous at any point ρ such that $\rho(A^*A) \neq 0$. Indeed, let $\mathcal{O}_1 = N(\rho^A : X_1, \dots, X_n, \epsilon)$ be a weak* neighborhood of ρ^A. Then, taking

$$\mathcal{O}_2 = N(\rho : A^*A, A^*X_1A, \dots, A^*X_nA, \delta), \tag{18}$$

and $\omega \in \mathcal{O}_2$, we have

$$|\rho(A^*X_iA) - \omega(A^*X_iA)| < \delta, \tag{19}$$

for $i = 1, \dots, n$, and

$$|\rho(A^*A) - \omega(A^*A)| < \delta. \tag{20}$$

By choosing $\delta < \rho(A^*A) \neq 0$, we also have $\omega(A^*A) \neq 0$, and thus

$$|\rho^A(X_i) - \omega^A(X_i)| < O(\delta) \leq \epsilon, \tag{21}$$

for an appropriate choice of δ. That is, $\omega^A \in \mathcal{O}_1$ for all $\omega \in \mathcal{O}_2$ and $\omega \to \omega^A$ is weak* continuous at ρ.

Specializing to the case where $\mathcal{R}_1 \subseteq \mathcal{R}_2'$, and $\mathcal{R}_{12} = \{\mathcal{R}_1 \cup \mathcal{R}_2\}''$, it is clear from the above that for any normal product state ω of $\mathcal{R}_{12}$ and for $A \in \mathcal{R}_1$, ω^A is again a normal product state. The same is true if ω is a convex combination of normal product states, or the weak* limit of such combinations. We summarize the results of this discussion in the following lemma:

Lemma 1. *For any separable state ω of $\mathcal{R}_{12}$ and any $A \in \mathcal{R}_1$, ω^A is again separable.*

Proposition 2. *Let $\mathcal{R}_1, \mathcal{R}_2$ be nonabelian von Neumann algebras such that $\mathcal{R}_1 \subseteq \mathcal{R}_2'$. If x is cyclic for $\mathcal{R}_1$, then ω_x is nonseparable across $\mathcal{R}_{12}$.*

Proof. From Lemma 2.1 of Summers & Werner (1995), there is a normal state ρ of $\mathcal{R}_{12}$ such that $\beta(\rho) = \sqrt{2}$. But since all normal states are in the (norm) closed convex hull of vector states (Kadison & Ringrose 1997, Thm. 7.1.12), and since β is norm continuous and convex, there is a vector $v \in \mathcal{S}$ such that $\beta(v) > 1$. By the continuity of β (on $\mathcal{S}$), there is an open neighborhood $\mathcal{O}$ of v in $\mathcal{S}$ such that $\beta(y) > 1$ for all $y \in \mathcal{O}$. Since x is cyclic for $\mathcal{R}_1$, there is an $A \in \mathcal{R}_1$ such that $Ax \in \mathcal{O}$. Thus, $\beta(Ax) > 1$ which entails that $\omega_{Ax} = (\omega_x)^A$ is a nonseparable state for $\mathcal{R}_{12}$. This, by the preceding lemma, entails that ω_x is nonseparable. □

Note that if $\mathcal{R}_1$ has at least one cyclic vector $x \in \mathcal{S}$, then $\mathcal{R}_1$ has a dense set of cyclic vectors in $\mathcal{S}$ (Dixmier & Maréchal 1971). Since each of the corresponding vector states is nonseparable across $\mathcal{R}_{12}$, Proposition 2 shows that if $\mathcal{R}_1$ has a cyclic vector, then the (open) set of vectors inducing nonseparable states across $\mathcal{R}_{12}$ is dense in $\mathcal{S}$. On the other hand, since the existence of a cyclic vector for $\mathcal{R}_1$ is not invariant under isomorphisms of $\mathcal{R}_{12}$, Proposition 2 does not entail that if $\mathcal{R}_1$ has a cyclic vector, then there is a norm dense set of nonseparable states in the entire normal state space of $\mathcal{R}_{12}$. (Contrast the analogous discussion preceding the proof of Proposition 1.) Indeed, if we let $\mathcal{R}_1 = \mathbf{B}(\mathbb{C}^2) \otimes I$, $\mathcal{R}_2 = I \otimes \mathbf{B}(\mathbb{C}^2)$, then any entangled state vector is cyclic for $\mathcal{R}_1$; but, the set of nonseparable states of $\mathbf{B}(\mathbb{C}^2) \otimes \mathbf{B}(\mathbb{C}^2)$ is *not* norm dense (Clifton & Halvorson 2000, Życzkowski *et al.* 1998). However, if in addition to $\mathcal{R}_1$ or $\mathcal{R}_2$ having a cyclic vector, $\mathcal{R}_{12}$ has a separating vector (as is often the case in quantum field theory), then all normal states of $\mathcal{R}_{12}$ are vector states (Kadison & Ringrose 1997, Thm. 7.2.3), and it follows that the nonseparable states *will* be norm dense in the entire normal state space of $\mathcal{R}_{12}$.

4 Applications to algebraic quantum field theory

Let (M, g) be a relativistic spacetime. The basic mathematical object of algebraic quantum field theory (see Haag 1992, Borchers 1996, Dimock 1980) is an association between precompact open subsets O of M and C^*-subalgebras $\mathcal{A}(O)$ of a unital C^*-algebra $\mathcal{A}$. (We assume that each $\mathcal{A}(O)$ contains the identity I of $\mathcal{A}$.) The motivation for this association is the idea that $\mathcal{A}(O)$ represents observables that can be measured in the region O. With this in mind, one assumes

1. *Isotony:* If $O_1 \subseteq O_2$, then $\mathcal{A}(O_1) \subseteq \mathcal{A}(O_2)$.

2. *Microcausality:* $\mathcal{A}(O') \subseteq \mathcal{A}(O)'$.

Here O' denotes the interior of the set of all points of M that are spacelike to every point in O.

In the case where (M, g) is Minkowski spacetime, it is assumed in addition that there is a faithful representation $\mathbf{x} \to \alpha_{\mathbf{x}}$ of the translation group of M in the group of automorphisms of $\mathcal{A}$ such that

3. *Translation Covariance:* $\alpha_{\mathbf{x}}(\mathcal{A}(O)) = \mathcal{A}(O + \mathbf{x})$.

4. *Weak Additivity:* For any $O \subseteq M$, $\mathcal{A}$ is the smallest C^*-algebra containing $\bigcup_{\mathbf{x} \in M} \mathcal{A}(O + \mathbf{x})$.

The class of physically relevant representations of $\mathcal{A}$ is decided by further desiderata such as — in the case of Minkowski spacetime — a unitary representation of the group of translation automorphisms which satisfies the spectrum condition. Relative to a fixed representation π, we let $\mathcal{R}_\pi(O)$ denote the von Neumann algebra $\pi(\mathcal{A}(O))''$ on the representation space $\mathcal{H}_\pi$. In what follows, we consider only nontrivial representations (i.e., $\dim \mathcal{H}_\pi > 1$), and we let $\mathcal{S}_\pi$ denote the set of unit vectors in $\mathcal{H}_\pi$.

Proposition 3. *Let $\{\mathcal{A}(O)\}$ be a net of local algebras over Minkowski spacetime. Let π be any representation in the local quasiequivalence class of some irreducible vacuum representation (e.g. superselection sectors in the sense of Doplicher-Haag-Roberts (1969) or Buchholz-Fredenhagen (1982)). If O_1, O_2 are any two open subsets of M such that $O_1 \subseteq O_2'$, then the set of vectors inducing Bell correlated states for $\mathcal{R}_\pi(O_1), \mathcal{R}_\pi(O_2)$ is open and dense in $\mathcal{S}_\pi$.*

Proof. Let O_3, O_4 be precompact open subsets of M such that $O_3 \subseteq O_1, O_4 \subseteq O_2$, and such that $O_3 + N \subseteq O_4'$ for some neighborhood N of the origin. In an irreducible vacuum representation ϕ, local algebras are of infinite type (Horuzhy 1988, Prop. 1.3.9), and since $O_3 + N \subseteq O_4'$, the Schlieder property holds for $\mathcal{R}_\phi(O_3), \mathcal{R}_\phi(O_4)$ (Schlieder 1969). If π is any representation in the local quasiequivalence class of ϕ, these properties hold for $\mathcal{R}_\pi(O_3), \mathcal{R}_\pi(O_4)$ as well. Thus, we may apply Proposition 1 to conclude that the set of vectors inducing Bell correlated states for $\mathcal{R}_\pi(O_3), \mathcal{R}_\pi(O_4)$ is dense in $\mathcal{S}_\pi$. Finally, note that any state Bell correlated for $\mathcal{R}_\pi(O_3), \mathcal{R}_\pi(O_4)$ is Bell correlated for $\mathcal{R}_\pi(O_1), \mathcal{R}_\pi(O_2)$. □

Proposition 4. *Let (M, g) be a globally hyperbolic spacetime, let $\{\mathcal{A}(O)\}$ be the net of local observable algebras associated with the free Klein-Gordon field (Dimock 1980), and let π be the GNS representation of some quasifree Hadamard state (Kay & Wald 1991). If O_1, O_2 are any two open subsets of M such that $O_1 \subseteq O_2'$, then the set of vectors inducing Bell correlated states for $\mathcal{R}_\pi(O_1), \mathcal{R}_\pi(O_2)$ is open and dense in $\mathcal{S}_\pi$.*

Proof. The regular diamonds (in the sense of Verch 1997) form a basis for the topology on M. Thus, we may choose regular diamonds O_3, O_4 such that $\overline{O_3} \subseteq O_1$ and $\overline{O_4} \subseteq O_2$. The nonfiniteness of the local algebras $\mathcal{R}_\pi(O_3), \mathcal{R}_\pi(O_4)$ is established in Verch (1997, Thm. 3.6.g), and the split property for these algebras is established in Verch (1997, Thm. 3.6.d). Since the split property entails the Schlieder property, it follows from Proposition 1 that the set of vectors inducing Bell correlated states for $\mathcal{R}_\pi(O_3), \mathcal{R}_\pi(O_4)$ [and thereby Bell correlated for $\mathcal{R}_\pi(O_1), \mathcal{R}_\pi(O_2)$] is dense in $\mathcal{S}_\pi$. □

There are many physically interesting states, such as the Minkowski vacuum itself, about which Propositions 3 and 4 are silent. However, Reeh-Schlieder type theorems entail that many of these physically interesting states are induced by vectors which are cyclic for local algebras, and thus it follows from Proposition 2 that these states are nonseparable across any spacelike separated pair of local algebras. In particular, although there is an upper bound on the Bell correlation of the Minkowski vacuum (in models with a mass gap) that decreases exponentially with spacelike separation (Summers & Werner 1995, Prop. 3.2), the vacuum state remains nonseparable (in our sense) at all distances. On the other hand, since nonseparability is only a *necessary* condition for Bell correlation, none of our results decide the question of whether the vacuum state always retains *some* Bell correlation across arbitrary spacelike separated regions.

Bibliography

Bell, J. S. (1964), 'On the Einstein-Podolsky-Rosen paradox', *Physics* **1**, 195–200.

Bell, J. S. (1987), *Speakable and Unspeakable in Quantum Mechanics*, Cambridge University Press, New York.

Borchers, H.-J. (1996), *Translation Group and Particle Representations in Quantum Field Theory*, Springer, New York.

Buchholz, D. & Fredenhagen, K. (1982), 'Locality and the structure of particle states', *Communications in Mathematical Physics* **84**, 1–54.

Clauser, J. F., Horne, M. A., Shimony, A. & Holt, R. A. (1969), 'Proposed experiment to test local hidden-variable theories', *Physical Review Letters* **26**, 880–4.

Clifton, R. K. & Halvorson, H. P. (2000), 'Bipartite mixed states of infinite-dimensional systems are generically nonseparable', *Physical Review A* **61**, 012108.

Clifton, R. K., Halvorson, H. P. & Kent, A. (2000), 'Non-local correlations are generic in infinite-dimensional bipartite systems', *Physical Review A* **61**, 042101. Chapter 10 of this volume.

Dell'Antonio, G. F. (1967), 'On the limit of sequences of normal states', *Communications in Pure and Applied Mathematics* **20**, 413–29.

Dimock, J. (1980), 'Algebras of local observables on a manifold', *Communications in Mathematical Physics* **77**, 219–28.

Dixmier, J. & Maréchal, O. (1971), 'Vecteurs totalisateurs d'une algèbre de von Neumann', *Communications in Mathematical Physics* **22**, 44–50.

Emch, G. G. (1972), *Algebraic Methods in Statistical Mechanics and Quantum Field Theory*, Wiley, New York.

Haag, R. (1992), *Local Quantum Physics*, 2nd ed., Springer, New York.

Horuzhy, S. S. (1988), *Introduction to Algebraic Quantum Field Theory*, Kluwer, Dordrecht.

Kadison, R. & Ringrose, J. (1997), *Fundamentals of the Theory of Operator Algebras*, American Mathematical Society, Providence, RI.

Kay, B. S. & Wald, R. M. (1991), 'Theorems on the uniqueness and thermal properties of stationary, nonsingular, quasifree states on spacetimes with a bifurcate Killing horizon', *Physics Reports* **207**, 49–136.

Landau, L. J. (1987*a*), 'On the non-classical structure of the vacuum', *Physics Letters A* **123**, 115–8.

Landau, L. J. (1987*b*), 'On the violation of Bell's inequality in quantum theory', *Physics Letters A* **120**, 54–6.

Popescu, S. (1995), 'Bell's inequalities and density matrices: Revealing "hidden" nonlocality', *Physical Review Letters* **74**, 2619–22.

Redhead, M. L. G. (1995), 'More ado about nothing', *Foundations of Physics* **25**, 123–37.

Schlieder, S. (1969), 'Einige Bemerkungen über Projektionsoperatoren', *Communications in Mathematical Physics* **13**, 216.

Summers, S. J. (1990), 'On the independence of local algebras in quantum field theory', *Reviews of Mathematical Physics* **2**, 201–47.

Summers, S. J. (1997), Bell's inequalities and algebraic structure, in S. Doplicher, R. Longo, J. E. Roberts & L. Zsido, eds, 'Operator Algebras and Quantum Field Theory', International Press, Cambridge, MA, pp. 633–46.

Summers, S. J. & Werner, R. F. (1995), 'On Bell's inequalities and algebraic invariants', *Letters in Mathematical Physics* **33**, 321–34.

Verch, R. (1997), 'Continuity of symplectically adjoint maps and the algebraic structure of Hadamard vacuum representations for quantum fields on curved spacetime', *Reviews of Mathematical Physics* **9**, 635–74.

Werner, R. F. (1989), 'Quantum states with Einstein-Podolsky-Rosen correlations admitting a hidden-variable model', *Physical Review A* **40**, 4277–81.

Życzkowski, K., Horodecki, P., Sanpera, A. & Lewenstein, M. (1998), 'Volume of the set of separable states', *Physical Review A* **58**, 883–92.

Chapter 7

Entanglement and open systems in algebraic quantum field theory

with Hans Halvorson

... despite its conservative way of dealing with physical principles, algebraic QFT leads to a radical change of paradigm. Instead of the Newtonian view of a space-time filled with a material content one enters the reality of Leibniz created by relation (in particular inclusions) between 'monads' ($\sim$ the hyperfinite type III_1 local von Neumann factors $\mathcal{A}(O)$ which as single algebras are nearly void of physical meaning).

— Bert Schroer (1998)

The authors are extremely grateful to Paul Busch (for helpful discussions about operations), Jeremy Butterfield (for helping us to clarify our critique of Redhead's discussion of the operational implications of cyclicity), Reinhard Werner (for filling in for us the argument for Eqn (10)), Fred Kronz (for urging us to include the final two sentences of note 1), Michael Redhead (for reminding us of the facts cited in note 14), and an anonymous referee (for prompting our inclusion of note 4). R. K. C. also wishes to thank All Souls College, Oxford for support under a Visiting Fellowship.

1 Introduction

In *PCT, Spin and Statistics, and All That*, Streater and Wightman claim that, as a consequence of the axioms of algebraic quantum field theory (AQFT), 'it is difficult to isolate a system described by fields from outside effects' (2000, p. 139). Haag makes a similar claim in *Local Quantum Physics*: 'From the previous chapters of this book it is evidently not obvious how to achieve a division of the world into parts to which one can assign individuality... Instead we used a division according to regions in space-time. This leads in general to open systems' (1992, p. 298). By a field system these authors mean that portion of a quantum field within a specified bounded open region O of spacetime, with its associated algebra of observables $\mathcal{A}(O)$ (constructed in the usual way, out of 'field operators' smeared with test-functions having support in O). The environment of a field system (so construed) is naturally taken to be the field in the region O', the spacelike complement of O. But then the claims above appear, at first sight, puzzling. After all, it is an axiom of AQFT that the observables in $\mathcal{A}(O')$ commute with those in $\mathcal{A}(O)$. And this implies — indeed, is *equivalent* to — the assertion that standard von Neumann measurements performed in O' *cannot* have 'outside effects' on the expectations of observables in O (Lüders 1951). What, then, could the above authors possibly mean by saying that the field in O must be regarded as an open system?

A similar puzzle is raised by a famous passage in which Einstein (1948) contrasts the picture of physical reality embodied in classical field theories with that which emerges when we try to take quantum theory to be complete:

> If one asks what is characteristic of the realm of physical ideas independently of the quantum theory, then above all the following attracts our attention: the concepts of physics refer to a real external world, i.e., ideas are posited of things that claim a 'real existence' independent of the perceiving subject (bodies, fields, etc.) ... it appears to be essential for this arrangement of the things in physics that, at a specific time, these things

> claim an existence independent of one another, insofar as these things 'lie in different parts of space'. Without such an assumption of the mutually independent existence (the 'being-thus') of spatially distant things, an assumption which originates in everyday thought, physical thought in the sense familiar to us would not be possible. Nor does one see how physical laws could be formulated and tested without such clean separation. ... For the relative independence of spatially distant things (A and B), this idea is characteristic: an external influence on A has no *immediate* effect on B; this is known as the 'principle of local action,' which is applied consistently in field theory. The complete suspension of this basic principle would make impossible the idea of the existence of (quasi-)closed systems and, thereby, the establishment of empirically testable laws in the sense familiar to us. (*ibid*, 321–22; Howard's 1989 translation)

There is a strong temptation to read Einstein's 'assumption of the mutually independent existence of spatially distant things' and his 'principle of local action' as anticipating, respectively, the distinction between separability and locality — or between nonlocal 'outcome-outcome' correlation and 'measurement-outcome' correlation — that some philosophers argue is crucial to unraveling the conceptual implications of Bell's theorem (see, e.g., Howard 1989). However, even in nonrelativistic quantum theory, there is no question of any nonlocal *measurement*-outcome correlation between distinct systems or degrees of freedom, whose observables are always represented as commuting. Making the reasonable assumption that Einstein knew this quite well, what is it about taking quantum theory at face value that he saw as a threat to securing the existence of physically closed systems?

What makes quantum systems open for Einstein, as well as for Streater and Wightman, and Haag, is that quantum systems can occupy entangled states in which they sustain nonclassical EPR correlations with other quantum systems outside their light cones. That is, while it is correct to read Einstein's discussion of the mutually independent existence of distant systems as an implicit critique of the way in which quantum theory typically represents their joint state as entangled, we believe it must be the *outcome-outcome* EPR correlations associated with entangled states that, in Einstein's view, pose a problem for the legitimate testing of the predictions of quantum theory. One could certainly doubt whether EPR correlations really pose any methodological problem, or whether they truly require the existence of physical (or 'causal') influences acting on a quantum system from outside. But the analogy with open systems in thermodynamics that Einstein and the others seem to be invoking is not entirely misplaced.

Consider the simplest toy universe consisting of two nonrelativistic quantum systems, represented by a tensor product of two-dimensional Hilbert spaces $\mathbb{C}^2_A \otimes \mathbb{C}^2_B$, where system A is the 'object' system, and B its 'environment'. Let x be any state vector for the composite system $A + B$, and $D_A(x)$ be the reduced density operator x determines for system A. Then the von Neumann entropy of A, $E_A(x) = -\mathrm{Tr}(D_A(x) \ln D_A(x))$ $(= E_B(x))$, varies with the degree to which A and B are entangled. If x is a product vector with no entanglement, $E_A(x) = 0$, whereas, at the opposite extreme, $E_A(x) = \ln 2$ when x is, say, a singlet or triplet state. The more A and B are entangled, the more 'disordered' A becomes, because it will then have more than one state available to it, and A's probabilities of occupying them will approach equality. In fact, exploiting an analogy to Carnot's heat cycle and the second law of thermodynamics (that it is impossible to construct a *perpetuum mobile*), Popescu & Rohrlich (1997) have shown that the general principle that it is impossible to create entanglement between pairs of systems by local operations on one member of each pair implies that the von Neumann entropy of either member provides the uniquely correct measure of their entanglement when they are in a pure state. Changes in their degree of entanglement, and hence in the entropy of either system A or B, can only come about in the presence of a nontrivial interaction Hamiltonian between them. But the fact remains that there is an intimate connection between a system's entanglement with its environment and the extent to which that system should be thought of as physically closed.

Returning to AQFT, Streater and Wightman, as well as Haag, all intend to make a far stronger claim about quantum field systems — a point that even applies to spacelike-separated regions of a *free* field, and might well have offended Einstein's physical sensibilities even more. The point is that quantum field systems are *unavoidably* and *intrinsically* open to entanglement. Streater and Wightman's comment is made in reference to the Reeh-Schlieder (1961) theorem, a consequence of the general axioms of AQFT. We shall show that this theorem entails severe *practical* obstacles to isolating field systems from entanglement with other field systems. Haag's comment goes deeper, and is related to the fact that the algebras associated with field systems localized in spacetime regions, are in all known models of the axioms, type III von Neumann algebras. We shall show that this feature of the local algebras imposes a fundamental limitation on isolating field systems from entanglement even *in principle*.

Think again of our toy nonrelativistic universe $A + B$, with Alice in possession of system A, and the state x entangled. Although there are no unitary operations Alice can perform on system A that will reduce its entropy, she can still try to destroy its entanglement with B by performing a standard von Neumann measurement on A. If $P_{\pm}$ are the eigenprojections of the observable Alice measures, and the initial density operator of $A + B$ is $D = P_x$ (where P_x is the projection onto the ray x generates), then the post-measurement joint state of $A + B$ will be given by the new density operator

$$D \to D' = (P_+ \otimes I)P_x(P_+ \otimes I) + (P_- \otimes I)P_x(P_- \otimes I). \tag{1}$$

Since the projections $P_{\pm}$ are one-dimensional, and x is entangled, there are nonzero vectors $a_x^{\pm} \in \mathbb{C}_A^2$ and $b_x^{\pm} \in \mathbb{C}_B^2$ such that $(P_{\pm} \otimes I)x = a_x^{\pm} \otimes b_x^{\pm}$, and a straightforward calculation reveals that D' may be re-expressed as

$$D' = \mathrm{Tr}[(P_+ \otimes I)P_x]P_+ \otimes P_{b_x^+} + \mathrm{Tr}[(P_- \otimes I)P_x]P_- \otimes P_{b_x^-}. \tag{2}$$

Thus, regardless of the initial state x, or the degree to which it was entangled, D' will always be a convex combination of product states, and there will no longer be any entanglement between

A and B. One might say that Alice's measurement operation on A has the effect of isolating A from any further EPR influences from B. Moreover, this result can be generalized. Given any finite or infinite dimension for the Hilbert spaces $\mathcal{H}_A$ and $\mathcal{H}_B$, there is always an operation Alice can perform on system A that will destroy its entanglement with B no matter what their initial state D was, pure or mixed. In fact, it suffices for Alice to measure any nondegenerate observable of A with a discrete spectrum. The final state D' will then be a convex combination of product states, each of which is a product density operator obtained by 'collapsing' D using some particular eigenprojection of the measured observable.[1]

The upshot is that if entanglement *does* pose a methodological threat, it can at least be brought under control in nonrelativistic quantum theory. Not so when we consider the analogous setup in quantum field theory, with Alice in the vicinity of one region A, and B any other spacelike-separated field system. We shall see that AQFT puts both practical and theoretical limits on Alice's ability to destroy entanglement between her field system and B. Again, while one can doubt whether this poses any real methodological problem for Alice — an issue to which we shall return in earnest later — we think it is ironic, considering Einstein's point of view, that such limits should be forced upon us once we make the transition to a fully *relativistic* formulation of quantum theory.

We begin in Section 2 by reviewing the formalism of AQFT, the concept of entanglement between spacelike-separated field systems, and the mathematical representation of an operation per-

[1] The fact that disentanglement of a state can always be achieved in this way does not conflict with the recently established result there can be no 'universal disentangling machine', i.e., no *unitary* evolution that maps an arbitrary $A + B$ state D to an unentangled state with the same reduced density operators as D (Mor 1999, Mor & Terno 1999). Also bear in mind that we have *not* required that a successful disentangling process leave the states of the entangled subsystems unchanged. Finally, though we have written of Alice's measurement 'collapsing' the density matrix D to D', we have *not* presupposed the projection postulate nor begged the question against no-collapse interpretations of quantum theory. What is at issue here is the destruction of entangling correlations between A and B, *not* between the compound system $M + A$, including Alice's measuring device M, and B.

formed within a local spacetime region on a field system. In Section 3, we connect the Reeh-Schlieder theorem with the practical difficulties involved in guaranteeing that a field system is disentangled from other field systems. The language of operations also turns out to be indispensable for clearing up some apparently paradoxical physical implications of the Reeh-Schlieder theorem that have been raised in the literature without being properly resolved. In Section 4, we discuss differences between type III von Neumann algebras and the standard type I von Neumann algebras employed in nonrelativistic quantum theory, emphasizing the radical implications type III algebras have for the ignorance interpretation of mixtures and entanglement. We end Section 4 by connecting the type III character of the algebra of a local field system with the inability, in principle, to perform local operations on the system that will destroy its entanglement with other spacelike-separated systems. We offer this result as one way to make precise the sense in which AQFT requires a radical change in paradigm — a change that, regrettably, has passed virtually unnoticed by philosophers of quantum theory.

2 AQFT, entanglement, and local operations

We first give a quick review of some of the mathematics needed to understand AQFT.

An abstract C^*-algebra is a Banach *-algebra, where the involution and norm are related by $|A^*A| = |A|^2$. Thus the algebra $\mathfrak{B}(\mathcal{H})$ of all bounded operators on a Hilbert space $\mathcal{H}$ is a C^*-algebra, with * taken to be the adjoint operation, and $|\cdot|$ the standard operator norm. Moreover, any *-*sub*algebra of $\mathfrak{B}(\mathcal{H})$ that is closed in the operator norm is a C^*-algebra, and, conversely, one can show that every abstract C^*-algebra has a concrete (faithful) representation as a norm-closed *-subalgebra of $\mathfrak{B}(\mathcal{H})$, for some appropriate Hilbert space $\mathcal{H}$ (Kadison & Ringrose 1997, Remark 4.5.7).

On the other hand, a von Neumann algebra is always taken to be a concrete collection of operators on some fixed Hilbert space $\mathcal{H}$. For F any set of operators on $\mathcal{H}$, let F' denote the commutant of F, the set of all operators on $\mathcal{H}$ that commute with *every* operator in

F. Observe that $F \subseteq F''$, that $F \subseteq G$ implies $G' \subseteq F'$, and (hence) that $A' = A'''$. $\mathcal{R}$ is called a von Neumann algebra exactly when $\mathcal{R}$ is a *-subalgebra of $\mathfrak{B}(\mathcal{H})$ that contains the identity and satisfies $\mathcal{R} = \mathcal{R}''$. This is equivalent, via von Neumann's famous double commutant theorem (Kadison & Ringrose 1997, Theorem 5.3.1), to the assertion that $\mathcal{R}$ is closed in the strong operator topology, where $Z_n \to Z$ strongly just in case $|(Z_n - Z)x| \to 0$ for all $x \in \mathcal{H}$.

If a sequence $\{Z_n\} \subseteq \mathcal{R}$ converges to $Z \in \mathcal{R}$ in norm, then since $|(Z_n - Z)x| \leq |Z_n - Z||x|$, the convergence is also strong, hence every von Neumann algebra is also a C^*-algebra. However, not every C^*-algebra of operators is a von Neumann algebra. For example, the C^*-algebra $\mathcal{C}$ of all compact operators on an infinite-dimensional Hilbert space $\mathcal{H}$ — that is, the norm closure of the *-subalgebra of all finite rank operators on $\mathcal{H}$ — does *not* contain the identity, nor does $\mathcal{C}$ satisfy $\mathcal{C} = \mathcal{C}''$. (Indeed, $\mathcal{C}'' = \mathfrak{B}(\mathcal{H})$, because only multiples of the identity commute with all finite-dimensional projections, and of course *every* operator commutes with all multiples of the identity.)

Finally, let S be any self-adjoint (i.e., *-closed) set of operators in $\mathfrak{B}(\mathcal{H})$. Then S' is a *-algebra containing the identity, and both S' $(= S''' = (S')'')$ and S'' $(= (S')' = (S')''' = (S'')'')$ are von Neumann algebras. If we suppose there is some other von Neumann algebra $\mathcal{R}$ such that $S \subseteq \mathcal{R}$, then $\mathcal{R}' \subseteq S'$, which in turn entails $S'' \subseteq \mathcal{R}'' = \mathcal{R}$. Thus S'' is actually the smallest von Neumann algebra containing S, i.e., the von Neumann algebra that S generates. For example, the von Neumann algebra generated by all finite rank operators is the whole of $\mathfrak{B}(\mathcal{H})$.

The basic mathematical object of AQFT on Minkowski spacetime M is an association $O \mapsto \mathcal{A}(O)$ between bounded open subsets O of M and C^*-subalgebras $\mathcal{A}(O)$ of an abstract C^*-algebra $\mathcal{A}$ (Horuzhy 1988, Haag 1992). The motivation for this association is that the self-adjoint elements of $\mathcal{A}(O)$ represent the physical magnitudes, or observables, of the field intrinsic to the region O. We shall see below how the elements of $\mathcal{A}(O)$ can also be used to represent mathematically the physical operations that can be performed within O, and often it is only this latter interpretation of $\mathcal{A}(O)$ that is emphasized (Haag 1992, p. 104). One naturally assumes

Isotony: If $O_1 \subseteq O_2$, then $\mathcal{A}(O_1) \subseteq \mathcal{A}(O_2)$.

As a consequence, the collection of all local algebras $\mathcal{A}(O)$ defines a net whose limit points can be used to define algebras associated with unbounded regions, and in particular $\mathcal{A}(M)$, which is identified with $\mathcal{A}$ itself.

One of the leading ideas in the algebraic approach to fields is that all of the physics of a particular field theory is encoded in the structure of its net of local algebras.[2] But there are some general assumptions about the net $\{\mathcal{A}(O) : O \subseteq M\}$ that all physically reasonable field theories are held to satisfy. First, one assumes

Microcausality: $\mathcal{A}(O') \subseteq \mathcal{A}(O)'$.

One also assumes that there is a faithful representation $\mathbf{x} \to \alpha_{\mathbf{x}}$ of the spacetime translation group of M in the group of automorphisms of $\mathcal{A}$, satisfying

Translation Covariance: $\alpha_{\mathbf{x}}(\mathcal{A}(O)) = \mathcal{A}(O + \mathbf{x})$.

Weak Additivity: For any $O \subseteq M$, $\mathcal{A}$ is the smallest C^*-algebra containing $\bigcup_{\mathbf{x} \in M} \mathcal{A}(O + \mathbf{x})$.

Finally, one assumes that there is some irreducible representation of the net $\{\mathcal{A}(O) : O \subseteq M\}$ in which these local algebras are identified with von Neumann algebras acting on a (nontrivial) Hilbert space $\mathcal{H}$, $\mathcal{A}$ is identified with a strongly dense subset of $\mathfrak{B}(\mathcal{H})$, and the following condition holds

Spectrum Condition: The generator of spacetime translations, the energy-momentum of the field, has a spectrum confined to the forward light-cone.

While the spectrum condition itself only makes sense relative to a representation (wherein one can speak, via Stone's theorem, of generators of the spacetime translation group of M — now concretely represented as a strongly continuous group of unitary operators $\{U(\mathbf{x})\}$ acting on $\mathcal{H}$), the requirement that the abstract net *have*

[2]In particular, though smearing any given field 'algebra' on M defines a unique net, the net underdetermines the field; see Borchers (1960).

a representation satisfying the spectrum condition does not require that one actually *pass* to such a representation to compute expectation values, cross-sections, etc. Indeed, Haag & Kastler (1964) have argued that there is a precise sense in which all concrete representations of a net are physically equivalent, including representations with and without a translationally invariant vacuum state vector Ω. Since we are not concerned with that argument here,[3] we shall henceforth take the 'Haag-Araki' approach of assuming that all the local algebras $\{\mathcal{A}(O) : O \subseteq M\}$ are von Neumann algebras acting on some $\mathcal{H}$, with $\mathcal{A}'' = \mathfrak{B}(\mathcal{H})$, and there is a translationally invariant vacuum state $\Omega \in \mathcal{H}$.[4]

We turn next to the concept of a state of the field. Generally, a physical state of a quantum system, represented by some von Neumann algebra $\mathcal{R} \subseteq \mathfrak{B}(\mathcal{H})$, is given by a normalized linear expectation functional τ on $\mathcal{R}$ that is both positive and countably additive. Positivity is the requirement that τ map any positive operator in $\mathcal{R}$ to a nonnegative expectation (a must, given that positive operators have nonnegative spectra), while countable additivity is the requirement that τ be additive over countable sums of mutually orthogonal projections in $\mathcal{R}$.[5] Every state on $\mathcal{R}$ extends to a state ρ on $\mathfrak{B}(\mathcal{H})$ which, in turn, can be represented by a density operator D_ρ on $\mathcal{H}$ via the standard formula $\rho(\cdot) = \mathrm{Tr}(D_\rho \cdot)$ (Kadi-

[3]See Arageorgis *et al.* (2002) for somewhat different criticisms of the Haag-Kastler argument.

[4] Since we do, after all, live in a heat bath at 3 degrees Kelvin, some might think it would be of more immediate *physical* interest if we investigated entanglement in finite temperature 'KMS' representations of the net $\{\mathcal{A}(O) : O \subseteq M\}$ that are 'disjoint' from the vacuum representation. However, aside from the fact that the vacuum representation is the simplest and most commonly discussed representation, we are interested here only in the conceptual foundations of particle physics, not quantum statistical mechanics. Moreover, much of value can be learned about the conceptual infrastructure of a theory by examining particular classes of its models — whether or not they are plausible candidates for describing our actual world. (In any case, we could hardly pretend to be discussing physics on a cosmological scale by looking at finite temperature representations, given that we would still be presupposing a flat spacetime background!)

[5]There are also non-countably additive or 'singular' states on $\mathcal{R}$ (Kadison & Ringrose 1997, p. 723), but whenever we use the term 'state' we shall mean *countably additive* state.

son & Ringrose 1997, p. 462). A pure state on $\mathfrak{B}(\mathcal{H})$, i.e., one that is not a nontrivial convex combination or mixture of other states of $\mathfrak{B}(\mathcal{H})$, is then represented by a vector $x \in \mathcal{H}$. We shall always use the notation ρ_x for the normalized state functional $\langle x, \cdot\, x\rangle / \|x\|^2$ $(= \mathrm{Tr}(P_x \cdot\,))$. If, furthermore, we consider the restriction $\rho_x|_{\mathcal{R}}$, the induced state on some von Neumann subalgebra $\mathcal{R} \subseteq \mathfrak{B}(\mathcal{H})$, we cannot in general expect it to be pure on $\mathcal{R}$ as well. For example, with $\mathcal{H} = \mathbb{C}^2_A \otimes \mathbb{C}^2_B$, $\mathcal{R} = \mathfrak{B}(\mathbb{C}^2_A) \otimes I$, and x entangled, we know that the induced state $\rho_x|_{\mathcal{R}}$, represented by $D_A(x) \in \mathfrak{B}(\mathbb{C}^2_A)$, is *always* mixed. Similarly, one cannot expect that a pure state ρ_x of the field algebra $\mathcal{A}'' = \mathfrak{B}(\mathcal{H})$ — which supplies a maximal specification of the state of the field *throughout* spacetime — will have a restriction to a local algebra $\rho_x|_{\mathcal{A}(O)}$ that is itself pure. In fact, we shall see later that the Reeh-Schlieder theorem entails that the vacuum state's restriction to any local algebra is always highly mixed.

There are two topologies on the state space of a von Neumann algebra $\mathcal{R}$ that we shall need to invoke.

One is the metric topology induced by the norm on linear functionals. The norm of a state ρ on $\mathcal{R}$ is defined by $\|\rho\| := \sup\{\, |\rho(Z)| : Z = Z^* \in \mathcal{R},\ \|Z\| \leq 1\}$. If two states, ρ_1 and ρ_2, are close to each other in norm, then they dictate close expectation values uniformly for *all* observables. In particular, if both ρ_1 and ρ_2 are vector states, i.e., they are induced by vectors $x_1, x_2 \in \mathcal{H}$ such that $\rho_1 = \rho_{x_1}|_{\mathcal{R}}$ and $\rho_2 = \rho_{x_2}|_{\mathcal{R}}$, then $\|x_1 - x_2\| \to 0$ implies $\|\rho_1 - \rho_2\| \to 0$.[6] More generally, whenever the trace norm distance between two density operators goes to zero, the norm distance between the states they induce on $\mathcal{R}$ goes to zero. Note also that since every state on $\mathfrak{B}(\mathcal{H})$ is given by a density operator, which in turn can be decomposed as an infinite convex combination of one-dimensional projections (with the infinite sum understood as trace norm convergence), it follows that every state on $\mathcal{R} \subseteq \mathfrak{B}(\mathcal{H})$ is the norm limit of convex combinations of vector states of $\mathcal{R}$ (cf. Kadison & Ringrose 1997, Thm. 7.1.12).

The other topology we shall invoke is the weak* topology: a net of states $\{\rho_i\}$ on $\mathcal{R}$ weak* converges to a state ρ just in case

[6]It is important not to conflate the terms 'vector state' and 'pure state', unless of course $\mathcal{R} = \mathfrak{B}(\mathcal{H})$ itself.

$\rho_i(Z) \to \rho(Z)$ for all $Z \in \mathcal{R}$. This convergence need not be uniform on all elements of $\mathcal{R}$, and is therefore weaker than the notion of approximation embodied by norm convergence. As it happens, any state on the whole of $\mathfrak{B}(\mathcal{H})$ that is the weak* limit of a set of states is also their norm limit. However, this is only true for type I von Neumann algebras (Connes & Størmer 1978, Cor. 9).

Next, we turn to defining entanglement in a field. Fix a state ρ on $\mathfrak{B}(\mathcal{H})$, and two mutually commuting subalgebras $\mathcal{R}_A, \mathcal{R}_B \subseteq \mathfrak{B}(\mathcal{H})$. To define what it means for ρ to be entangled across the algebras, we need only consider the restriction $\rho|_{\mathcal{R}_{AB}}$ to the von Neumann algebra they generate, i.e., $\mathcal{R}_{AB} = (\mathcal{R}_A \cup \mathcal{R}_B)''$, and of course we need a definition that also applies when $\rho|_{\mathcal{R}_{AB}}$ is mixed. A state ω on $\mathcal{R}_{AB}$ is called a *product state* just in case there are states ω_A of $\mathcal{R}_A$ and ω_B of $\mathcal{R}_B$ such that $\omega(XY) = \omega_A(X)\omega_B(Y)$ for all $X \in \mathcal{R}_A$, $Y \in \mathcal{R}_B$. Clearly, product states, or convex combinations of product states, possess only classical correlations. Moreover, if one can *approximate* a state with convex combinations of product states, its correlations do not significantly depart from those characteristic of a classical statistical theory. Therefore, we define ρ to be *entangled* across $(\mathcal{R}_A, \mathcal{R}_B)$ just in case $\rho|_{\mathcal{R}_{AB}}$ is *not* a weak* limit of convex combinations of product states of $\mathcal{R}_{AB}$ (see Chap. 6, page 171). Notice that we chose weak* convergence rather than convergence in norm, hence we obtain a strong notion of entanglement. In the case $\mathcal{H} = \mathcal{H}_A \otimes \mathcal{H}_B$, $\mathcal{R}_A = \mathfrak{B}(\mathcal{H}_A) \otimes I$, and $\mathcal{R}_B = I \otimes \mathfrak{B}(\mathcal{H}_B)$, the definition obviously coincides with the usual notion of entanglement for a pure state (convex combinations and approximations being irrelevant in that case), and also coincides with the definition of entanglement (usually called 'nonseparability') for a mixed density operator that is standard in quantum information theory (Werner 1989, Clifton *et al.* 2000). Further evidence that the definition captures an essentially nonclassical feature of correlations is given by the fact that $\mathcal{R}_{AB}$ will possess an entangled state in the sense defined above if and *only if* both $\mathcal{R}_A$ and $\mathcal{R}_B$ are nonabelian (Bacciagaluppi 1993, Thm. 7; Summers & Werner 1995, Lemma 2.1). Returning to AQFT, it is therefore reasonable to say that a global state of the field ρ on $\mathcal{A}'' = \mathfrak{B}(\mathcal{H})$ is entangled across a pair of spacelike-separated regions (O_A, O_B) just in case $\rho|_{\mathcal{A}_{AB}}$, ρ's restriction to $\mathcal{A}_{AB} = [\mathcal{A}(O_A) \cup \mathcal{A}(O_B)]''$, falls outside the weak* closure of the convex hull of $\mathcal{A}_{AB}$'s product states.

2.1 Operations, local operations, and entanglement

Our next task is to review the mathematical representation of operations, highlight some subtleties in their physical interpretation, and then discuss what is meant by *local* operations on a system. We then end this section by showing that local operations performed in either of two spacelike-separated regions (O_A, O_B) cannot create entanglement in a state across the regions.

The most general transformation of the state of a quantum system with Hilbert space $\mathcal{H}$ is described by an *operation* on $\mathfrak{B}(\mathcal{H})$, defined to be a positive, weak* continuous, linear map $T : \mathfrak{B}(\mathcal{H}) \to \mathfrak{B}(\mathcal{H})$ satisfying $0 \leq T(I) \leq I$ (Haag & Kastler 1964, Davies 1976, Kraus 1983, Busch *et al.* 1995, Werner 1987). (The weak* topology on a von Neumann algebra $\mathcal{R}$ is defined in complete analogy to the weak* topology on its state space, viz., $\{Z_n\} \subseteq \mathcal{R}$ weak* converges to $Z \in \mathcal{R}$ just in case $\rho(Z_n) \to \rho(Z)$ for all states ρ of $\mathcal{R}$.) Any such T induces a map $\rho \to \rho^T$ from the state space of $\mathfrak{B}(\mathcal{H})$ into itself or 0, where, for all $Z \in \mathfrak{B}(\mathcal{H})$,

$$\rho^T(Z) := \begin{cases} \rho(T(Z))/\rho(T(I)) & \text{if } \rho(T(I)) \neq 0 \\ 0 & \text{otherwise.} \end{cases} \tag{3}$$

The number $\rho(T(I))$ is the probability that an ensemble in state ρ will respond 'Yes' to the question represented by the positive operator $T(I)$. An operation T is called *selective* if $T(I) < I$, and nonselective if $T(I) = I$.

The final state after a selective operation on an ensemble of identically prepared systems is obtained by ignoring those members of the ensemble that fail to respond 'Yes' to $T(I)$. Thus a selective operation involves performing a physical operation on an ensemble followed by a *purely conceptual* operation in which one makes a selection of a subensemble based on the outcome of the physical operation (assigning 'state' 0 to the remainder). Nonselective operations, by contrast, always elicit a 'Yes' response from any state, hence the final state is not obtained by selection but purely as a result of the physical interaction between object system and the device that effects the operation. (We shall shortly discuss some actual physical examples to make this general description of operations concrete.)

An operation T, which quantum information theorists call a superoperator (acting, as it does, on operators to produce operators), 'can describe any combination of unitary operations, interactions with an ancillary quantum system or with the environment, quantum measurement, classical communication, and subsequent quantum operations conditioned on measurement results' (Bennett *et al.* 1999). Interestingly, a superoperator itself can always be represented in terms of operators, as a consequence of the Kraus representation theorem (Kraus 1983, p. 42): For any operation $T : \mathfrak{B}(\mathcal{H}) \to \mathfrak{B}(\mathcal{H})$, there exists a (not necessarily unique) countable collection of Kraus operators $\{K_i\} \subseteq \mathfrak{B}(\mathcal{H})$ such that

$$T(\cdot) = \sum_i K_i^*(\cdot)K_i, \qquad \text{with } 0 \leq \sum_i K_i^* K_i \leq I\,, \tag{4}$$

where both sums, if infinite, are to be understood in terms of weak* convergence. It is not difficult to show that the sum $\sum_i K_i K_i^*$ must also weak* converge, hence we can let T^* denote the operation conjugate to T whose Kraus operators are $\{K_i^*\}$. It then follows (using the linearity and cyclicity of the trace) that if a state ρ is represented by a density operator D on $\mathcal{H}$, ρ^T will be represented by the density operator $T^*(D)$. If the mapping $\rho \to \rho^T$, or equivalently, $D \to T^*(D)$, maps pure states to pure states, then the operation T is called a *pure operation*, and this corresponds to it being representable by a *single* Kraus operator.

More generally, the Kraus representation shows that a general operation is always equivalent to mixing the results of separating an initial ensemble into subensembles to which one applies pure (possibly selective) operations, represented by the individual Kraus operators. To see this, let T be an arbitrary operation performed on a state ρ, where $\rho^T \neq 0$, and suppose T is represented by Kraus operators $\{K_i\}$. Let ρ^{K_i} denote the result of applying to ρ the pure operation given by the mapping $T_i(\cdot) = K_i^*(\cdot)K_i$, and (for convenience) define $\lambda_i = \rho(T_i(I))/\rho(T(I))$. Then, at least when there are finitely many Kraus operators, it is easy to see that T itself maps ρ to the convex combination $\rho^T = \sum_i \lambda_i \rho^{K_i}$. In the infinite case, this sum converges not just weak* but *in norm*, and it is a useful exercise in the topologies we have introduced to see why. Letting ρ_n^T denote the partial sum $\sum_{i=1}^n \lambda_i \rho^{K_i}$, we need to establish that

$$\lim_{n\to\infty}\left[\sup\{\,|\rho^T(Z) - \rho_n^T(Z)| : Z = Z^* \in \mathfrak{B}(\mathcal{H}),\ \|Z\| \leq 1\}\right] = 0. \quad (5)$$

For *any* $Z \in \mathfrak{B}(\mathcal{H})$, we have

$$|\rho^T(Z) - \rho_n^T(Z)| \;=\; \rho(T(I))^{-1}\left|\sum_{i=n+1}^{\infty} \rho(K_i^* Z K_i)\right|. \quad (6)$$

However, $\rho(K_i^*(\,\cdot\,)K_i)$, being a positive linear functional, has a norm that may be computed by its action on the identity (Kadison & Ringrose 1997, Thm. 4.3.2). Therefore, $|\rho(K_i^* Z K_i)| \leq \|Z\|\rho(K_i^* K_i)$, and we obtain

$$|\rho^T(Z) - \rho_n^T(Z)| \;\leq\; \rho(T(I))^{-1}\|Z\| \sum_{i=n+1}^{\infty} \rho(K_i^* K_i). \quad (7)$$

However, since $\sum_i K_i^* K_i$ weak* converges, this last summation is the tail set of a convergent series. Therefore, when $\|Z\| \leq 1$, the right-hand side of (7) goes to zero independently of Z.

To get a concrete idea of how operations work physically, and to highlight two important interpretational pitfalls, let us again consider our toy universe, with $\mathcal{H} = \mathbb{C}_A^2 \otimes \mathbb{C}_B^2$ and x an entangled state. Recall that Alice disentangled x by measuring an observable of A with eigenprojections $P_\pm$. Her measurement corresponds to applying the nonselective operation T with Kraus operators $K_1 = P_+ \otimes I$ and $K_2 = P_- \otimes I$, resulting in the final state $T^*(P_x) = T(P_x) = D'$, as given in (1). If Alice were to further 'apply' the pure selective operation T' represented by the single Kraus operator $P_+ \otimes I$, the final state of her ensemble, as is apparent from (2), would be the product state $D'' = P_+ \otimes P_{b_x^+}$. But, as we have emphasized, this corresponds to a conceptual operation in which Alice just throws away all members of the original ensemble that yielded measurement outcome -1.

On the other hand, it is essential not to lose sight of the issue that troubled Einstein. *Whatever* outcome Alice selects for, she will then be in a position to assert that certain B observables — those that have either b_x^+ or b_x^- as an eigenvector, depending on the outcome she favors — have a sharp value in the ensemble she is left with. But prior to Alice performing the first operation T, such an assertion would have contradicted the orthodox interpretation of the entangled superposition x. If, contra Bohr, one were to view this change in B's state as a *real physical* change brought about by one of the operations Alice performs, surely the innocuous conceptual operation T' could not be the culprit — it must have been T which forced B to 'choose' between the alternatives $b_x^\pm$. Unfortunately, this clear distinction between the physical operation T and conceptual operation T' is not reflected well in the formalism of operations. For we could equally well have represented Alice's final product state $D'' = P_+ \otimes P_{b_x^+}$, not as the result of successively applying the operations T and T', but as the outcome of applying the single composite operation $T' \circ T$, which is just the mapping T'. And *this* T' now needs to be understood, not purely as a conceptual operation, but as also involving a physical operation, with possibly real nonlocal effects on B, depending on one's view of the EPR paradox.[7]

There is a second pitfall that concerns interpreting the result of *mixing* subensembles, as opposed to singling out a particular subensemble. Consider an alternative method available to Alice for disentangling a state x. For concreteness, let us suppose that x is the singlet state $1/\sqrt{2}(a^+ \otimes b^- - a^- \otimes b^+)$. Alice applies the nonselective operation with Kraus representation

$$T(\cdot) = \frac{1}{2}(\sigma_a \otimes I)(\cdot)(\sigma_a \otimes I) + \frac{1}{2}(I \otimes I)(\cdot)(I \otimes I), \tag{8}$$

[7]In particular, keep in mind that you are taking the first step on the road to conceding the incompleteness of quantum theory if you attribute the change in the state of B brought about by T' in this case to a mere change in Alice's *knowledge* about B's state.

where σ_a is the spin observable with eigenstates $a^{\pm}$. Since $\sigma_a \otimes I$ maps x to the triplet state $1/\sqrt{2}(a^+ \otimes b^- + a^- \otimes b^+)$, T^* $(= T)$ will map P_x to an equal mixture of the singlet and triplet, which admits the following convex decomposition into product states

$$D' = \frac{1}{2}P_{a^+\otimes b^-} + \frac{1}{2}P_{a^-\otimes b^+}. \tag{9}$$

Has Alice truly disentangled A from B? Technically, Yes. Yet all Alice has done, physically, is to separate the initial A ensemble into two subensembles in equal proportion, left the second subensemble alone while performing a (pure, nonselective) unitary operation $\sigma_a \otimes I$ on the first that maps all its $A + B$ pairs to the triplet state, and then remixed the ensembles. Thus, notwithstanding the above decomposition of the final density matrix D', Alice *knows quite well* that she is in possession of an ensemble of A systems each of which is entangled either via the singlet or triplet state with the corresponding B systems. This will of course be recognized as one aspect of the problem with the ignorance interpretation of mixtures. We have two different ways to decompose D' — as an equal mixture of the singlet and triplet or of two product states — but which is the correct way to understand how the ensemble is *actually* constituted? The definition of entanglement is just not sensitive to the answer.[8] Nevertheless, we are inclined to think the destruction of the singlet's entanglement that Alice achieves by applying the operation in (8) is an artifact of her mixing process, in which she is represented as simply forgetting about the history of the A systems. And this is the view we shall take when we consider similar possibilities for destroying entanglement between field systems in AQFT.

In the two examples considered above, Alice applies operations whose Kraus operators lie in the subalgebra $\mathfrak{B}(\mathcal{H}_A) \otimes I$ associated with system A. In the case of a nonselective operation, this is clearly sufficient for her operation not to have any effect on the expecta-

[8] It is exactly this insensitivity that is at the heart of the recent dispute over whether NMR quantum computing is correctly understood as implementing genuine *quantum* computing that cannot be simulated classically (Braunstein *et al.* 1999, Laflamme 1998).

tions of the observables of system B. However, it is also necessary. The point is quite general.

Let us define a nonselective operation T to be (*pace* Einstein!) *local* to the subsystem represented by a von Neumann subalgebra $\mathcal{R} \subseteq \mathfrak{B}(\mathcal{H})$ just in case $\rho^T|_{\mathcal{R}'} = \rho|_{\mathcal{R}'}$ for all states ρ. Thus, we require that T leave the expectations of observables outside of $\mathcal{R}$, as well as those in its center $\mathcal{R} \cap \mathcal{R}'$, unchanged. Since distinct states of $\mathcal{R}'$ cannot agree on all expectation values, this means T must act like the identity operation on $\mathcal{R}'$. Now fix an arbitrary element $Y \in \mathcal{R}'$, and suppose T is represented by Kraus operators $\{K_i\}$. A straightforward calculation reveals that

$$\sum_i [Y, K_i]^* [Y, K_i] = T(Y^2) - T(Y)Y \quad YT(Y) + YT(I)Y. \tag{10}$$

Since $T(I) = I$, and T leaves the elements of $\mathcal{R}'$ fixed, the right-hand side of (10) reduces to zero. Thus each of the terms in the sum on the left-hand side, which are positive operators, must individually be zero. Since Y was an arbitrary element of $\mathcal{R}'$, it follows that $\{K_i\} \subseteq (\mathcal{R}')' = \mathcal{R}$. So we see that nonselective operations local to $\mathcal{R}$ *must* be represented by Kraus operators taken from the subalgebra $\mathcal{R}$.

As for selective operations, we have already seen that they *can* 'change' the global statistics of a state ρ outside the subalgebra $\mathcal{R}$, particularly when ρ is entangled. However, a natural extension of the definition of local operation on $\mathcal{R}$ to cover the case when T is selective is to require that $T(Y) = T(I)Y$ for all $Y \in \mathcal{R}'$. This implies $\rho^T(Y) = \rho(T(I)Y)/\rho(T(I))$, and so guarantees that T will leave the statistics of any observable in $\mathcal{R}'$ the same *modulo* whatever correlations that observable might have had in the initial state with the Yes/No question represented by the positive operator $T(I)$. Further motivation is provided by the fact this definition is equivalent to requiring that T factor across the algebras $(\mathcal{R}, \mathcal{R}')$, in the sense that $T(XY) = T(X)Y$ for all $X \in \mathcal{R}$, $Y \in \mathcal{R}'$ (Werner 1987, Lemma). If there exist product states across $(\mathcal{R}, \mathcal{R}')$ (an assumption we shall later see does *not* usually hold when $\mathcal{R}$ is a local algebra in AQFT), this guarantees that any local selective operation on $\mathcal{R}$, when the global state is an entirely uncorrelated product state, will leave the statistics of that state on $\mathcal{R}'$ unchanged. Finally, observe that $T(Y) = T(I)Y$ for all $Y \in \mathcal{R}'$ implies that the right-hand side of (10) again reduces to zero. Thus it follows (as before) that selective local operations on $\mathcal{R}$ must also be represented by Kraus operators taken from the subalgebra $\mathcal{R}$.

Applying these considerations to field theory, any local operation on the field system within a region O, whether or not the operation is selective, is represented by a family of Kraus operators taken from $\mathcal{A}(O)$. In particular, each individual element of $\mathcal{A}(O)$ represents a pure operation that can be performed within O (cf. Haag & Kastler 1964, p. 850). We now need to argue that local operations performed by two experimenters in spacelike-separated regions cannot create entanglement in a state across the regions where it had none before. This point, well known by quantum information theorists working in nonrelativistic quantum theory, in fact applies quite generally to any two commuting von Neumann algebras $\mathcal{R}_A$ and $\mathcal{R}_B$.

Suppose that a state ρ is not entangled across $(\mathcal{R}_A, \mathcal{R}_B)$, local operations T_A and T_B are applied to ρ, and the result is nonzero (i.e., some members of the initial ensemble are not discarded). Since the Kraus operators of these operations commute, it is easy to check that $(\rho^{T_A})^{T_B} = (\rho^{T_B})^{T_A}$, so it does not matter in which order we take the operations. It is sufficient to show that ρ^{T_A} will again be unentangled, for then we can just repeat the same argument to obtain that neither can $(\rho^{T_A})^{T_B}$ be entangled. Next, recall that a general operation T_A will just produce a mixture over the results of applying a countable collection of pure operations to ρ; more precisely, the result will be the norm, and hence weak*, limit of finite convex combinations of the results of applying pure operations to ρ. If the states that result from ρ under those pure operations are themselves not entangled, ρ^{T_A} itself could not be either, because the set of unentangled states is by definition convex and weak* closed. Without loss of generality, then, we may assume that the local operation T_A is pure and, hence, given by $T_A(\cdot) = K^*(\cdot)K$, for some *single* Kraus operator $K \in \mathcal{R}_A$. As before, we shall denote the resulting state ρ^{T_A} by ρ^K ($\equiv \rho(K^* \cdot K)/\rho(K^*K)$).

Next, suppose that ω is any product state on $\mathcal{R}_{AB}$ with restrictions to $\mathcal{R}_A$ and $\mathcal{R}_B$ given by ω_A and ω_B, and such that $\omega^K \neq 0$. Then, for any $X \in \mathcal{R}_A, Y \in \mathcal{R}_B$,

$$\omega^K(XY) = \frac{\omega(K^*(XY)K)}{\omega(K^*K)} \tag{11}$$

$$= \frac{\omega(K^*XKY)}{\omega(K^*K)} \tag{12}$$

$$= \frac{\omega_A(K^*XK)}{\omega_A(K^*K)}\omega_B(Y) = \omega_A^K(X)\omega_B(Y). \tag{13}$$

It follows that K maps product states of $\mathcal{R}_{AB}$ to product states (or to zero). Suppose, instead, that ω is a convex combination of states on $\mathcal{R}_{AB}$, i.e., $\omega = \sum_{i=1}^n \lambda_i \omega_i$. Then, setting $\lambda_i^K = \omega_i(K^*K)/\omega(K^*K)$, it is easy to see that $\omega^K = \sum_{i=1}^n \lambda_i^K \omega_i^K$, hence K preserves convex combinations of states on $\mathcal{R}_{AB}$ as well. It is also not difficult to see that the mapping $\omega \mapsto \omega^K$ is weak* continuous at any point where $\omega^K \neq 0$ (cf. Sec. 3 of Chap. 6).

Returning to our original state ρ, our hypothesis is that it is not entangled. Thus, there is a net of states $\{\omega_n\}$ on $\mathcal{R}_{AB}$, each of which is a convex combination of product states, such that $\omega_n \to \rho|_{\mathcal{R}_{AB}}$ in the weak* topology. It follows from the above considerations that $\omega_n^K \to \rho^K|_{\mathcal{R}_{AB}}$, where each of the states $\{\omega_n^K\}$ is again a convex combination of product states. Hence, by definition, $\rho^K|_{\mathcal{R}_{AB}}$ is not entangled either.

In summary, we have shown:

> *If $\mathcal{R}_A$ and $\mathcal{R}_B$ are any two commuting von Neumann algebras, and ρ is any unentangled state across $(\mathcal{R}_A, \mathcal{R}_B)$, then operations on ρ, local to either or both of A and B, cannot produce an entangled state.*

3 The operational implications of the Reeh-Schlieder theorem

Again, let $\mathcal{R} \subseteq \mathfrak{B}(\mathcal{H})$ be any von Neumann algebra. A vector $x \in \mathcal{H}$ is called *cyclic* for $\mathcal{R}$ if the norm closure of the set $\{Ax : A \in \mathcal{R}\}$ is the *whole* of $\mathcal{H}$. In AQFT, the Reeh-Schlieder (RS) theorem connects this formal property of cyclicity to the physical property of a field state having bounded energy.[9] A pure global state x of the field has bounded energy just in case $E([0,r])x = x$ for some $r < \infty$, where E is the spectral measure for the global Hamiltonian of the field. In other words, the probability in state x that the field's energy is confined to the bounded interval $[0,r]$ is unity. In particular, the vacuum Ω is an eigenstate of the Hamiltonian with eigenvalue 0, and hence trivially has bounded energy. The RS theorem implies that

> *If x has bounded energy, then x is cyclic for any local algebra $\mathcal{A}(O)$.*

Our first order of business is to explain Streater and Wightman's comment that the RS theorem entails 'it is difficult to isolate a system described by fields from outside effects' (2000, p. 139).

A vector x is called *separating* for a von Neumann algebra $\mathcal{R}$ if $Ax = 0$ implies $A = 0$ whenever $A \in \mathcal{R}$. It is an elementary result of von Neumann algebra theory that x is cyclic for $\mathcal{R}$ if and only if x is separating for $\mathcal{R}'$ (Kadison & Ringrose 1997, Cor. 5.5.12). To illustrate with a simple example, take $\mathcal{H} = \mathcal{H}_A \otimes \mathcal{H}_B$. If $\dim \mathcal{H}_A \geq \dim \mathcal{H}_B$, then there is a vector $x \in \mathcal{H}$ that has Schmidt decomposition $\sum_i c_i a_i \otimes b_i$ where $|c_i|^2 \neq 0$ for $i = 1, \ldots, \dim \mathcal{H}_B$. If we act on such an x by an operator in the subalgebra $I \otimes \mathfrak{B}(\mathcal{H}_B)$, of form $I \otimes B$, then $(I \otimes B)x = 0$ only if B itself maps all the basis vectors $\{b_i\}$ to zero, i.e., $I \otimes B = 0$. Thus such vectors are separating for $I \otimes \mathfrak{B}(\mathcal{H}_B)$, and therefore cyclic for $\mathfrak{B}(\mathcal{H}_A) \otimes I$. Conversely, it is easy to convince oneself that $\mathfrak{B}(\mathcal{H}_A) \otimes I$ possesses a cyclic vector — equivalently, $I \otimes \mathfrak{B}(\mathcal{H}_B)$ has a separating vector — *only if* $\dim \mathcal{H}_A \geq \dim \mathcal{H}_B$. So, to take another example, each of the A and B subalgebras have a cyclic and a separating vector just in case $\mathcal{H}_A$ and $\mathcal{H}_B$ have the same dimension (cf. the proof of Clifton *et al.* 1998, Thm. 4).

---

[9] More generally, the connection is between cyclicity and field states that are 'analytic' in the energy.

Consider, now, a local algebra $\mathcal{A}(O)$ with $O' \neq \emptyset$, and a field state x with bounded energy. The RS theorem tells us that x is cyclic for $\mathcal{A}(O')$, and therefore, separating for $\mathcal{A}(O')'$. But by microcausality, $\mathcal{A}(O) \subseteq \mathcal{A}(O')'$, hence x must be separating for the subalgebra $\mathcal{A}(O)$ as well. Thus it is an immediate corollary to the RS theorem that

> *If x has bounded energy, then x is separating for any local algebra $\mathcal{A}(O)$ with $O' \neq \emptyset$.*

It is this corollary that prompted Streater and Wightman's remark. But what has it got to do with thinking of the field system $\mathcal{A}(O)$ as isolated? For a start, we can now show that the local restriction $\rho_x|_{\mathcal{A}(O)}$ of a state with bounded energy is always a highly 'noisy' mixed state. Recall that a state ω on a von Neumann algebra $\mathcal{R}$ is said to be a *component* of another state ρ if there is a third state τ such that $\rho = \lambda\omega + (1-\lambda)\tau$ with $\lambda \in (0,1)$ (van Fraassen 1991, p. 161). We are going to show that $\rho_x|_{\mathcal{A}(O)}$ has a *norm* dense set of components in the state space of $\mathcal{A}(O)$.

Once again, the point is quite general. Let $\mathcal{R}$ be any von Neumann algebra, x be separating for $\mathcal{R}$, and let ω be an arbitrary state of $\mathcal{R}$. We must find a sequence $\{\omega_n\}$ of states of $\mathcal{R}$ such that each ω_n is a component of $\rho_x|_{\mathcal{R}}$ and $\|\omega_n - \omega\| \to 0$. Since $\mathcal{R}$ has a separating vector, it follows that every state of $\mathcal{R}$ is a vector state (Kadison & Ringrose 1997, Thm. 7.2.3).[10] In particular, there is a nonzero vector $y \in \mathcal{H}$ such that $\omega = \omega_y$. Since x is separating for $\mathcal{R}$, x is cyclic for $\mathcal{R}'$, therefore we may choose a sequence of operators $\{A_n\} \subseteq \mathcal{R}'$ so that $A_n x \to y$. Since $\|A_n x - y\| \to 0$, $\|\omega_{A_n x} - \omega_y\| \to 0$ (see page 189). We claim now that each $\omega_{A_n x}$ is a component of $\rho_x|_{\mathcal{R}}$. Indeed, for any positive element $B^*B \in \mathcal{R}$, we have:

$$\begin{aligned} \langle A_n x, B^*BA_n x\rangle &= \langle x, A_n^*A_nB^*Bx\rangle = \langle Bx, A_n^*A_nBx\rangle & (14)\\ &\leq \|A_n^*A_n\|\langle Bx, Bx\rangle = \|A_n\|^2\langle x, B^*Bx\rangle. & (15)\end{aligned}$$

[10] That this should be so is not as surprising as it sounds. Again, if $\mathcal{H} = \mathcal{H}_A \otimes \mathcal{H}_B$, and $\dim \mathcal{H}_A \geq \dim \mathcal{H}_B$, then as we have seen, the B subalgebra possesses a separating vector. But it is also easy to see, in this case, that every state on $I \otimes \mathfrak{B}(\mathcal{H}_B)$ is the reduced density operator obtained from a pure state on $\mathfrak{B}(\mathcal{H})$ determined by a vector in $\mathcal{H}$.

Thus,

$$\omega_{A_n x}(B^*B) \;=\; \frac{\langle A_n x, B^*BA_n x\rangle}{\|A_n x\|^2} \;\leq\; \frac{\|A_n\|^2}{\|A_n x\|^2}\,\rho_x(B^*B). \qquad (16)$$

If we now take $\lambda = \|A_n x\|^2/\|A_n\|^2 \in (0,1)$, and consider the linear functional τ on $\mathcal{R}$ given by $\tau = (1-\lambda)^{-1}(\rho_x|_{\mathcal{R}} - \lambda\omega_{A_n x})$, then (16) implies that τ is a state (in particular, positive), and we see that $\rho_x|_{\mathcal{R}} = \lambda\omega_{A_n x} + (1-\lambda)\tau$ as required.[11]

So bounded energy states are, locally, highly mixed. And such states are far from special — they lie norm dense in the pure state space of $\mathfrak{B}(\mathcal{H})$. To see this, recall that it is part of the spectral theorem for the global Hamiltonian that $E([0,n])$ converges strongly to the identity as $n \to \infty$. Thus we may approximate any vector $y \in \mathcal{H}$ by the sequence of bounded energy states $\{E([0,n])y/\|E([0,n])y\|\}_{n=0}^{\infty}$. Since there are so many bounded energy states of the field, that are locally so 'noisy', Streater and Wightman's comment is entirely warranted. But somewhat more can be said. As we saw with our toy example in Section 1, when a local subsystem of a global system in a pure state is itself in a mixed state, this is a sign of that subsystem's entanglement with its environment. And there is entanglement lurking in bounded energy states too. But, first, we want to take a closer look at the operational implications of local cyclicity.

If a vector x is cyclic for $\mathcal{R}$, then for any $y \in \mathcal{H}$, there is a sequence $A_n \in \mathcal{R}$ such that $A_n x \to y$. Thus for any $\epsilon > 0$ there is an $A \in \mathcal{R}$ such that $\|\rho_{Ax} - \rho_y\| < \epsilon$. However, ρ_{Ax} is just the state one gets by applying the pure operation given by the Kraus operator $K = A/\|A\| \in \mathcal{R}$ to ρ_x. It follows that if x is cyclic for $\mathcal{R}$, one can get arbitrarily close in norm to any other pure state of $\mathfrak{B}(\mathcal{H})$ by applying an appropriate pure local operation in $\mathcal{R}$ to ρ_x. In particular, pure operations on the vacuum Ω within a local region O, no matter how small, can prepare essentially any global state of the field. As Haag emphasizes, to do this the operation must 'judiciously exploit

[11] This result also holds more generally for states ρ of $\mathcal{R}$ that are faithful, i.e., $\rho(Z) = 0$ entails $Z = 0$ for any positive $Z \in \mathcal{R}$; see the proof of Theorem 2.1 of Summers & Werner (1988).

the small but nonvanishing long distance correlations which exist in the vacuum' (1992, p. 102). This, as Redhead (1995) has argued by analogy to the singlet state, is made possible by the fact that the vacuum is highly entangled (cf. Clifton *et al.* 1998).

3.1 Physical versus conceptual operations

The first puzzle we need to sort out is that it looks as though entirely *physical* operations in O can change the global state, in particular the vacuum Ω, to any desired state![12]

Redhead's analysis of the cyclicity of the singlet state $x = 1/\sqrt{2}(a^+ \otimes b^- - a^- \otimes b^+)$ for the subalgebra $\mathfrak{B}(\mathbb{C}^2_A) \otimes I$ is designed to remove this puzzle (*ibid*, 128).[13] Redhead writes:

> ... we want to distinguish clearly two senses of the term 'operation'. Firstly there are physical operations such as making measurements, selecting subensembles according to the outcome of measurements, and mixing ensembles with probabilistic weights, and secondly there are the mathematical operations of producing superpositions of states by taking linear combinations of pure states produced by appropriate selective measurement procedures. These superpositions are of course quite different from the mixed states whose preparation we have listed as a physical operation. (1995, pp. 128–9)

Note that, in stark contrast to our discussion in the previous section, Redhead counts selecting subensembles and mixing as physical operations; it is only the operation of superposition that warrants the adjective 'mathematical'. When he explains the cyclicity of the singlet state, Redhead first notes (*ibid*, 129) that the four basis states

$$a^+ \otimes b^-, a^- \otimes b^-, a^- \otimes b^+, a^+ \otimes b^-, \tag{17}$$

[12]For example, Segal & Goodman (1965) have called this 'bizarre' and 'physically quite surprising', sentiments echoed recently by Fleming (2000) who calls it 'amazing!', and Fleming & Butterfield (1999, p. 161) who think it is 'hard to square with naïve, or even educated, intuitions about localization!'

[13]Note that in this simple 2×2-dimensional case, Redhead could equally well have chosen *any* entangled state, since they are all separating for $I \otimes \mathfrak{B}(\mathbb{C}^2_B)$.

are easily obtained by the physical operations of applying projections and unitary transformations to the singlet state, and exploiting the fact that the singlet strictly correlates σ_a with σ_b. He goes on:

> But *any* state for the joint system is some linear combination of these four states, so by the *mathematical* operation of linear combination, we can see how to generate an arbitrary state in $\mathcal{H}_1 \otimes \mathcal{H}_2$ from physical operations performed on particle one. But all the operations we have described can be represented in the algebra of operators on $\mathcal{H}_1$ (extended to $\mathcal{H}_1 \otimes \mathcal{H}_2$). (*ibid*, 129)

Now, while Redhead's explanation of why it is mathematically possible for x to be cyclic is perfectly correct, he actually misses the mark when it comes to the physical interpretation of cyclicity. The point is that superposition *of states* is a red herring. Certainly a superposition of the states in (17) could not be prepared by physical operations confined to the A system. But, as Redhead himself notes in the final sentence above, one can get the same *effect* as superposing those states by acting on x with an operator of form $A \otimes I$ in the subalgebra $\mathfrak{B}(\mathbb{C}_A^2) \otimes I$ — an operator that is itself a 'superposition' of other operators in that algebra. What Redhead neglects to point out is that the action of this operator on x *does have a local physical interpretation*: as we have seen, it is a Kraus operator that represents the outcome of a generalized positive operator-valued measurement on the A system. The key to the puzzle is, rather, that this positive operator-valued measurement will generally have to be *selective*. For one certainly could never, with nonselective operations on A alone, get as close as one likes to any state vector in $\mathbb{C}_A^2 \otimes \mathbb{C}_B^2$ (otherwise all state vectors would induce the same state on $I \otimes \mathfrak{B}(\mathbb{C}_B^2)$!).

We conclude that the correct way to view the physical content of cyclicity is that changes in the global state are partly due to an experimenter's ability to perform a generalized measurement on A, and partly due (*pace* Redhead) to the purely conceptual operation of selecting a subensemble based on the outcome of the experimenter's measurement together with the consequent 'change' in the state of B via the EPR correlations between A and B.[14]

One encounters the same interpretational pitfall concerning the cyclicity of the vacuum in relation to localized states in AQFT. A global state of the field is said to be *localized* in O if its expectations on the algebra $\mathcal{A}(O')$ agree with vacuum expectation values (Haag 1992, 102). Thus localized states are 'excitations' of the vacuum confined to O. In particular, $U\Omega$ is a localized state whenever U is a unitary operator taken from $\mathcal{A}(O)$ (since unitary operations are nonselective). But every element of a C^*-algebra is a finite linear combination of unitary operators (Kadison & Ringrose 1997, Thm. 4.1.7). Since Ω is cyclic for $\mathcal{A}(O)$, this means we must be able to approximate any global state by linear superpositions of vectors describing states localized in O — even approximate states that are localized in regions spacelike separated from O! Haag, rightly cautious, calls this a '(superficial) paradox' (1992, p. 254; parenthesis his), but he neglects to put his finger on its resolution: while unitary operations are nonselective, a local operation in $\mathcal{A}(O)$ given by a Kraus operator that is a linear combination of local unitary operators will generally be *selective*.[15]

The (common) point of the previous two paragraphs is perhaps best summarized as follows. Both Redhead and Haag would agree that unitary Kraus operators in $\mathcal{A}(O)$ give rise to purely physical operations in the local region O. But there are many Kraus operators in $\mathcal{A}(O)$ that do not represent purely physical operations in O

---

[14] In fairness to Redhead, we would like to add that in his first book (1989, p. 58) he includes an exceptionally clear discussion of the difference between nonselective and selective measurements. In particular, while we have dubbed the latter 'conceptual' operations, he uses the term 'mental', without implying anything mystical is involved. As he puts it, while a nonselective operation *can* have a physical component — like the physical action of throwing some subensemble of particles into a box for further examination — it is the *decision* to focus on a particular subensemble to the exclusion of the rest that is not dictated by the physics.

[15] Haag *does* make the interesting point that only a proper subset of the state space of a field can be approximated if we restrict ourselves to local operations that involve a physically reasonable expenditure of energy. But we do not share the view of Schroer (1999) that this point by itself reconciles the RS theorem with 'common sense'.

insofar as they are selective. Since every Kraus operator is a linear superposition of unitary operators, it follows that 'superposition of local operations' does not preserve (pure) physicality (so to speak). Redhead is right that the key to diffusing the paradox is in noting that superpositions are involved — but it is essential to understand these superpositions as occurring locally in $\mathcal{A}(O)$, not in the Hilbert space.

3.2 Cyclicity and entanglement

Our next order of business is to supply the rigorous argument behind Redhead's intuition about the connection between cyclicity and entanglement. The point, again, is quite general (cf. Chap. 6):

> *For any two commuting nonabelian von Neumann algebras $\mathcal{R}_A$ and $\mathcal{R}_B$, and any state vector x cyclic for $\mathcal{R}_A$ (or $\mathcal{R}_B$), ρ_x is entangled across the algebras.*

For suppose, in order to extract a contradiction, that ρ_x is *not* entangled. Then as we have seen, operations on ρ_x that are local to $\mathcal{R}_A$ cannot turn that state into an entangled state across $(\mathcal{R}_A, \mathcal{R}_B)$. Yet, by the cyclicity of x, we know that we can apply pure operations to ρ_x, that are local to $\mathcal{R}_A$ (or $\mathcal{R}_B$), and approximate in norm (and hence weak* approximate) any other vector state of $\mathcal{R}_{AB}$. It follows that no vector state of $\mathcal{R}_{AB}$ is entangled across $(\mathcal{R}_A, \mathcal{R}_B)$, and the same goes for all its mixed states (which lie in the norm closed convex hull of the vector states). But this means that $\mathcal{R}_{AB}$ would possess *no* entangled states at all — in flat contradiction with the fact that neither $\mathcal{R}_A$ nor $\mathcal{R}_B$ is abelian (see page 190 and following).

Returning to the context of AQFT, if we now consider *any* two spacelike separated field systems, $\mathcal{A}(O_A)$ and $\mathcal{A}(O_B)$, then the argument we just gave establishes that the dense set of field states bounded in the energy will *all* be entangled across the regions (O_A, O_B).[16] However, by itself this result does not imply that Alice cannot destroy a bounded energy state x's entanglement across

[16]Note that the fact that $\mathcal{A}(O_A)$ and $\mathcal{A}(O_B)$ are nonabelian is *itself* a consequence of the RS theorem. For if, say, $\mathcal{A}(O_A)$ were abelian, then since by the RS theorem that algebra possesses a cyclic vector, it must be a maximal abelian subalgebra of

(O_A, O_B) by performing local operations in O_A. In fact, Borchers (1965, Cor. 7) has shown that any state of the field induced by a vector of form Ax, for any nontrivial $A \in \mathcal{A}(O_A)$, *never* has bounded energy.[17] So it might seem that all Alice needs to do is perform any pure operation within O_A and the resulting state, because it is no longer subject to the RS theorem, need no longer be entangled across (O_A, O_B).

However, the RS theorem gives only a sufficient, *not* a necessary, condition for a state x of the field to be cyclic for $\mathcal{A}(O_A)$. And notwithstanding that no pure operation Alice performs can preserve boundedness in the energy, *almost all* the pure operations she could perform *will* preserve the state's cyclicity! The reason is, once again, quite general.

Again let $\mathcal{R}_A$ and $\mathcal{R}_B$ be two commuting nonabelian von Neumann algebras, suppose x is cyclic for $\mathcal{R}_A$, and consider the state induced by the vector Ax where $A \in \mathcal{R}_A$. Now every element in a von Neumann algebra is the strong limit of invertible elements in the algebra (Dixmier & Maréchal 1971, Prop. 1). Therefore, there is a sequence of invertible operators $\{\tilde{A}_n\} \subseteq \mathcal{R}_A$ such that $\tilde{A}_n x \to Ax$, i.e., $\|\rho_{\tilde{A}_n x} - \rho_{Ax}\| \to 0$. Notice, however, that since each $\tilde{A}_n$ is invertible, each vector $\tilde{A}_n x$ is again cyclic for $\mathcal{R}_A$, because we can 'cycle back' to x by applying to $\tilde{A}_n x$ the inverse operator $\tilde{A}_n^{-1} \in \mathcal{R}_A$, and from there we know, by hypothesis, that we can cycle with elements of $\mathcal{R}_A$ arbitrarily close to any other vector in $\mathcal{H}$. It follows that, even though Alice may *think* she has applied the pure operation given by some Kraus operator $A/\|A\|$ to x, she could well have *actually* applied an invertible Kraus operation given by one of the operators $\tilde{A}_n/\|\tilde{A}_n\|$ in a strong neighborhood of $A/\|A\|$. And if she actually did this, then she certainly would *not* disentangle x, because she would not have succeeded in destroying the *cyclicity* of the field state for her local algebra.

$\mathfrak{B}(\mathcal{H})$ (Kadison & Ringrose 1997, Cor. 7.2.16). The same conclusion would have to follow for any subregion $\tilde{O}_A \subset O_A$ whose closure is a proper subset of O_A. And this, by isotony, would lead to the absurd conclusion that $\mathcal{A}(\tilde{O}_A) = \mathcal{A}(O_A)$, which is readily shown to be inconsistent with the axioms of AQFT (Horuzhy 1988, Lemma 1.3.10).

[17] Nor will the state be 'analytic' in the energy (see note 9).

We could, of course, give Alice the freedom to employ more general mixing operations in O_A. But as we saw in the last section, it is far from clear whether a mixing operation should count as a successful disentanglement when all the states that are mixed by her operation are themselves entangled — or at least not *known* by Alice to be disentangled (given her practical inability to specify exactly which Kraus operations go into the pure operations of her mixing process).

Besides this, there is a more fundamental practical limitation facing Alice, even if we allow her any local operation she chooses. If, as we have seen, we can approximate the result of acting on x with any given operator A in von Neumann algebra $\mathcal{R}$ by acting on x with an invertible operator that preserves x's cyclicity, then the set of all such 'invertible actions' on x must itself produce a dense set of vector states, given that $\{Ax : A \in \mathcal{R}\}$ is dense. It follows that if a von Neumann algebra possesses even just one cyclic vector, it must possess a dense set of them (Dixmier & Maréchal 1971, Lemma 4; cf. Clifton *et al.* 1998).

Now consider, again, the general situation of two commuting nonabelian algebras $\mathcal{R}_A$ and $\mathcal{R}_B$, where either algebra possesses a cyclic vector, and hence a dense set of such. If, in addition, the algebra $\mathcal{R}_{AB} = (\mathcal{R}_A \cup \mathcal{R}_B)''$ possesses a separating vector, then *all* states of that algebra are vector states, a *norm* dense set of which are therefore entangled across $(\mathcal{R}_A, \mathcal{R}_B)$. And since the entangled states of $\mathcal{R}_{AB}$ are open in the weak* topology, they must be open in the (stronger) norm topology too — so we are dealing with a truly generic set of states. Thus, it follows — quite independently of the RS theorem — that

> **Generic Result:** *If $\mathcal{R}_A$ and $\mathcal{R}_B$ are commuting nonabelian von Neumann algebras either of which possesses a cyclic vector, and $\mathcal{R}_{AB}$ possesses a separating vector, then the generic state of $\mathcal{R}_{AB}$ is entangled across $(\mathcal{R}_A, \mathcal{R}_B)$.*

The role that the RS theorem plays is to guarantee that the antecedent conditions of this Generic Result are satisfied whenever we consider spacelike-separated regions (and corresponding algebras) satisfying $(O_A \cup O_B)' \neq \emptyset$. This is a very weak requirement, which is satisfied, for example, when we assume both regions are bounded in spacetime. In that case, in order to be *certain* that her local operation in O_A (pure or mixed) produced a disentangled state, Alice would need the extraordinary ability to distinguish the state of $\mathcal{A}_{AB}$ which results from her operation from the generic set of states of $\mathcal{A}_{AB}$ that are entangled!

Finally, while we noted in our introduction the irony that limitations on disentanglement arise precisely when one considers *relativistic* quantum theory, the practical limitations we have just identified — as opposed to the *intrinsic* limits on disentanglement which are the subject of the next section — are not characteristic of AQFT alone. In particular, the existence of locally cyclic states does not depend on field theory. As we have seen, both the A and B subalgebras of $\mathfrak{B}(\mathcal{H}_A) \otimes \mathfrak{B}(\mathcal{H}_B)$ possess a cyclic vector just in case $\dim \mathcal{H}_A = \dim \mathcal{H}_B$. Indeed, operator algebraists so often find themselves dealing with von Neumann algebras that, together with their commutants, possess a cyclic vector, that such algebras are said by them to be in 'standard form'. So we should not think that local cyclicity is somehow peculiar to the states of local quantum fields.

Neither is it the case that our Generic Result above finds its only application in quantum *field* theory. For example, consider the infinite-by-infinite state space $\mathcal{H}_A \otimes \mathcal{H}_B$ of any two nonrelativistic particles, ignoring their spin degrees of freedom. Take the tensor product with a third auxiliary infinite-dimensional Hilbert space $\mathcal{H}_A \otimes \mathcal{H}_B \otimes \mathcal{H}_C$. Then obviously $\infty = \dim \mathcal{H}_C \geq \dim(\mathcal{H}_A \otimes \mathcal{H}_B) = \infty$, whence the C subalgebra possesses a cyclic vector, which is therefore separating for the $A + B$ algebra. On the same dimensional grounds, both the A and B subalgebras possess cyclic vectors of their own. So our Generic Result applies immediately yielding the conclusion that a typical state of $A + B$ is entangled.

Nor should we think of local cyclicity or the applicability of our Generic Result as peculiar to standard *local* quantum field theory. After noting that the local cyclicity of the vacuum in AQFT was a 'great, counterintuitive, surprise' when it was first proved, Fleming (2000, p. 4) proposes, instead, to build up local algebras associated with bounded open spatial sets within hyperplanes from raising and lowering operators associated with nonlocal Newton-Wigner position eigenstates — a proposal that goes back at least as far as Segal (1964). Fleming then observes, as did Segal (1964, p. 143), that the resulting vacuum state is *not* entangled, nor cyclic for any such local algebra. Nevertheless, as Segal points out, each Segal-Fleming local algebra will be isomorphic to the algebra $\mathfrak{B}(\mathcal{H})$ of all bounded operators on an *infinite*-dimensional Hilbert space $\mathcal{H}$, and algebras associated with spacelike-separated regions in the same hyperplane commute. It follows that if we take any two spacelike-separated bounded open regions O_A and O_B lying in the same hyperplane, then $[\mathcal{A}(O_A) \cup \mathcal{A}(O_B)]''$ is naturally isomorphic to $\mathfrak{B}(\mathcal{H}_A) \otimes \mathfrak{B}(\mathcal{H}_B)$ (Horuzhy 1988, Lemma 1.3.28), and the result of the previous paragraph applies. So Fleming's 'victory' over the RS theorem of standard local quantum field theory rings hollow. Even though the Newton-Wigner vacuum is not itself entangled or locally cyclic across the regions (O_A, O_B), it is indistinguishable from globally pure states of the Newton-Wigner field that are![18]

On the other hand, generic entanglement is certainly not to be expected in every quantum-theoretic context. For example, if we ignore external degrees of freedom, and just consider the spins of two particles with joint state space $\mathcal{H}_A \otimes \mathcal{H}_B$, where both spaces are nontrivial and *finite*-dimensional, then the Generic Result no longer applies. Taking the product with a third auxiliary Hilbert space H_C does not work, because in order for the $A + B$ subalgebra to have a separating vector we would need $\dim H_C \geq \dim H_A \dim H_B$, but for either the A or B subalgebras to possess a cyclic vector we would *also* need that either $\dim H_A \geq \dim H_B \dim H_C$ or $\dim H_B \geq \dim H_A \dim H_C$ — both of which contradict the fact $\mathcal{H}_A$ and $\mathcal{H}_B$ are nontrivial and finite-dimensional.[19]

The point is that while the conditions for generic entanglement may or may not obtain in *any* quantum-theoretical context — de-

---

[18]For further critical discussion of the Segal-Fleming approach to quantum fields, see Halvorson (2001).

[19]In fact, it can be shown that the spins of any pair of particles are *not* generically entangled, unless of course we ignore their mixed spin states; see Clifton & Halvorson (2000) for further discussion.

pending on the observables and dimensions of the state spaces involved — the beauty of the RS theorem is that it allows us to deduce that generic entanglement between bounded open spacetime regions *must* obtain just by making some very general and natural assumptions about what should count as a physically reasonable relativistic quantum field theory.

4 Type III von Neumann algebras and intrinsic entanglement

Though it is not known to follow from the general axioms of AQFT (cf. Kadison 1963), all known concrete models of the axioms are such that the local algebras associated with bounded open regions in M are type III factors (Horuzhy 1988, pp. 29, 35; Haag 1992, Sec. V.6). We start by reviewing what precisely is meant by the designation 'type III factor'.

A von Neumann algebra $\mathcal{R}$ is a factor just in case its center $\mathcal{R} \cap \mathcal{R}'$ consists only of multiples of the identity. It is easy to verify that this is equivalent to $(\mathcal{R} \cup \mathcal{R}')'' = \mathfrak{B}(\mathcal{H})$. Thus, $\mathcal{R}$ induces a 'factorization' of the total Hilbert space algebra $\mathfrak{B}(\mathcal{H})$ into two subalgebras which together generate that algebra.

To understand what 'type III' means, a few further definitions need to be absorbed. A *partial isometry* V is an operator on a Hilbert space $\mathcal{H}$ that maps some closed subspace $C \subseteq \mathcal{H}$ isometrically onto another closed subspace $C' \subseteq \mathcal{H}$, and maps $C^{\perp}$ to zero. (Think of V as a 'hybrid' unitary/projection operator.) Given the set of projections in a von Neumann algebra $\mathcal{R}$, we can define the following equivalence relation on this set: $P \sim Q$ just in case there is a partial isometry $V \in \mathcal{R}$ that maps the range of P onto the range of Q.[20] For example, any two infinite-dimensional projections in $\mathfrak{B}(\mathcal{H})$ are equivalent (when $\mathcal{H}$ is separable), including projections one of whose range is properly contained in the other (cf. Kadison

[20]It is important to notice that this definition of equivalence is relative to the particular von Neumann algebra $\mathcal{R}$ that the projections are considered to be members of.

& Ringrose 1997, Cor. 6.3.5). A nonzero projection $P \in \mathcal{R}$ is called *abelian* if the von Neumann algebra $P\mathcal{R}P$ acting on the subspace $P\mathcal{H}$ (with identity P) is abelian. One can show that the abelian projections in a factor $\mathcal{R}$ are exactly the atoms in its projection lattice (Kadison & Ringrose 1997, Prop. 6.4.2). For example, the atoms of the projection lattice of $\mathfrak{B}(\mathcal{H})$ are all its one-dimensional projections, and they are all (trivially) abelian, whereas it is clear that higher-dimensional projections are not. Finally, a projection $P \in \mathcal{R}$ is called *infinite* (relative to $\mathcal{R}$) when it is equivalent to another projection $Q \in \mathcal{R}$ such that $Q < P$, i.e., Q projects onto a proper subspace of the range of P. One can also show that any abelian projection in a von Neumann algebra is *finite*, i.e., not infinite (Kadison & Ringrose 1997, Prop. 6.4.2).

A von Neumann factor $\mathcal{R}$ is said to be *type* I just in case it possesses an abelian projection. For example, $\mathfrak{B}(\mathcal{H})$ for any Hilbert space $\mathcal{H}$ is type I; and, indeed, every type I factor is isomorphic to $\mathfrak{B}(\mathcal{H})$ for some Hilbert space $\mathcal{H}$ (Kadison & Ringrose 1997, Thm. 6.6.1). On the other hand, a factor is type III if all its nonzero projections are infinite and equivalent. In particular, this entails that the algebra itself is not abelian, nor could it even possess an abelian projection — which would have to be finite. And since a type III factor contains no abelian projections, its projection lattice has no atoms. Another fact about type III algebras (acting on a separable Hilbert space) is that they *always* possess a vector that is both cyclic and separating (Sakai 1971, Cor. 2.9.28). Therefore we know that type III algebras will always possess a dense set of cyclic vectors, and that all their states will be vector states. *Notwithstanding this,* type III algebras possess *no* pure states, as a consequence of the fact that they lack atoms.

To get some feeling for why this is the case — and for the general connection between the failure of the projection lattice of an algebra to possess atoms and its failure to possess pure states — let $\mathcal{R}$ be any non-atomic von Neumann algebra possessing a separating vector (so all of its states are vector states), and let ρ_x be any state of $\mathcal{R}$. We shall need two further definitions. The *support projection* S_x of ρ_x in $\mathcal{R}$ is defined to be the meet of all projections $P \in \mathcal{R}$

such that $\rho_x(P) = 1$. (So S_x is the smallest projection in $\mathcal{R}$ that ρ_x 'makes true'.) The *left-ideal* $\mathcal{I}_x$ of ρ_x in $\mathcal{R}$ is defined to be the set of all $A \in \mathcal{R}$ such that $\rho_x(A^*A) = 0$. Now since S_x is not an atom, there is some nonzero $P \in \mathcal{R}$ such that $P < S_x$. Choose any vector y in the range of P (noting it follows that $S_y \leq P$). We shall first show that $\mathcal{I}_x$ is a proper subset of $\mathcal{I}_y$. So let $A \in \mathcal{I}_x$. Clearly this is equivalent to saying that $Ax = 0$, or that x lies in the range of $N(A)$, the projection onto the null-space of A. $N(A)$ itself lies $\mathcal{R}$ (Kadison & Ringrose 1997, Lemma 5.1.5 and Prop. 2.5.13), thus, $\rho_x(N(A)) = 1$, and accordingly $S_x \leq N(A)$. But since $S_y \leq P < S_x$, we also have $\rho_y(N(A)) = 1$. Thus, y too lies in the range of $N(A)$, i.e., $Ay = 0$, and therefore $A \in \mathcal{I}_y$. To see that the inclusion $\mathcal{I}_x \subseteq \mathcal{I}_y$ is proper, note that since $\langle y, S_y y\rangle = 1$, $\langle y, [I - S_y]^2 y\rangle = 0$, and thus $I - S_y \in \mathcal{I}_y$. However, certainly $I - S_y \notin \mathcal{I}_x$, for the contrary would entail that $\langle x, S_y x\rangle = 1$, in other words, $S_x \leq S_y \leq P < S_x$, which is a contradiction. We can now see, finally, that ρ_x is not pure. For, a pure state of a von Neumann algebra $\mathcal{R}$ determines a maximal left-ideal in $\mathcal{R}$ (Kadison & Ringrose 1997, Thm. 10.2.10), whereas we have just shown (under thc assumption that $\mathcal{R}$ is non-atomic) that $\mathcal{I}_x \subset \mathcal{I}_y$.

The fact that every state of a type III algebra $\mathcal{R}$ is mixed throws an entirely new wrench into the works of the ignorance interpretation of mixtures.[21] Not only is there no preferred way to pick out components of a mixture, but the components of states of $\mathcal{R}$ are always mixed states. Thus, it is impossible to understand the physical preparation of such a mixture in terms of mixing pure states — the states of $\mathcal{R}$ are always irreducibly or what we shall call *intrinsically* mixed. Note, however, that while the states of type III factors fit this description, so do the states of certain *abelian* von Neumann algebras. For example, the 'multiplication' algebra $\mathcal{M} \subseteq \mathfrak{B}(L_2(\mathbb{R}))$ of all bounded functions of the position operator for a single particle lacks atomic projections because position has no eigenvectors. Moreover, all the states of $\mathcal{M}$ are vector states, because any state vector that corresponds to a wavefunction whose support is the whole of $\mathbb{R}$ is

[21] To our knowledge, van Aken (1985) is the only philosopher of quantum theory to have noticed this.

separating for $\mathcal{M}$. Thus the previous paragraph's argument applies equally well to $\mathcal{M}$.

Of course no properly *quantum* system has an abelian algebra of observables, and, as we have already noted, systems with abelian algebras are never entangled with other systems (Bacciagaluppi 1993, Thm. 3). This makes the failure of a type III factor $\mathcal{R}$ to have pure states importantly different from that failure in the case of an abelian algebra. Because $\mathcal{R}$ is *non*abelian, and taking the commutant preserves type (Kadison & Ringrose 1997, Thm. 9.1.3) so that $\mathcal{R}'$ will also be nonabelian, one suspects that any pure state of $(\mathcal{R} \cup \mathcal{R}')'' = \mathfrak{B}(\mathcal{H})$ — which must restrict to an intrinsically mixed state on both subalgebras $\mathcal{R}$ and $\mathcal{R}'$ — has to be *intrinsically entangled* across $(\mathcal{R}, \mathcal{R}')$. And that intuition is exactly right. Indeed, one can show that there are not even any *product* states across $(\mathcal{R}, \mathcal{R}')$ (Summers 1990, p. 213). And, of course, if there are no unentangled states across $(\mathcal{R}, \mathcal{R}')$, then the infamous distinction, some have argued is important to preserve, between so-called 'improper' mixtures that arise by restricting an entangled state to a subsystem, and 'proper' mixtures that do not, becomes *irrelevant*.

Even more interesting is the fact that in all known models of AQFT, the local algebras are 'type III_1' (cf. Haag 1992, p. 267). It would take us too far afield to explain the standard subclassification of factors presupposed by the subscript '1'. We wish only to draw attention to an equivalent characterization of type III_1 algebras established by Connes & Størmer (1978, Cor. 6): A factor $\mathcal{R}$ acting standardly on a (separable) Hilbert space is type III_1 just in case for *any two* states ρ, ω of $\mathfrak{B}(\mathcal{H})$, and any $\epsilon > 0$, there are unitary operators $U \in \mathcal{R}$, $U' \in \mathcal{R}'$ such that $\|\rho - \omega^{UU'}\| < \epsilon$. Notice that this result immediately implies that there are no unentangled states across $(\mathcal{R}, \mathcal{R}')$; for, if some ω were not entangled, it would be impossible to act on this state with local unitary operations in $\mathcal{R}$ and $\mathcal{R}'$ and get arbitrarily close to the states that *are* entangled across $(\mathcal{R}, \mathcal{R}')$. Furthermore — and this is the interesting fact — the Connes-Størmer characterization immediately implies the impossibility of distinguishing in any reasonable way between the different degrees of entanglement that states might have across $(\mathcal{R}, \mathcal{R}')$. For it is a standard assumption in quantum information theory that all reasonable measures of entanglement must be *invariant* under unitary operations on the separate entangled systems (cf. Vedral *et al.* 1997), and presumably such a measure should assign close degrees of entanglement to states that are close to each other in norm. In light of the Connes-Størmer characterization, however, imposition of both these requirements forces triviality on any proposed measure of entanglement across $(\mathcal{R}, \mathcal{R}')$.[22]

The above considerations have particularly strong physical implications when we consider local algebras associated with *diamond regions* in M, i.e., regions given by the intersection of the timelike future of a given spacetime point p with the timelike past of another point in p's future. When $\diamondsuit \subseteq M$ is a diamond, it can be shown in many models of AQFT, including for noninteracting fields, that $\mathcal{A}(\diamondsuit') = \mathcal{A}(\diamondsuit)'$ (Haag 1992, Sec. III.4.2). Thus every global state of the field is *intrinsically* entangled across $(\mathcal{A}(\diamondsuit), \mathcal{A}(\diamondsuit'))$, and it is never possible to think of the field system in a diamond region $\diamondsuit$ as disentangled from its spacelike complement. Though he does not use the language of entanglement, this is precisely the reason for Haag's remark that field systems are always open. In particular, Alice would have *no hope whatsoever* of using local operations in $\diamondsuit$ to disentangle that region's state from that of the rest of the world.

Suppose, however, that Alice has only the more limited goal of disentangling a state of the field across some isolated pair of *strictly* spacelike-separated regions (O_A, O_B), i.e., regions which remain spacelike separated when either is displaced by an arbitrarily small amount. It is also known that in many models of AQFT the local algebras possess the *split property*: for any bounded open $O \subseteq M$, and any larger region $\tilde{O}$ whose interior contains the closure of O, there is a type I factor $\mathcal{N}$ such that $\mathcal{A}(O) \subset \mathcal{N} \subset \mathcal{A}(\tilde{O})$ (Buch-

[22]Of course, the standard von Neumann entropy measure we discussed in Section 1 is norm continuous, and, because of the unitary invariance of the trace, this measure is invariant under unitary operations on the component systems. But in the case of a type III factor $\mathcal{R}$, that measure, as we should expect, is *not* available. Indeed, the state of a system described by $\mathcal{R}$ cannot be represented by any density operator *in* $\mathcal{R}$ because $\mathcal{R}$ cannot contain compact operators, like density operators, whose spectral projections are all finite!

holz 1974, Werner 1987). This implies that the von Neumann algebra generated by a pair of algebras for strictly spacelike-separated regions is isomorphic to their tensor product and, as a consequence, that there *are* product states across $(\mathcal{A}(O_A), \mathcal{A}(O_B))$ (cf. Summers 1990, pp. 239–40). Since, therefore, not every state of $\mathcal{A}_{AB}$ is entangled, we might hope that whatever the global field state is, Alice could *at least in principle* perform an operation in O_A on that state that disentangles it across (O_A, O_B). However, we are now going to use the fact that $\mathcal{A}(O_A)$ lacks abelian projections to show that a norm dense set of entangled states of $\mathcal{A}_{AB}$ cannot be disentangled by any pure local operation performed in $\mathcal{A}(O_A)$.

Let ρ_x be any one of the norm dense set of entangled states of $\mathcal{A}_{AB}$ induced by a vector $x \in \mathcal{H}$ cyclic for $\mathcal{A}(O_B)$, and let $K \in \mathcal{A}(O_A)$ be an arbitrary Kraus operator. (Observe that $\rho_x^K \neq 0$ because x is separating for $\mathcal{A}(O_B)'$ — which includes $\mathcal{A}(O_A)$ — and $K^*K \in \mathcal{A}(O_A)$ is positive.) Suppose for reductio ad absurdum that ω_x^K is not entangled. Let Ky, with $y \in \mathcal{H}$, be any nonzero vector in the range of K. Then, since x is cyclic for $\mathcal{A}(O_B)$, we have, for some sequence $\{B_i\} \subseteq \mathcal{A}(O_B)$, $Ky = K(\lim B_i x) = \lim(B_i Kx)$, which entails $\|(\omega_x^K)^{B_i/\|B_i\|} - \omega_{Ky}\| \to 0$. Since ω_x^K is not entangled across $(\mathcal{A}(O_A), \mathcal{A}(O_B))$, and the local pure operations on $\mathcal{A}(O_B)$ given by the Kraus operators $B_i/\|B_i\|$ cannot create entanglement, we see that ω_{Ky} is the norm (hence weak*) limit of a sequence of unentangled states and, as such, is not itself entangled either. Since y was arbitrary, it follows that every nonzero vector in the range of K induces an unentangled state across $(\mathcal{A}(O_A), \mathcal{A}(O_B))$. Obviously, the same conclusion follows for any nonzero vector in the range of $R(K)$ — the projection onto the range of K — since the range of the latter lies dense in that of the former.

Next, consider the von Neumann algebra

$$\mathcal{C}_{AB} \equiv [R(K)\mathcal{A}(O_A)R(K) \cup R(K)\mathcal{A}(O_B)R(K)]'' \qquad (18)$$

acting on the Hilbert space $R(K)\mathcal{H}$. Since $K \in \mathcal{A}(O_A)$, $R(K) \in \mathcal{A}(O_A)$ (Kadison & Ringrose 1997, p. 309), and thus the subalgebra $R(K)\mathcal{A}(O_A)R(K)$ cannot be abelian — on pain of contradicting the fact that $\mathcal{A}(O_A)$ has no abelian projections. And neither is

$R(K)\mathcal{A}(O_B)R(K)$ abelian. For since $\mathcal{A}(O_B)$ itself is nonabelian, there are $Y_1, Y_2 \in \mathcal{A}(O_B)$ such that $[Y_1, Y_2] \neq 0$. And because our regions (O_A, O_B) are strictly spacelike-separated, they have the Schlieder property: $0 \neq A \in \mathcal{A}(O_A), 0 \neq B \in \mathcal{A}(O_B)$ implies $AB \neq 0$ (Summers 1990, Thm. 6.7). Therefore,

$$[R(K)Y_1R(K), R(K)Y_2R(K)] = [Y_1, Y_2]R(K) \neq 0. \qquad (19)$$

So we see that neither algebra occurring in $\mathcal{C}_{AB}$ is abelian; yet they commute, and so there must be at least one entangled state across those algebras (see page 190). But this conflicts with the conclusion of the preceding paragraph! For the vector states of $\mathcal{C}_{AB}$ are precisely those induced by the vectors in the range of $R(K)$, and we concluded above that these all induce unentangled states across $(\mathcal{A}(O_A), \mathcal{A}(O_B))$. Therefore, by restriction, they all induce unentangled states across the algebra $\mathcal{C}_{AB}$. But if none of $\mathcal{C}_{AB}$'s vector states are entangled, it can possess *no* entangled states at all.

The above argument still goes through under the weaker assumption that Alice applies any mixed *projective* operation, i.e., any operation T corresponding to a standard von Neumann measurement associated with a mutually orthogonal set $\{P_i\} \in \mathcal{A}(O_A)$ of projection operators. For suppose, again for reductio ad absurdum, that $\rho_x^T = \sum_i \lambda_i \rho_x^{P_i}$ is not entangled across the regions. Then, since entanglement cannot be created by a further application to ρ_x^T of the local projective operation given by (say) $T_1(\cdot) = P_1(\cdot)P_1$, it follows that $(\rho_x^T)^{T_1} = (\rho_x^{T_1 \circ T}) = \rho_x^{P_1}$ must again be unentangled, and the above reasoning to a contradiction goes through *mutatis mutandis* with $K = P_1$. This is to be contrasted to the nonrelativistic case we considered in Section 1, where Alice *was* able to disentangle an arbitrary state of $\mathfrak{B}(\mathcal{H}_A) \otimes \mathfrak{B}(\mathcal{H}_B)$ by a nonselective projective operation on A. And a moment's reflection will reveal that that was possible precisely because of the availability of abelian projections in the algebra of her subsystem A.

We have not, of course, shown that the above argument covers *arbitrary* mixing operations Alice might perform in O_A; in particular, positive operator-valued mixings, where the Kraus operators $\{K_i\}$ of a local operation T in O_A do not have mutually orthogonal ranges. However, although it would be interesting to know how far the result could be pushed, we have already expressed our reservations about whether arbitrary mixing operations should count as disentangling when none of the pure operations of which they are composed could possibly produce disentanglement on their own.

In summary:

> *There are many regions of spacetime within which no local operations can be performed that will disentangle that region's state from that of its spacelike complement, and within which no pure or projective operation on any one of a norm dense set of states can yield disentanglement from the state of any other strictly spacelike-separated region.*

Clearly the advantage of the formalism of AQFT is that it allows us to see clearly just how much more deeply entrenched entanglement is in *relativistic* quantum theory. At the very least, this should serve as a strong note of caution to those who would quickly assert that quantum nonlocality cannot peacefully exist with relativity!

4.1 Neutralizing the methodological worry

What then becomes of Einsteinian worries about the possibility of doing science in such a highly entangled world? As we shall now explain, *for all practical purposes* the split property of local algebras neutralizes Einstein's main methodological worry.[23]

Let us suppose Alice wants to prepare some state ρ on $\mathcal{A}(O_A)$ for subsequent testing. By the split property, there is a type I factor $\mathcal{N}$ satisfying $\mathcal{A}(O_A) \subset \mathcal{N} \subset \mathcal{A}(\tilde{O}_A)$ for any super-region $\tilde{O}_A$ that contains the closure of O_A. Since ρ is a vector state (when we assume $(O_A)' \neq \emptyset$), its vector representative defines a state on $\mathcal{N}$ that extends ρ and is, therefore, represented by some density operator D_ρ in the type I algebra $\mathcal{N}$. Now D_ρ is a convex combination $\sum_i \lambda_i P_i$ of mutually orthogonal atomic projections in $\mathcal{N}$ satisfying $\sum_i P_i = I$ with $\sum_i \lambda_i = 1$. But each such projection is equivalent, in the type

[23]The following arguments are essentially just an amplification of the reasoning in Werner (1987) and Summers (1990, Thm. 3.13).

III algebra $\mathcal{A}(\tilde{O}_A)$, to the identity operator. Thus, for each i, there is a partial isometry $V_i \in \mathcal{A}(\tilde{O}_A)$ satisfying $V_i V_i^* = P_i$ and $V_i^* V_i = I$.

Next, consider the nonselective operation T on $\mathcal{A}(\tilde{O}_A)$ given by Kraus operators $K_i = \sqrt{\lambda_i} V_i$. We claim that $T(X) = \rho(X)I$ for all $X \in \mathcal{A}(O_A)$. Indeed, because each P_i is abelian in $\mathcal{N} \supseteq \mathcal{A}(O_A)$, the operator $P_i X P_i$ acting on $P_i \mathcal{H}$ can only be some multiple, c_i, of the identity operator P_i on $P_i \mathcal{H}$, and taking the trace of both sides of the equation

$$P_i X P_i = c_i P_i \tag{20}$$

immediately reveals that $c_i = \mathrm{Tr}(P_i X)$. Moreover, acting on the left of (20) with V_i^* and on the right with V_i, we obtain $V_i^* X V_i = \mathrm{Tr}(P_i X) I$, which yields the desired conclusion when multiplied by λ_i and summed over i.

Finally, since $T(X) = \rho(X)I$ for all $X \in \mathcal{A}(O_A)$, obviously $\omega^T = \rho$ for all initial states ω of $\mathcal{A}(O_A)$. Thus, once we allow Alice to perform an operation like T that is *approximately* local to $\mathcal{A}(O_A)$ (choosing $\tilde{O}_A$ to approximate O_A as close as we like), she has the freedom to prepare any state of $\mathcal{A}(O_A)$ that she pleases!

Notice that, ironically, testing the theory is actually *easier* here than in nonrelativistic quantum theory. For we were able to exploit above the type III character of $\mathcal{A}(\tilde{O}_A)$ to show that Alice can always prepare her desired state on $\mathcal{A}(O_A)$ *nonselectively*, i.e., without ever having to sacrifice any members of her ensemble! Also observe that the result of her preparing operation T, because it is local to $\mathcal{A}(\tilde{O}_A)$, will always produce a product state across (O_A, O_B) when $O_B \subseteq (\tilde{O}_A)'$. That is, for any initial state ω across the regions, and all $X \in \mathcal{A}(O_A)$ and $Y \in \mathcal{A}(O_B)$, we have

$$\omega^T(XY) = \omega(T(X)Y) = \omega(\rho(X)Y) = \rho(X)\omega(Y). \tag{21}$$

So as soon as we allow Alice to perform *approximately* local operations on her field system, she *can* isolate it from entanglement with other strictly spacelike-separated field systems, while simultaneously preparing its state as she likes and with relative ease. *God is subtle, but not malicious.*

Bibliography

Arageorgis, A., Earman, J. & Ruetsche, L. (2002), 'Weyling the time away: the non-unitary implementability of quantum field dynamics on curved spacetime', *Studies in History and Philosophy of Modern Physics* **33**(2), 151–84.

Bacciagaluppi, G. (1993), Separation theorems and Bell inequalities in algebraic quantum mechanics, in P. Busch, P. Lahti & P. Mittelstaedt, eds, 'Symposium on the Foundations of Modern Physics 1993', World Scientific, Singapore, pp. 29–37.

Bennett, C. H., DiVincenzo, D. P., Fuchs, C. A., Mor, T., Rains, E., Schor, P. W., Smolin, J. A. & Wooters, W. K. (1999), 'Quantum nonlocality without entanglement', *Physical Review A* **59**, 1070–91.

Borchers, H.-J. (1960), 'Über die Mannigfaltigkeit der interpolierenden Felder zu einer kausalen *S*-Matrix', *Nuovo Cimento (10)* **15**, 784–94.

Borchers, H.-J. (1965), 'On the vacuum state in quantum field theory, II.', *Communications in Mathematical Physics* **1**, 57–79.

Braunstein, S. L., Caves, C. M., Jozsa, R., Linden, N., Popescu, S. & Schack, R. (1999), 'Separability of very noisy mixed states and implications for NMR quantum computing', *Physical Review Letters* **83**, 1054–7.

Buchholz, D. (1974), 'Product states for local algebras', *Communications in Mathematical Physics* **36**, 287–304.

Busch, P., Grabowski, M. & Lahti, P. (1995), *Operational Quantum Physics*, Springer, New York.

Clifton, R. K. & Halvorson, H. P. (2000), 'Bipartite mixed states of infinite-dimensional systems are generically nonseparable', *Physical Review A* **61**, 012108.

Clifton, R. K., Feldman, D., Halvorson, H. P., Redhead, M. L. G. & Wilce, A. (1998), 'Superentangled states', *Physical Review A* **58**, 135–45.

Clifton, R. K., Halvorson, H. P. & Kent, A. (2000), 'Non-local correlations are generic in infinite-dimensional bipartite systems', *Physical Review A* **61**, 042101. Chapter 10 of this volume.

Connes, A. & Størmer, E. (1978), 'Homogeneity of the state space of factors of type III_1', *Journal of Functional Analysis* **28**, 187–96.

Davies, E. B. (1976), *Quantum Theory of Open Systems*, Academic Press, London.

Dixmier, J. & Maréchal, O. (1971), 'Vecteurs totalisateurs d'une algèbre de von Neumann', *Communications in Mathematical Physics* **22**, 44–50.

Einstein, A. (1948), 'Quantenmechanik und Wirklichkeit', *Dialectica* **2**, 320–4.

Fleming, G. (2000), 'Reeh-Schlieder meets Newton-Wigner', *Philosophy of Science* **67**, S495–S515.

Fleming, G. & Butterfield, J. (1999), Strange positions, in J. Butterfield & C. Pagonis, eds, 'From Physics to Philosophy', Cambridge University Press, New York.

Haag, R. (1992), *Local Quantum Physics*, 2nd ed., Springer, New York.

Haag, R. & Kastler, D. (1964), 'An algebraic approach to quantum field theory', *Journal of Mathematical Physics* **5**, 848–61.

Halvorson, H. P. (2001), 'Reeh-Schlieder defeats Newton-Wigner: On alternative localization schemes in relativistic quantum field theory', *Philosophy of Science* **68**, 111–33.

Horuzhy, S. S. (1988), *Introduction to Algebraic Quantum Field Theory*, Kluwer, Dordrecht.

Howard, D. (1989), Holism, separability, and the metaphysical implications of the Bell experiments, in J. T. Cushing & E. McMullin, eds, 'The Philosophical Consequences of Bell's Theorem', Notre Dame University Press, Notre Dame, pp. 224–53.

Kadison, R. V. (1963), 'Remarks on the type of von Neumann algebras of local observables in quantum field theory', *Journal of Mathematical Physics* **4**, 1511–6.

Kadison, R. V. & Ringrose, J. (1997), *Fundamentals of the Theory of Operator Algebras*, American Mathematical Society, Providence, RI.

Kraus, K. (1983), *States, Effects, and Operations*, Springer, Berlin.

Laflamme, R. (1998), Review of 'Separability of very noisy mixed states and implications for NMR quantum computing' by Braunstein et al., in 'Quick Reviews in Quantum Computation and Information'. http://quantum-computing.lanl.gov/qcreviews/qc.

Lüders, G. (1951), 'Über die Zustandsänderung durch den Messprozess', *Annalen der Physik* **8**, 322–8.

Mor, T. (1999), 'Disentangling quantum states while preserving all local properties', *Physical Review Letters* **83**, 1451–4.

Mor, T. & Terno, D. R. (1999), 'Sufficient conditions for a disentanglement', *Physical Review A* **60**, 4341–3.

Popescu, S. & Rohrlich, D. (1997), 'Thermodynamics and the measure of entanglement', *Physical Review A* **56**, R3319–21.

Redhead, M. L. G. (1989), *Incompleteness, Nonlocality, and Realism*, 2nd ed., Clarendon Press, Oxford.

Redhead, M. L. G. (1995), 'More ado about nothing', *Foundations of Physics* **25**, 123–37.

Reeh, H. & Schlieder, S. (1961), 'Bemerkungen zur Unitäräquivalenz von Lorentzinvarianten Feldern', *Nuovo Cimento* **22**, 1051–68.

Sakai, S. (1971), *C^*-algebras and W^*-algebras*, Springer, New York.

Schroer, B. (1999), 'Basic quantum theory and measurement from the viewpoint of local quantum physics'. E-print: quant-ph/9904072.

Segal, I. E. (1964), Quantum fields and analysis in the solution manifolds of differential equations, in W. T. Martin & I. E. Segal, eds, 'Proceedings of a Conference on the Theory and Applications of Analysis in Function Space', MIT Press, Cambridge, MA, pp. 129–53.

Segal, I. E. & Goodman, R. W. (1965), 'Anti-locality of certain Lorentz-invariant operators', *Journal of Mathematics and Mechanics* **14**, 629–38.

Streater, R. F. & Wightman, A. S. (2000), *PCT, Spin and Statistics, and All That*, 3rd ed., Addison-Wesley, New York.

Summers, S. J. (1990), 'On the independence of local algebras in quantum field theory', *Reviews of Mathematical Physics* **2**, 201–47.

Summers, S. J. & Werner, R. F. (1988), 'Maximal violation of Bell's inequalities for algebras of observables in tangent spacetime regions', *Annales de l'Institut Henri Poincaré* **49**, 215–43.

Summers, S. J. & Werner, R. F. (1995), 'On Bell's inequalities and algebraic invariants', *Letters in Mathematical Physics* **33**, 321–34.

van Aken, J. (1985), 'Analysis of quantum probability theory', *Journal of Philosophical Logic* **14**, 267–96.

van Fraassen, B. C. (1991), *Quantum Mechanics: An Empiricist View*, Clarendon Press, Oxford.

Vedral, V., Plenio, M. B., Rippin, M. A. & Knight, P. L. (1997), 'Quantifying entanglement', *Physical Review Letters* **78**, 2275–9.

Werner, R. F. (1987), 'Local preparability of states and the split property', *Letters in Mathematical Physics* **13**, 325–9.

Werner, R. F. (1989), 'Quantum states with Einstein-Podolsky-Rosen correlations admitting a hidden-variable model', *Physical Review A* **40**, 4277–81.

Part III

The Concept of a Particle

Chapter 8

No place for particles in relativistic quantum theories?

with Hans Halvorson

1 Introduction

It is a widespread belief, at least within the physics community, that there is no relativistic quantum theory of (localizable) particles; and, thus, that the only relativistic quantum theory is a theory of fields. This belief has received much support in recent years in the form of rigorous no-go theorems by Malament (1996) and Hegerfeldt (1998*a*,*b*). In particular, Hegerfeldt shows that in a generic quantum theory (relativistic or non-relativistic), if there are states with localized particles, and if there is a lower bound on the system's energy, then superluminal spreading of the wavefunction must occur. Similarly, Malament shows the inconsistency of a few intuitive desiderata for a relativistic quantum mechanics of (localizable) par-

---

Thanks to Jeff Barrett and David Malament for helpful comments on an earlier version of this paper.

ticles. Thus, it appears that quantum theory engenders a fundamental conflict between relativistic causality and localizability.

What is thc philosophical lesson of this conflict between relativistic causality and localizability? One the one hand, if we believe that the assumptions of Malament's theorem must hold for any empirically adequate theory, then it follows that our world cannot be correctly described by a particle theory. On the other hand, if we believe that our world *can* be correctly described by a particle theory, then one (or more) of Malament's assumptions must be false. Malament clearly endorses the first response; that is, he argues that his theorem entails that there is no relativistic quantum mechanics of localizable particles (insofar as any relativistic theory precludes act-outcome correlations at spacelike separation). Others, however, have argued that the assumptions of Malament's theorem need not hold for any relativistic, quantum-mechanical theory (cf. Fleming & Butterfield 1999), or that we cannot judge the truth of the assumptions until we resolve the interpretive issues of elementary quantum mechanics (cf. Barrett 2002).

We do not think that these objections to the soundness of Malament's argument are cogent. However, there are other tacit assumptions of Malament's theorem that some might be tempted to question. For example, Malament's theorem depends on the assumption that there is no preferred inertial reference frame, which some believe to have very little empirical support (cf. Cushing 1996). Furthermore, Malament's theorem establishes only that there is no relativistic quantum mechanics in which particles can be completely localized in spatial regions with sharp boundaries; it leaves open the possibility that there might be a relativistic quantum mechanics of 'unsharply' localized particles.

In this paper, we present two new no-go theorems which show that these tacit assumptions of Malament's theorem are not needed to sustain an argument against localizable particles. First, we derive a no-go theorem against localizable particles that does not assume the equivalence of all inertial frames (Theorem 1). Second, we derive a no-go theorem that shows that there is no relativistic quantum mechanics of unsharply localized particles (Theorem 2).

However, it would be a mistake to think that these results show — or, are intended to show — that a field ontology, rather than a particle ontology, is appropriate for relativistic quantum theories. While these results show that there is no position observable that satisfies relativistic constraints, quantum field theories — both relativistic *and* non-relativistic — already reject the notion of a position observable in favor of localized field observables. Thus, our first two results have nothing to say about the possibility that relativistic quantum field theory (RQFT) might permit a 'particle interpretation', in which localized particles are supervenient on the underlying localized field observables. To exclude this latter possibility, we formulate (in Section 6) a necessary condition for a quantum field theory to permit a particle interpretation, and we then show that this condition fails in any relativistic theory (Theorem 3).

Presumably, any empirically adequate theory must be able to reproduce the predictions of special relativity and of quantum mechanics. Therefore, our no-go results show that the existence of localizable particles is, strictly speaking, ruled out by the empirical data. However, in Section 7 we defuse this counterintuitive consequence by showing that RQFT itself explains how the illusion of localizable particles can arise, and how 'particle talk' — although strictly fictional — can still be useful.

2 Malament's theorem

Malament's theorem shows the inconsistency of a few intuitive desiderata for a relativistic quantum mechanics of (localizable) particles. It strengthens previous results (e.g., Schlieder 1971) by showing that the assumption of 'no superluminal wavepacket spreading' can be replaced by the weaker assumption of 'microcausality', and by making it clear that Lorentz invariance is not needed to derive a conflict between relativistic causality and localizability.

In order to present Malament's result, we assume that our background spacetime M is an affine space, with a foliation S into spatial hyperplanes. (For ease, we can think of an affine space as a vector space, so long as we do not assign any physical significance to the origin.) This will permit us to consider a wide range of relativistic (e.g., Minkowski) as well as nonrelativistic (e.g., Galilean) spacetimes. The pure states of our quantum-mechanical system are given by rays in some Hilbert space $\mathcal{H}$. We assume that there is a mapping $\Delta \mapsto E_\Delta$ of *bounded* subsets of hyperplanes in M into projections on $\mathcal{H}$. We think of E_Δ as representing the proposition that the particle is localized in Δ; or, from a more operational point of view, E_Δ represents the proposition that a position measurement is certain to find the particle within Δ. We also assume that there is a strongly continuous representation $\mathbf{a} \mapsto U(\mathbf{a})$ of the translation group of M in the unitary operators on $\mathcal{H}$. Here strong continuity means that for any unit vector $\psi \in \mathcal{H}$, $\langle \psi, U(\mathbf{a})\psi \rangle \to 1$ as $\mathbf{a} \to 0$; and it is equivalent (via Stone's theorem) to the assumption that there are energy and momentum observables for the particle. If all of the preceding conditions hold, we say that the triple $(\mathcal{H}, \Delta \mapsto E_\Delta, \mathbf{a} \mapsto U(\mathbf{a}))$ is a *localization system* over M.

The following conditions should hold for any localization system — either relativistic or nonrelativistic — that describes a single particle.

Localizability: If Δ and Δ' are disjoint subsets of a single hyperplane, then $E_\Delta E_{\Delta'} = 0$.

Translation covariance: For any Δ and for any translation $\mathbf{a}$ of M, $U(\mathbf{a})E_\Delta U(\mathbf{a})^* = E_{\Delta+\mathbf{a}}$.

Energy bounded below: For any timelike translation $\mathbf{a}$ of M, the generator $H(\mathbf{a})$ of the one-parameter group $\{U(t\mathbf{a}) : t \in \mathbb{R}\}$ has a spectrum bounded from below.

We recall briefly the motivation for each of these conditions. 'Localizability' says that the particle cannot be detected in two disjoint spatial sets at a given time. 'Translation covariance' gives us a connection between the symmetries of the spacetime M and the symmetries of the quantum-mechanical system. In particular, if we displace the particle by a spatial translation $\mathbf{a}$, then the original wavefunction ψ will transform to some wavefunction $\psi_\mathbf{a}$. Since the statistics for the displaced detection experiment should be identical to the original statistics, we have $\langle\psi, E_\Delta\psi\rangle = \langle\psi_{\mathbf{a}}, E_{\Delta+\mathbf{a}}\psi_{\mathbf{a}}\rangle$. By Wigner's theorem, however, the symmetry is implemented by some unitary operator $U(\mathbf{a})$. Thus, $U(\mathbf{a})\psi = \psi_{\mathbf{a}}$, and $U(\mathbf{a})E_\Delta U(\mathbf{a})^* = E_{\Delta+\mathbf{a}}$. In the case of time translations, the covariance condition entails that the particle has unitary dynamics. (This might seem to beg the question against a collapse interpretation of quantum mechanics; we dispel this worry at the end of this section.) Finally, the 'energy bounded below' condition asserts that, relative to any free-falling observer, the particle has a lowest possible energy state. If it were to fail, we could extract an arbitrarily large amount of energy from the particle as it drops down through lower and lower states of energy.

We now turn to the 'specifically relativistic' assumptions needed for Malament's theorem. The special theory of relativity entails that there is a finite upper bound on the speed at which (detectable) physical disturbances can propagate through space. Thus, if Δ and Δ' are distant regions of space, then there is a positive lower bound on the amount of time it should take for a particle localized in Δ to travel to Δ'. We can formulate this requirement precisely by saying that for any timelike translation $\mathbf{a}$, there is an $\epsilon > 0$ such that, for every state ψ, if $\langle\psi, E_\Delta\psi\rangle = 1$ then $\langle\psi, E_{\Delta'+t\mathbf{a}}\psi\rangle = 0$ whenever $0 \leq t < \epsilon$. This is equivalent to the following assumption.

Strong causality: If Δ and Δ' are disjoint subsets of a single hyperplane, and if the distance between Δ and Δ' is nonzero, then for any timelike translation $\mathbf{a}$, there is an $\epsilon > 0$ such that $E_\Delta E_{\Delta'+t\mathbf{a}} = 0$ whenever $0 \leq t < \epsilon$.

(Note that strong causality entails localizability.) Although strong causality is a reasonable condition for relativistic theories, Malament's theorem requires only the following weaker assumption (which he himself calls 'locality').

Microcausality: If Δ and Δ' are disjoint subsets of a single hyperplane, and if the distance between Δ and Δ' is nonzero, then for any timelike translation $\mathbf{a}$, there is an $\epsilon > 0$ such that $[E_\Delta, E_{\Delta'+t\mathbf{a}}] = 0$ whenever $0 \leq t < \epsilon$.

If E_Δ can be measured within Δ, microcausality is equivalent to the assumption that a measurement within Δ cannot influence the statistics of measurements performed in regions that are spacelike to Δ (see Malament 1996, p. 5). Conversely, a failure of microcausality would entail the possibility of act-outcome correlations at spacelike separation. Note that both strong and weak causality make sense for nonrelativistic spacetimes (as well as for relativistic spacetimes); though, of course, we should not expect either causality condition to hold in the nonrelativistic case.

Theorem (Malament). *Let* $(\mathcal{H}, \Delta \mapsto E_\Delta, \mathbf{a} \mapsto U(\mathbf{a}))$ *be a localization system over Minkowski spacetime that satisfies:*

1. *Localizability*
2. *Translation covariance*
3. *Energy bounded below*
4. *Microcausality*

Then $E_\Delta = 0$ *for all* Δ.

Thus, in every state, there is no chance that the particle will be detected in any local region of space. As Malament claims, this serves as a *reductio ad absurdum* of any relativistic quantum mechanics of a single (localizable) particle.

2.1 The soundness of Malament's argument

Several authors have claimed that Malament's theorem is not sufficient to rule out a relativistic quantum mechanics of localizable particles. In particular, these authors argue that it is not reasonable to expect the conditions of Malament's theorem to hold for any relativistic, quantum-mechanical theory of particles. For example, Dickson (1998) argues that a 'quantum' theory does not need a position *operator* (equivalently, a system of localizing projections) in order to treat position as a physical quantity; Barrett (2002) argues that time-translation covariance is suspect; and Fleming & Butterfield (1999) argue that the microcausality assumption is not warranted by special relativity. We now show, however, that none of these arguments is decisive against the assumptions of Malament's theorem.

Dickson (1998, p. 214) cites the Bohmian interpretation of the Dirac equation as a counterexample to the claim that any 'quantum' theory must represent position by an operator. In order to see what Dickson might mean by this, recall that the Dirac equation admits both positive and negative energy solutions. If $\mathcal{H}$ denotes the Hilbert space of all (both positive and negative energy) solutions, then we may define the 'standard position operator' Q by setting $Q\psi(\mathbf{x}) = \mathbf{x} \cdot \psi(\mathbf{x})$ (Thaller 1992, p. 7). If, however, we restrict to the Hilbert space $\mathcal{H}_{\text{pos}} \subset \mathcal{H}$ of positive energy solutions, then the probability density given by the Dirac wavefunction does not correspond to a self-adjoint position operator (Thaller 1992, p. 32). According to Holland (1993, p. 502), this lack of a position operator on $\mathcal{H}_{\text{pos}}$ precludes a Bohmian interpretation of $\psi(\mathbf{x})$ as a probability amplitude for finding the particle in an elementary volume $d^3\mathbf{x}$ around $\mathbf{x}$.

Since the Bohmian interpretation of the Dirac equation uses all states (both positive and negative energy), and the corresponding position observable Q, it is not clear what Dickson means by saying that the Bohmian interpretation of the Dirac equation dispenses with a position observable. Moreover, since the energy is not bounded below in $\mathcal{H}$, this would not in any case give us a counterexample to Malament's theorem. However, Dickson could have developed his argument by appealing to the positive energy subspace $\mathcal{H}_{\text{pos}}$. In this case, we *can* talk about positions despite the fact that we do not have a position observable in the usual sense. In particular, we shall show in Section 5 that, for talk about positions, it suffices to have a family of 'unsharp' localization observables. (And, yet, we shall show that relativistic quantum theories do not permit even this attenuated notion of localization.)

Barrett (2002) argues that the significance of Malament's theorem cannot be assessed until we have solved the measurement problem:

> If we might have to violate the apparently weak and obvious assumptions that go into proving Malament's theorem in order to get a satisfactory solution to the mea-

> surement problem, then all bets are off concerning the applicability of the theorem to the detectable entities that inhabit our world. (Barrett 2002, p. 16)

In particular, a solution to the measurement problem may require that we abandon unitary dynamics. But if we abandon unitary dynamics, then the translation covariance condition does not hold, and we need not accept the conclusion that there is no relativistic quantum mechanics of (localizable) particles.

Unfortunately, it is not clear that we could avoid the upshot of Malament's theorem by moving to a collapse theory. Existing (nonrelativistic) collapse theories take the empirical predictions of quantum theory seriously. That is, the 'statistical algorithm' of quantum mechanics is assumed to be at least approximately correct; and collapse is introduced only to ensure that we obtain determinate properties at the end of a measurement. However, in the present case, Malament's theorem shows that the statistical algorithm of any quantum theory predicts that if there are local particle detections, then act-outcome correlations are possible at spacelike separation. Thus, if a collapse theory is to stay close to these predictions, it too would face a conflict between localizability and relativistic causality.

Perhaps, then, Barrett is suggesting that the price of accommodating localizable particles might be a complete abandonment of unitary dynamics, *even at the level of a single particle*. In other words, we may be forced to adopt a collapse theory *without* having any underlying (unitary) quantum theory. But even if this is correct, it wouldn't count against Malament's theorem, which was intended to show that there is no relativistic *quantum* theory of localizable particles. Furthermore, noting that Malament's theorem requires unitary dynamics is one thing; it would be quite another thing to provide a model in which there *are* localizable particles — at the price of nonunitary dynamics — but which is also capable of reproducing the well-confirmed quantum interference effects at the micro-level. Until we have such a model, pinning our hopes for localizable particles on a failure of unitary dynamics is little more than wishful thinking.

Like Barrett, Fleming (Fleming & Butterfield 1999, p. 158ff) disagrees with the reasonableness of Malament's assumptions. Unlike Barrett, however, Fleming provides a concrete model in which there are localizable particles (viz., using the Newton-Wigner position operator as a localizing observable) and in which Malament's microcausality assumption fails. Nonetheless, Fleming argues that this failure of microcausality is perfectly consistent with relativistic causality.

According to Fleming, the property 'localized in Δ' (represented by E_Δ) need not be detectable within Δ. As a result, $[E_\Delta, E_{\Delta'}] \neq 0$ does not entail that it is possible to send a signal from Δ to Δ'. However, by claiming that local *beables* need not be local *observables*, Fleming undercuts the primary utility of the notion of localization, which is to indicate those physical quantities that are operationally accessible in a given region of spacetime. Indeed, it is not clear what motivation there could be — aside from indicating what is locally measurable — for assigning observables to spatial regions. If E_Δ is *not* measurable in Δ, then why should we say that 'E_Δ is localized in Δ'? Why not say instead that 'E_Δ is localized in Δ'' (where $\Delta' \neq \Delta$)? Does either statement have any empirical consequences and, if so, how do their empirical consequences differ? Until these questions are answered, we maintain that local beables are always local observables; and a failure of microcausality *would* entail the possibility of act-outcome correlations at spacelike separation. (For a more detailed argument along these lines, see Halvorson 2001, Section 6.)

2.2 Tacit assumptions of Malament's theorem

The objections to the four assumptions of Malament's theorem are unconvincing. By any reasonable understanding of special relativity and of quantum theory, these assumptions should hold for any theory that is capable of reproducing the predictions of both theories. Nonetheless, we anticipate that further objections could be directed against the more or less tacit assumptions of Malament's theorem.

As we noted earlier, Malament's theorem does not make use of the full structure of Minkowski spacetime (e.g., Lorentz invariance). However, the following example shows that the theorem fails if there is a preferred inertial reference frame.

Example 1. Let $M = \mathbb{R}^1 \oplus \mathbb{R}^3$ be full Newtonian spacetime with a distinguished timelike direction **a**. To any set of the form $\{(t, x) : x \in \Delta\}$, with $t \in \mathbb{R}$, and Δ a bounded open subset of $\mathbb{R}^3$, we assign the spectral projection E_Δ of the position operator for a particle in three dimensions. Thus, the conclusion of Malament's theorem is false, while both the microcausality and localizability conditions hold. Let $P_0 = 0$, and for $i = 1, 2, 3$, let $P_i = -i(d/dx_i)$. For any four-vector **b**, let $U(\mathbf{b}) = \exp\{i(\mathbf{b} \cdot \mathbf{P})\}$, where

$$\mathbf{b} \cdot \mathbf{P} = b_0 P_0 + b_1 P_1 + b_2 P_2 + b_3 P_3. \tag{1}$$

Thus, translation covariance holds, and since the energy is identically zero, the energy condition trivially holds. (Note, however, that if M is *not* regarded as having a distinguished timelike direction, then this example violates the energy condition.) □

A brief inspection of Malament's proof shows that the following assumption on the affine space M is sufficient for his theorem to go through.

No absolute velocity: Let **a** be a spacelike translation of M. Then there is a pair $(\mathbf{b}, \mathbf{c})$ of timelike translations of M such that $\mathbf{a} = \mathbf{b} - \mathbf{c}$.

Despite the fact that "no absolute velocity" is a feature of both Galilean and Minkowski spacetimes, there are some who claim that the existence of a (undetectable) preferred reference frame is perfectly consistent with all current empirical evidence (cf. Cushing 1996). What is more, the existence of a preferred frame is an absolutely essential feature of a number of "realistic" interpretations of quantum theory (cf. Maudlin 1994, Chap. 7). Thus, this tacit assumption of Malament's theorem could be a source of contention for those wishing to maintain the possibility of a relativistic quantum mechanics of localizable particles.

Second, some might wonder whether Malament's result is an artifact of special relativity, and whether a notion of localizable particles might be restored in the context of general relativity. Indeed, it is not difficult to see that Malament's result does *not* automatically generalize to arbitrary relativistic spacetimes.

To see this, suppose that M is an arbitrary globally hyperbolic manifold. (That is, M is a manifold that permits at least one foliation $\mathcal{S}$ into spacelike hypersurfaces). Although M will not typically have a translation group, we assume that M has a transitive Lie group G of diffeomorphisms. (Just as a manifold is locally isomorphic to $\mathbb{R}^n$, a Lie group is locally isomorphic to a group of translations.) We require that G has a representation $g \mapsto U(g)$ in the unitary operators on $\mathcal{H}$; and, the translation covariance condition now says that $E_{g(\Delta)} = U(g)E_\Delta U(g)^*$ for all $g \in G$. The following example then shows that Malament's theorem fails even for the very simple case where M is a two-dimensional cylinder.

Example 2. Let $M = \mathbb{R} \oplus S^1$, where S^1 is the one-dimensional unit circle, and let G denote the Lie group of timelike translations and rotations of M. It is not difficult to construct a unitary representation of G that satisfies the energy bounded below condition. (We can use the Hilbert space of square-integrable functions from S^1 into $\mathbb{C}$, and the procedure for constructing the unitary representation is directly analogous to the case of a single particle moving on a line.) Fix a spacelike hypersurface Σ, and let μ denote the normalized rotation-invariant measure on Σ. For each open subset Δ of Σ, let $E_\Delta = I$ if $\mu(\Delta) \geq 2/3$, and let $E_\Delta = 0$ if $\mu(\Delta) < 2/3$. Then localizability holds, since for any pair (Δ, Δ') of disjoint open subsets of Σ, either $\mu(\Delta) < 2/3$ or $\mu(\Delta') < 2/3$. □

Nonetheless, Examples 1 and 2 hardly serve as physically interesting counterexamples to a strengthened version of Malament's theorem. In particular, in Example 1 the energy is identically zero, and therefore the probability for finding the particle in a given region of space remains constant over time. In Example 2, the particle is localized in every region of space with volume greater than 2/3, and the particle is never localized in a region of space with volume less than 2/3. In the following two sections, then, we will formulate explicit conditions to rule out such pathologies, and we will use these conditions to derive a strengthened version of Malament's theorem that applies to generic spacetimes.

3 Hegerfeldt's theorem

Hegerfeldt's (1998*a*, 1998*b*) recent results on localization apply to arbitrary (globally hyperbolic) spacetimes, and they do not make use of the 'no absolute velocity' condition. Thus, we will suppose henceforth that M is a globally hyperbolic spacetime, and we will fix a foliation $\mathcal{S}$ of M, as well as a unique isomorphism between any two hypersurfaces in this foliation. If $\Sigma \in \mathcal{S}$, we will write $\Sigma + t$ for the hypersurface that results from 'moving Σ forward in time by t units'; and if Δ is a subset of Σ, we will use $\Delta + t$ to denote the corresponding subset of $\Sigma + t$. We assume that there is a representation $t \mapsto U_t$ of the time-translation group $\mathbb{R}$ in the unitary operators on $\mathcal{H}$, and we will say that the localization system $(\mathcal{H}, \Delta \mapsto E_\Delta, t \mapsto U_t)$ satisfies *time-translation covariance* just in case $U_t E_\Delta U_{-t} = E_{\Delta+t}$ for all Δ and all $t \in \mathbb{R}$.

Hegerfeldt's result is based on the following root lemma.

Lemma 1 (Hegerfeldt). *Suppose that $U_t = e^{itH}$, where H is a self-adjoint operator with spectrum bounded from below. Let A be a positive operator (e.g., a projection operator). Then for any state ψ, either*

$$\langle U_t\psi, AU_t\psi\rangle \neq 0\,, \qquad \text{for almost all } \; t \in \mathbb{R},$$

or

$$\langle U_t\psi, AU_t\psi\rangle = 0\,, \qquad \text{for all } \; t \in \mathbb{R}\,.$$

Hegerfeldt claims that this lemma has the following consequence for localization:

> If there exist particle states which are strictly localized in some finite region at $t = 0$ and later move towards infinity, then finite propagation speed cannot hold for localization of particles. (Hegerfeldt 1998*a*, p. 243)

Hegerfeldt's argument for this conclusion is as follows:

> Now, if the particle or system is strictly localized in Δ at $t = 0$ it is, a fortiori, also strictly localized in any larger

> region Δ' containing Δ. If the boundaries of Δ' and Δ have a finite distance and *if finite propagation speed holds* then the probability to find the system in Δ' must also be 1 for sufficiently small times, e.g. $0 \leq t < \epsilon$. But then [Lemma 1], with $A \equiv I - E_{\Delta'}$, states that the system stays in Δ' for *all* times. Now, we can make Δ' smaller and let it approach Δ. Thus we conclude that if a particle or system is at time $t = 0$ strictly localized in a region Δ, then finite propagation speed implies that it stays in Δ for all times and therefore prohibits motion to infinity. (Hegerfeldt 1998*a*, pp. 242–3; notation adapted, but italics in original)

Let us attempt now to put this argument into a more precise form.

First, Hegerfeldt claims that the following is a consequence of 'finite propagation speed': If $\Delta \subseteq \Delta'$, and if the boundaries of Δ and Δ' have a finite distance, then a state initially localized in Δ will continue to be localized in Δ' for some finite amount of time. We can capture this precisely by means of the following condition.

No instantaneous wavepacket spreading (NIWS): If $\Delta \subseteq \Delta'$, and the boundaries of Δ and Δ' have a finite distance, then there is an $\epsilon > 0$ such that $E_\Delta \leq E_{\Delta'+t}$ whenever $0 \leq t < \epsilon$.

(Note that NIWS plus localizability entails strong causality.) In the argument, Hegerfeldt also assumes that if a particle is localized in every one of a family of sets that 'approaches' Δ, then it is localized in Δ. We can capture this assumption in the following condition.

Monotonicity: If $\{\Delta_n : n \in \mathbb{N}\}$ is a downward nested family of subsets of Σ such that $\bigcap_n \Delta_n = \Delta$, then $\bigwedge_n E_{\Delta_n} = E_\Delta$.

Using this assumption, Hegerfeldt argues that if NIWS holds, and if a particle is initially localized in some finite region Δ, then it will remain in Δ for all subsequent times. In other words, if $E_\Delta\psi = \psi$, then $E_\Delta U_t\psi = U_t\psi$ for all $t \geq 0$. We can now translate this into the following rigorous no-go theorem.

Theorem (Hegerfeldt). *Suppose that the localization system $(\mathcal{H}, \Delta \mapsto E_\Delta, t \mapsto U_t)$ satisfies:*

1. *Monotonicity*
2. *Time-translation covariance*
3. *Energy bounded below*
4. *No instantaneous wavepacket spreading*

Then $U_t E_\Delta U_{-t} = E_\Delta$ *for all* $\Delta \subset \Sigma$ *and all* $t \in \mathbb{R}$.

(The proof of this and all subsequent theorems can be found in the appendix.)

Thus, conditions 1–4 can be satisfied only if the particle has trivial dynamics.

The following Lemma then shows how to derive Malament's conclusion from Hegerfeldt's theorem.

Lemma 2. *Let* M *be an affine space. Suppose that the localization system* $(\mathcal{H}, \Delta \mapsto E_\Delta, \mathbf{a} \mapsto U(\mathbf{a}))$ *satisfies localizability, time-translation covariance, and no absolute velocity. For any bounded spatial set* Δ, *if* $U(\mathbf{a})E_\Delta U(\mathbf{a})^* = E_\Delta$ *for all timelike translations* $\mathbf{a}$ *of* M, *then* $E_\Delta = 0$.

Thus, if we add "no absolute velocity" to the assumptions of Hegerfeldt's theorem, then it follows that $E_\Delta = 0$ for all bounded Δ. However, NIWS is a stronger causality assumption than microcausality. In fact, while NIWS plus localizability entails strong causality (and hence microcausality), the following example shows that NIWS is not entailed by the conjunction of strong causality, monotonicity, time-translation covariance, and energy bounded below.

Example 3. Let Q, P denote the standard position and momentum operators on $\mathcal{H} = L_2(\mathbb{R})$, and let $H = P^2/2m$ for some $m > 0$. Let $\Delta \mapsto E^Q_\Delta$ denote the spectral measure for Q. Fix some bounded subset Δ_0 of $\mathbb{R}$, and let $E_\Delta = E^Q_\Delta \otimes E^Q_{\Delta_0}$ (a projection operator on $\mathcal{H} \otimes \mathcal{H}$) for all Borel subsets Δ of $\mathbb{R}$. Thus, $\Delta \mapsto E_\Delta$ is a (nonnormalized) projection-valued measure. Let $U_t = I \otimes e^{itH}$, and let $E_{\Delta+t} = U_t E_\Delta U_{-t}$ for all $t \in \mathbb{R}$. It is clear that monotonicity, time-translation covariance, and energy bounded below hold. To see that strong causality holds, let Δ and Δ' be disjoint subsets of a single hyperplane Σ. Then,

$$E_{\Delta}U_tE_{\Delta'}U_{-t} = E^Q_{\Delta}E^Q_{\Delta'} \otimes E^Q_{\Delta_0}E^Q_{\Delta_0+t} = 0 \otimes E^Q_{\Delta}E^Q_{\Delta_0+t} = 0, \quad (2)$$

for all $t \in \mathbb{R}$. On the other hand, $U_tE_{\Delta}U_{-t} \neq E_{\Delta}$ for any nonempty Δ and for any $t \neq 0$. Thus, it follows from Hegerfeldt's theorem that NIWS fails. □

Thus, we could not recapture the full strength of Malament's theorem simply by adding 'no absolute velocity' to the conditions of Hegerfeldt's theorem.

4 Doing without 'no absolute velocity'

Example 3 shows that Hegerfeldt's theorem fails if NIWS is replaced by strong causality (or by microcausality). On the other hand, Example 3 is hardly a physically interesting counterexample to a strengthened version of Hegerfeldt's theorem. In particular, if Σ is a fixed spatial hypersurface, and if $\{\Delta_n : n \in \mathbb{N}\}$ is a covering of Σ by bounded sets (i.e., $\bigcup_n \Delta_n = \Sigma$), then $\bigvee_n E_{\Delta_n} = I \otimes E_{\Delta_0} \neq I \otimes I$. Thus, it is not certain that the particle will be detected *somewhere or other* in space. In fact, if $\{\Delta_n : n \in \mathbb{N}\}$ is a covering of Σ and $\{\Pi_n : n \in \mathbb{N}\}$ is a covering of $\Sigma + t$, then

$$\bigvee_{n\in\mathbb{N}} E_{\Delta_n} = I \otimes E_{\Delta_0} \neq I \otimes E_{\Delta_0+t} = \bigvee_{n\in\mathbb{N}} E_{\Pi_n}. \quad (3)$$

Thus, the total probability for finding the particle somewhere or other in space can change over time.

It would be completely reasonable to require that $\bigvee_n E_{\Delta_n} = I$ whenever $\{\Delta_n : n \in \mathbb{N}\}$ is a covering of Σ. This would be the case, for example, if the mapping $\Delta \mapsto E_{\Delta}$ (restricted to subsets of Σ) were the spectral measure of some position operator. However, we propose that — at the very least — any physically interesting model should satisfy the following weaker condition.

Probability conservation: If $\{\Delta_n : n \in \mathbb{N}\}$ is a covering of Σ, and $\{\Pi_n : n \in \mathbb{N}\}$ is a covering of $\Sigma + t$, then $\bigvee_n E_{\Delta_n} = \bigvee_n E_{\Pi_n}$.

Probability conservation guarantees that there is a well-defined total probability for finding the particle somewhere or other in space, and this probability remains constant over time. In particular, if both $\{\Delta_n : n \in \mathbb{N}\}$ and $\{\Pi_n : n \in \mathbb{N}\}$ consist of pairwise disjoint sets, then the localizability condition entails that $\bigvee_n E_{\Delta_n} = \sum_n E_{\Delta_n}$ and $\bigvee_n E_{\Pi_n} = \sum_n E_{\Pi_n}$. In this case, probability conservation is equivalent to

$$\sum_{n\in\mathbb{N}} \mathrm{Prob}^{\psi}(E_{\Delta_n}) = \sum_{n\in\mathbb{N}} \mathrm{Prob}^{\psi}(E_{\Pi_n}), \tag{4}$$

for any state ψ. Note, finally, that probability conservation is neutral with respect to relativistic and nonrelativistic models.[1]

Theorem 1. *Suppose that the localization system* $(\mathcal{H}, \Delta \mapsto E_\Delta, t \mapsto U_t)$ *satisfies:*

1. *Localizability*
2. *Probability conservation*
3. *Time-translation covariance*
4. *Energy bounded below*
5. *Microcausality*

Then $U_t E_\Delta U_{-t} = E_\Delta$ *for all* Δ *and all* $t \in \mathbb{R}$.

If M is an affine space, and if we add 'no absolute velocity' as a sixth condition in this theorem, then it follows that $E_\Delta = 0$ for all Δ (see Lemma 2). Thus, modulo the probability conservation condition, Theorem 1 recaptures the full strength of Malament's theorem. Moreover, we can now trace the difficulties with localization to microcausality *alone*: there are localizable particles only if it is possible to have act-outcome correlations at spacelike separation.

[1]Probability conservation would fail if a particle could escape to infinity in a finite amount of time (cf. Earman 1986, p. 33). However, a particle can escape to infinity only if there is an infinite potential well, and this would violate the energy condition. Thus, given the energy condition, probability conservation should also hold for nonrelativistic particle theories.

We now give examples to show that each condition in Theorem 1 is indispensable; that is, no four of the conditions suffices to entail the conclusion. (Example 1 shows that conditions 1–5 can be simultaneously satisfied.) Suppose for simplicity that M is two-dimensional. (All examples work in the four-dimensional case as well.) Let Q, P be the standard position and momentum operators on $L_2(\mathbb{R})$, and let $H = P^2/2m$. Let Σ be a spatial hypersurface in M, and suppose that a coordinatization of Σ has been fixed, so that there is a natural association between each bounded open subset Δ of Σ and a corresponding spectral projection E_Δ of Q.

(1+2+3+4) (a) Consider the standard localization system for a single nonrelativistic particle. That is, let Σ be a fixed spatial hyperplane, and let $\Delta \mapsto E_\Delta$ (with domain the Borel subsets of Σ) be the spectral measure for Q. For $\Sigma + t$, set $E_{\Delta+t} = U_t E_\Delta U_{-t}$, where $U_t = e^{itH}$. (b) The Newton-Wigner approach to relativistic QM uses the standard localization system for a nonrelativistic particle, only replacing the nonrelativistic Hamiltonian $P^2/2m$ with the relativistic Hamiltonian $(P^2 + m^2 I)^{1/2}$, whose spectrum is also bounded from below.

(1+2+3+5) (a) For a mathematically simple (but physically uninteresting) example, take the first example above and replace the Hamiltonian $P^2/2m$ with P. In this case, microcausality trivially holds, since $U_t E_\Delta U_{-t}$ is just a shifted spectral projection of Q. (b) For a physically interesting example, consider the relativistic quantum theory of a single spin-1/2 electron (see Section 2.1). Due to the negative energy solutions of the Dirac equation, the spectrum of the Hamiltonian is not bounded from below.

(1+2+4+5) Consider the standard localization system for a nonrelativistic particle, but set $E_{\Delta+t} = E_\Delta$ for all $t \in \mathbb{R}$. Thus, we escape the conclusion of trivial dynamics, but only by disconnecting the (nontrivial) unitary dynamics from the (trivial) association of projections with spatial regions.

(1+3+4+5) (a) Let Δ_0 be some bounded open subset of Σ, and let E_{Δ_0}

be the corresponding spectral projection of Q. When $\Delta \neq \Delta_0$, let $E_\Delta = 0$. Let $U_t = e^{itH}$, and let $E_{\Delta+t} = U_t E_\Delta U_{-t}$ for all Δ. This example is physically uninteresting, since the particle cannot be localized in any region besides Δ_0, including proper supersets of Δ_0. (b) See Example 3.

(2+3+4+5) Let Δ_0 be some bounded open subset of Σ, and let E_{Δ_0} be the corresponding spectral projection of Q. When $\Delta \neq \Delta_0$, let $E_\Delta = I$. Let $U_t = e^{itH}$, and let $E_{\Delta+t} = U_t E_\Delta U_{-t}$ for all Δ. Thus, the particle is always localized in every region other than Δ_0, and is sometimes localized in Δ_0 as well.

5 Are there unsharply localizable particles?

We have argued that attempts to undermine the four explicit assumptions of Malament's theorem are unsuccessful. We have also now shown that the 'no absolute velocity' condition is not necessary to rule out localizable particles. However, there is one further question that might arise concerning the soundness of Malament's argument. In particular, some might argue that it is possible to have a quantum-mechanical particle theory in the absence of a family $\{E_\Delta\}$ of localizing projections. What is more, one might argue that localizing projections represent an unphysical idealization — viz., that a 'particle' can be completely contained in a finite region of space with a sharp boundary, when in fact it would require an infinite amount of energy to prepare a particle in such a state. Thus, there remains a possibility that relativistic causality can be reconciled with 'unsharp' localizability.

To see how we can define 'particle talk' without having projection operators, consider again the relativistic theory of a single spin-1/2 electron (where we now restrict to the subspace $\mathcal{H}_{\text{pos}}$ of positive energy solutions of the Dirac equation). In order to treat the '$\mathbf{x}$' of the Dirac wavefunction as an observable, it would be sufficient to define a probability amplitude and density for the particle to be found at $\mathbf{x}$; and these can be obtained from the Dirac wavefunction itself. That is, for a subset Δ of Σ, we set

$$\text{Prob}^{\psi}(\mathbf{x} \in \Delta) = \int_{\Delta} |\psi(\mathbf{x})|^2 d\mathbf{x}. \tag{5}$$

Now let $\Delta \mapsto E_\Delta$ be the spectral measure for the standard position operator on the Hilbert space $\mathcal{H}$ (which includes both positive and negative energy solutions). That is, E_Δ multiplies a wavefunction by the characteristic function of Δ. Let F denote the orthogonal projection of $\mathcal{H}$ onto $\mathcal{H}_{\text{pos}}$. Then,

$$\int_{\Delta} |\psi(\mathbf{x})|^2 d\mathbf{x} = \langle \psi, E_\Delta \psi \rangle = \langle \psi, F E_\Delta \psi \rangle, \tag{6}$$

for any $\psi \in \mathcal{H}_{\text{pos}}$. Thus, we can apply the standard recipe to the operator FE_Δ (defined on $\mathcal{H}_{\text{pos}}$) to compute the probability that the particle will be found within Δ. However, FE_Δ does *not* define a projection operator on $\mathcal{H}_{\text{pos}}$. (In fact, it can be shown that FE_Δ does not have any eigenvectors with eigenvalue 1.) Thus, we do not need a family of *projection* operators in order to define probabilities for localization.

Now, in general, to define the probability that a particle will be found in Δ, we need only assume that there is an operator A_Δ such that $\langle \psi, A_\Delta \psi \rangle \in [0,1]$ for any unit vector ψ. Such operators are called *effects*, and include the projection operators as a proper subclass. Thus, we say that the triple $(\mathcal{H}, \Delta \mapsto A_\Delta, \mathbf{a} \mapsto U(\mathbf{a}))$ is an *unsharp localization system* over M just in case $\Delta \mapsto A_\Delta$ is a mapping from subsets of hyperplanes in M to effects on $\mathcal{H}$, and $\mathbf{a} \mapsto U(\mathbf{a})$ is a continuous representation of the translation group of M in unitary operators on $\mathcal{H}$. (We assume for the present that M is again an affine space.)

Most of the conditions from the previous sections can be applied, with minor changes, to unsharp localization systems. In particular, since the energy bounded below condition refers only to the unitary representation, it can be carried over intact; and translation covariance also generalizes straightforwardly. However, we will need to take more care with microcausality and with localizability.

If E and F are projection operators, $[E, F] = 0$ just in case for any state, the statistics of a measurement of F are not affected by a nonselective measurement of E and vice versa (cf. Malament 1996, p. 5). This fact, along with the assumption that E_Δ is measurable in Δ, motivates the microcausality assumption. For the case of an association of arbitrary effects with spatial regions, Busch (1999, Proposition 2) has shown that $[A_\Delta, A_{\Delta'}] = 0$ just in case for any state, the statistics for a measurement of A_Δ are not affected by a nonselective measurement of $A_{\Delta'}$ and vice versa. Thus, we may carry over the microcausality assumption intact, again seen as enforcing a prohibition against act-outcome correlations at spacelike separation.

The localizability condition is motivated by the idea that a particle cannot be simultaneously localized (with certainty) in two disjoint regions of space. In other words, if Δ and Δ' are disjoint subsets of a single hyperplane, then $\langle\psi, E_\Delta\psi\rangle = 1$ entails that $\langle\psi, E_{\Delta'}\psi\rangle = 0$. It is not difficult to see that this last condition is equivalent to the assumption that $E_\Delta + E_{\Delta'} \leq I$. That is,

$$\langle\psi, (E_\Delta + E_{\Delta'})\psi\rangle \leq \langle\psi, I\psi\rangle\,, \tag{7}$$

for any state ψ. Now, it is an accidental feature of projection operators (as opposed to arbitrary effects) that $E_\Delta + E_{\Delta'} \leq I$ is equivalent to $E_\Delta E_{\Delta'} = 0$. Thus, the appropriate generalization of localizability to unsharp localization systems is the following condition.

Localizability: If Δ and Δ' are disjoint subsets of a single hyperplane, then $A_\Delta + A_{\Delta'} \leq I$.

That is, the probability for finding the particle in Δ, plus the probability for finding the particle in some disjoint region Δ', never totals more than 1. It would, in fact, be reasonable to require a slightly stronger condition, viz., the probability of finding a particle in Δ plus the probability of finding a particle in Δ' equals the probability of finding a particle in $\Delta \cup \Delta'$. If this is true for all states ψ, we have:

Additivity: If Δ and Δ' are disjoint subsets of a single hyperplane, then $A_\Delta + A_{\Delta'} = A_{\Delta\cup\Delta'}$.

With just these mild constraints, Busch (1999) was able to derive the following no-go result.

Theorem (Busch). *Suppose that the unsharp localization system* $(\mathcal{H}, \Delta \mapsto A_\Delta, \mathbf{a} \mapsto U(\mathbf{a}))$ *satisfies localizability, translation covariance, energy bounded below, microcausality, and no absolute velocity. Then, for all* Δ, A_Δ *has no eigenvector with eigenvalue* 1.

Thus, it is not possible for a particle to be localized with certainty in any bounded region Δ. Busch's theorem, however, leaves it an open question whether there are (nontrivial) 'strongly unsharp' localization systems that satisfy microcausality. The following result shows that there are not.

Theorem 2. *Suppose that the unsharp localization system* $(\mathcal{H}, \Delta \mapsto A_\Delta, \mathbf{a} \mapsto U(\mathbf{a}))$ *satisfies:*

1. *Additivity*
2. *Translation covariance*
3. *Energy bounded below*
4. *Microcausality*
5. *No absolute velocity*

Then $A_\Delta = 0$ *for all* Δ.

Theorem 2 shows that invoking the notion of unsharp localization does nothing to resolve the tension between relativistic causality and localizability. For example, we can now conclude that the positive energy Dirac theory violates microcausality.[2]

Unfortunately, Theorem 2 does not generalize to arbitrary globally hyperbolic spacetimes, as the following example shows.

Example 4. Let M be the cylinder spacetime from Example 2. Let G denote the group of timelike translations and rotations of M, and let $g \mapsto U(g)$ be a positive energy representation of G in the unitary

[2]For any unit vector $\psi \in \mathcal{H}_{\text{pos}}$, there is a bounded set Δ such that $\int_\Delta |\psi|^2 d\mathbf{x} \neq 0$ and therefore $A_\Delta \neq 0$. On the other hand, additivity, translation covariance, energy bounded below, and no absolute velocity hold. Thus, it follows from Theorem 2 that microcausality fails.

operators on a Hilbert space $\mathcal{H}$. For any $\Sigma \in \mathcal{S}$, let μ denote the normalized rotation-invariant measure on Σ, and let $A_\Delta = \mu(\Delta)I$. Then, conditions 1–5 of Theorem 2 are satisfied, but the conclusion of the theorem is false. □

The previous counterexample can be excluded if we require there to be a fixed positive constant δ such that, for each Δ, there is a state ψ with $\langle \psi, A_\Delta \psi \rangle \geq \delta$. In fact, with this condition added, Theorem 2 holds for any globally hyperbolic spacetime. (The proof is an easy modification of the proof we give in Section 8.) However, it is not clear what physical motivation there could be for requiring this further condition. Note also that Example 4 has trivial dynamics; i.e., $U_t A_\Delta U_{-t} = A_\Delta$ for all Δ. We conjecture that every counterexample to a generalized version of Theorem 2 will have trivial dynamics.

Theorem 2 strongly supports the conclusion that there is no relativistic quantum mechanics of a single (localizable) particle; and that the only consistent combination of special relativity and quantum mechanics is in the context of quantum field theory. However, neither Theorem 1 nor Theorem 2 says anything about the ontology of relativistic quantum field theory itself; they leave open the possibility that relativistic quantum field theory might permit an ontology of localizable particles. To eliminate this latter possibility, we will now proceed to present a more general result which shows that there are no localizable particles in *any* relativistic quantum theory.

6 Are there localizable particles in RQFT?

The localizability assumption is motivated by the idea that a 'particle' cannot be detected in two disjoint spatial regions at once. However, in the case of a many-particle system, it is certainly possible for there to be particles in disjoint spatial regions. Thus, the localizability condition does not apply to many-particle systems; and Theorems 1 and 2 cannot be used to rule out a relativistic quantum mechanics of $n > 1$ localizable particles.

Still, one might argue that we could use E_Δ to represent the proposition that a measurement is certain to find that *all* n particles lie within Δ, in which case localizability should hold. Note, however, that when we alter the interpretation of the localization operators $\{E_\Delta\}$, we must alter our interpretation of the conclusion. In particular, the conclusion now shows only that it is not possible for all n particles to be localized in a bounded region of space. This leaves open the possibility that there are localizable particles, but that they are governed by some sort of 'exclusion principle' that prohibits them all from clustering in a bounded spacetime region.

Furthermore, Theorems 1 and 2 only show that it is impossible to define *position operators* that obey appropriate relativistic constraints. But it does not immediately follow from this that we lack any notion of localization in relativistic quantum theories. Indeed,

> ... a position operator is inconsistent with relativity. This compels us to find another way of modeling localization of events. In field theory, we model localization by making the observables dependent on position in spacetime. (Ticiatti 1999, p. 11)

However, it is not a peculiar feature of *relativistic* quantum field theory that it lacks a position operator: Any quantum field theory (either relativistic or nonrelativistic) will model localization by making the observables dependent on position in spacetime. Moreover, in the case of nonrelativistic QFT, these 'localized' observables suffice to provide us with a concept of localizable particles. In particular, for each spatial region Δ, there is a 'number operator' N_Δ whose eigenvalues give the number of particles within the region Δ. Thus, we have no difficultly in talking about the particle content in a given region of space despite the absence of any position operator.

Abstractly, a number operator N on $\mathcal{H}$ is any operator with eigenvalues contained in $\{0, 1, 2, \dots\}$. In order to describe the number of particles locally, we require an association $\Delta \mapsto N_\Delta$ of subsets of spatial hyperplanes in M to number operators on $\mathcal{H}$, where N_Δ represents the number of particles in the spatial region Δ. If $\mathbf{a} \mapsto U(\mathbf{a})$ is a unitary representation of the translation group, we say that the triple $(\mathcal{H}, \Delta \mapsto N_\Delta, \mathbf{a} \mapsto U(\mathbf{a}))$ is a *system of local number operators* over M.

Note that a localization system $(\mathcal{H}, \Delta \mapsto E_\Delta, \mathbf{a} \mapsto U(\mathbf{a}))$ is a special case of a system of local number operators where the eigenvalues of each N_Δ are restricted to $\{0, 1\}$. Furthermore, if we loosen our assumption that number operators have a discrete spectrum, and instead require only that they have a spectrum contained in $[0, \infty)$, then we can also include unsharp localization systems within the general category of systems of local number operators. Thus, a system of local number operators is the *minimal* requirement for a concept of localizable particles in any quantum theory.

In addition to the natural analogues of the energy bounded below condition, translation covariance, and microcausality, we will be interested in the following two requirements on a system of local number operators:[3]

Additivity: If Δ and Δ' are disjoint subsets of a single hyperplane, then $N_\Delta + N_{\Delta'} = N_{\Delta \cup \Delta'}$.

Number conservation: If $\{\Delta_n : n \in \mathbb{N}\}$ is a disjoint covering of Σ, then the sum $\sum_n N_{\Delta_n}$ converges to a densely defined, self-adjoint operator N on $\mathcal{H}$ (independent of the chosen covering), and $U(\mathbf{a})NU(\mathbf{a})^* = N$ for any timelike translation $\mathbf{a}$ of M.

Additivity asserts that, when Δ and Δ' are disjoint, the expectation value (in any state ψ) for the number of particles in $\Delta \cup \Delta'$ is the sum of the expectations for the number of particles in Δ and the number of particles in Δ'. In the pure case, it asserts that the number of particles in $\Delta \cup \Delta'$ is the sum of the number of particles in Δ and the number of particles in Δ'. The 'number conservation' condition tells us that there is a well-defined total number of particles (at a given

[3]Due to the unboundedness of number operators, we would need to take some care in giving technically correct versions of the following conditions. In particular, the additivity condition should technically include the clause that N_Δ and $N_{\Delta'}$ have a common dense domain, and the operator $N_{\Delta \cup \Delta'}$ should be thought of as the self-adjoint closure of $N_\Delta + N_{\Delta'}$. In the number conservation condition, the sum $N = \sum_n N_{\Delta_n}$ can be made rigorous by exploiting the correspondence between self-adjoint operators and 'quadratic forms' on $\mathcal{H}$. In particular, we can think of N as deriving from the upper bound of quadratic forms corresponding to finite sums.

time), and that the total number of particles does not change over time. This condition holds for any noninteracting model of QFT.

It is a well-known consequence of the Reeh-Schlieder theorem that relativistic quantum field theories do not admit systems of local number operators (cf. Redhead 1995). We will now derive the same conclusion from strictly weaker assumptions. In particular, we show that microcausality is the *only* specifically relativistic assumption needed for this result. The relativistic spectrum condition — which requires that the spectrum of the four-momentum lie in the forward light cone, and which is used in the proof of the Reeh-Schlieder theorem — plays no role in our proof.[4]

Theorem 3. *Suppose that the system* $(\mathcal{H}, \Delta \mapsto N_\Delta, \mathbf{a} \mapsto U(\mathbf{a}))$ *of local number operators satisfies:*

1. *Additivity*
2. *Translation covariance*
3. *Energy bounded below*
4. *Number conservation*
5. *Microcausality*
6. *No absolute velocity*

Then $N_\Delta = 0$ *for all* Δ.

Thus, in every state, there are no particles in any local region. This serves as a *reductio ad absurdum* for any notion of localizable particles in a relativistic quantum theory.

Unfortunately, Theorem 3 is not the strongest result we could hope for, since 'number conservation' can only be expected to hold in the (trivial) case of noninteracting fields. However, we would

[4]Microcausality is not only sufficient, but also necessary for the proof that there are no local number operators. The Reeh-Schlieder theorem entails the cyclicity of the vacuum state. But the cyclicity of the vacuum state alone does not entail that there are no local number operators; we must also assume microcausality (cf. Requardt 1982; Halvorson 2001).

need a more general approach in order to deal with interacting relativistic quantum fields, because (due to Haag's theorem; cf. Streater & Wightman 2000, p. 163) their dynamics are not unitarily implementable on a fixed Hilbert space. On the other hand it would be wrong to think of this as indicating a limitation on the generality of our conclusion: Haag's theorem also entails that interacting models of RQFT have no number operators — either global or local.[5] Still, it would be interesting to recover this conclusion (perhaps working in a more general algebraic setting) without using the full strength of Haag's assumptions.

7 Particle talk without particle ontology

The results of the previous sections show that, insofar as we can expect any relativistic quantum theory to satisfy a few basic conditions, these theories do not admit (localizable) particles into their ontology. We also considered and rejected several arguments which attempt to show that one (or more) of these conditions can be jettisoned without doing violence to the theory of relativity or to quantum mechanics. Thus, we have yet to find a good reason to reject one of the premises on which our argument against localizable particles is based. However, Segal (1964) and Barrett (2002) claim that we have independent grounds for rejecting the conclusion; that is, we have good reasons for believing that there *are* localizable particles.

The argument for localizable particles appears to be very simple: Our experience shows us that objects (particles) occupy finite regions of space. But the reply to this argument is just as simple: These experiences are illusory! Although no object is strictly localized in a bounded region of space, an object can be well-enough localized to give the appearance to us (finite observers) that it is

[5]If a total number operator exists in a representation of the canonical commutation relations, then that representation is quasi-equivalent to a free-field (Fock) representation (Chaiken 1968). However, Haag's theorem entails that in relativistic theories, representations with nontrivial interactions are *not* quasi-equivalent to a free-field representation.

strictly localized. In fact, relativistic quantum field theory *itself* shows how the 'illusion' of localizable particles can arise, and how talk about localizable particles can be a useful fiction.

In order to assess the possibility of 'approximately localized' objects in relativistic quantum field theory, we shall now pursue the investigation in the framework of algebraic quantum field theory.[6] Here, one assumes that there is a correspondence $O \mapsto \mathcal{R}(O)$ between bounded open subsets of M and subalgebras of observables on some Hilbert space $\mathcal{H}$. Observables in $\mathcal{R}(O)$ are considered to be 'localized' (i.e., measurable) in O. Thus, if O and O' are spacelike separated, we require that $[A, B] = 0$ for any $A \in \mathcal{R}(O)$ and $B \in \mathcal{R}(O')$. Furthermore, we assume that there is a continuous representation $\mathbf{a} \mapsto U(\mathbf{a})$ of the translation group of M in unitary operators on $\mathcal{H}$, and that there is a unique 'vacuum' state $\Omega \in \mathcal{H}$ such that $U(\mathbf{a})\Omega = \Omega$ for all $\mathbf{a}$. This latter condition entails that the vacuum appears the same to all observers, and that it is the unique state of lowest energy.

In this context, a particle detector can be represented by an effect C such that $\langle \Omega, C\Omega \rangle = 0$. That is, C should register no particles in the vacuum state. However, the Reeh-Schlieder theorem entails that no positive local observable can have zero expectation value in the vacuum state. Thus, we again see that (strictly speaking) it is impossible to detect particles by means of local measurements; instead, we will have to think of particle detections as 'approximately local' measurements.

If we think of an observable as representing a measurement procedure (or, more precisely, an equivalence class of measurement procedures), then the norm distance $\|C - C'\|$ between two observables gives a quantitative measure of the physical similarity between the corresponding procedures. (In particular, if $\|C - C'\| < \delta$, then the expectation values of C and C' never differ by more than δ.)[7] Moreover, in the case of real-world measurements, the existence

[6]For general information on algebraic quantum field theory, see Haag (1992) and Buchholz (2000). For specific information on particle detectors and 'almost local' observables, see Chapter 6 of Haag (1992) and Section 4 of Buchholz (2000).

[7]Recall that $\|C - C'\| = \sup\{\|(C - C')\psi\| : \psi \in \mathcal{H}, \|\psi\| = 1\}$. It follows then from the Cauchy-Schwartz inequality that $|\langle \psi, (C - C')\psi \rangle| \leq \|C - C'\|$ for any unit

of measurement errors and environmental noise make it impossible for us to determine precisely which measurement procedure we have performed. Thus, practically speaking, we can at best determine a neighborhood of observables corresponding to a concrete measurement procedure.

In the case of present interest, what we actually measure is always a local observable — i.e., an element of $\mathcal{R}(O)$, where O is bounded. However, given a fixed error bound δ, if an observable C is within norm distance δ from some local observable $C' \in \mathcal{R}(O)$, then a measurement of C' will be practically indistinguishable from a measurement of C. Thus, if we let

$$\mathcal{R}_\delta(O) = \{C : \exists C' \in \mathcal{R}(O) \text{ such that } \|C - C'\| < \delta\}, \qquad (8)$$

denote the family of observables 'almost localized' in O, then 'FAPP' (i.e., 'for all practical purposes') we can locally measure any observable from $\mathcal{R}_\delta(O)$. That is, measurement of an element from $\mathcal{R}_\delta(O)$ can be simulated to a high degree of accuracy by local measurement of an element from $\mathcal{R}(O)$. However, for any local region O, and for any $\delta > 0$, $\mathcal{R}_\delta(O)$ *does* contain (nontrivial) effects that annihilate the vacuum.[8] Thus, particle detections can always be simulated by purely local measurements; and the appearance of (fairly well) localized objects can be explained without the supposition that there are localizable particles in the strict sense.

However, it may not be easy to pacify Segal and Barrett with a FAPP solution to the problem of localization. Both appear to think that the absence of localizable particles (in the strict sense) is not simply contrary to our manifest experience, but would undermine the very possibility of objective empirical science. For example, Segal claims that,

> ... it is an elementary fact, *without which experimentation of the usual sort would not be possible*, that particles are

vector ψ.

[8]Suppose that $A \in \mathcal{R}(O)$, and let $A(\mathbf{x}) = U(\mathbf{x})AU(\mathbf{x})^*$. If f is a test function on M whose Fourier transform is supported in the complement of the forward light cone, then $L = \int f(\mathbf{x})A(\mathbf{x})d\mathbf{x}$ is almost localized in O and $\langle\Omega, L\Omega\rangle = 0$ (cf. Buchholz 2000, p. 7).

> indeed localized in space at a given time. (Segal 1964, p. 145; our italics)

Furthermore, 'particles would not be observable without their localization in space at a particular time' (1964, p. 139). In other words, experimentation involves observations of particles, and these observations can occur only if particles are localized in space. Unfortunately, Segal does not give any argument for these claims. It seems to us, however, that the moral we should draw from the no-go theorems is that Segal's account of observation is false. In particular, it is not (strictly speaking) true that we observe particles. Rather, there are 'observation events', and these observation events are consistent (to a good degree of accuracy) with the supposition that they are brought about by localizable particles.

Like Segal, Barrett (2002) claims that we will have trouble explaining how empirical science can work if there are no localizable particles. In particular, Barrett claims that empirical science requires that we be able to keep an account of our measurement results so that we can compare these results with the predictions of our theories. Furthermore, we identify measurement records by means of their location in space. Thus, if there were no localized objects, then there would be no identifiable measurement records, and '... it would be difficult to account for the possibility of empirical science at all' (Barrett 2002, p. 3).

However, it is not clear what the difficulty here is supposed to be. On the one hand, we have seen that relativistic quantum field theory does predict that the appearances will be FAPP consistent with the supposition that there are localized objects. So, for example, we could distinguish two record tokens at a given time if there were two disjoint regions O and O' and particle detector observables $C \in \mathcal{R}_\delta(O)$ and $C' \in \mathcal{R}_\delta(O')$ (approximated by observables *strictly* localized in O and O respectively) such that $\langle \psi, C\psi \rangle \approx 1$ and $\langle \psi, C'\psi \rangle \approx 1$. Now, it may be that Barrett is also worried about how, given a field ontology, we could assign any sort of trans-temporal identity to our record tokens. But this problem, however important philosophically, is distinct from the problem of localization. Indeed, it also arises in the context of nonrelativistic quantum field theory, where there is *no* problem with describing localizable particles. Finally, Barrett might object that once we supply a quantum-theoretical model of a particle detector itself, then the superposition principle will prevent the field and detector from getting into a state where there is a fact of the matter as to whether, 'a particle has been detected in the region O'. But this is simply a restatement of the standard quantum measurement problem that infects *all* quantum theories — and we have made no pretense of solving that here.

8 Conclusion

Malament claims that his theorem justifies the belief that,

> ... in the attempt to reconcile quantum mechanics with relativity theory ... one is driven to a field theory; all talk about 'particles' has to be understood, at least in principle, as talk about the properties of, and interactions among, quantized fields. (Malament 1996, p. 1)

In order to buttress Malament's argument for this claim, we provided two further results (Theorems 1 and 2) which show that the conclusion continues to hold for generic spacetimes, as well as for unsharp localization observables. We then went on to show that RQFT does not permit an ontology of localizable particles; and so, strictly speaking, our talk about localizable particles is a fiction. Nonetheless, RQFT does permit *talk* about particles — albeit, if we understand this talk as really being about the properties of, and interactions among, quantized fields. Indeed, modulo the standard quantum measurement problem, RQFT has no trouble explaining the appearance of macroscopically well-localized objects, and shows that our talk of particles, though a *façon de parler*, has a legitimate role to play in empirically testing the theory.

Appendix: Proofs of theorems

Theorem (Hegerfeldt). *Suppose that the localization system* $(\mathcal{H}, \Delta \mapsto E_\Delta, t \mapsto U_t)$ *satisfies monotonicity, time-translation covariance, energy* *bounded below, and NIWS. Then $U_t E_\Delta U_{-t} = E_\Delta$ for all $\Delta \subset \Sigma$ and all $t \in \mathbb{R}$.*

Proof. The formal proof corresponds directly to Hegerfeldt's informal proof. Thus, let Δ be a subset of some spatial hypersurface Σ. If $E_\Delta = 0$ then obviously $U_t E_\Delta U_{-t} = E_\Delta$ for all $t \in \mathbb{R}$. So, suppose that $E_\Delta \neq 0$, and let ψ be a unit vector such that $E_\Delta \psi = \psi$. Since Σ is a manifold, and since $\Delta \neq \Sigma$, there is a family $\{\Delta_n : n \in \mathbb{N}\}$ of subsets of Σ such that, for each $n \in \mathbb{N}$, the distance between the boundaries of Δ_n and Δ is nonzero, and such that $\bigcap_n \Delta_n = \Delta$. Fix $n \in \mathbb{N}$. By NIWS and time-translation covariance, there is an $\epsilon_n > 0$ such that $E_{\Delta_n} U_t \psi = U_t \psi$ whenever $0 \leq t < \epsilon_n$. That is, $\langle U_t \psi, E_{\Delta_n} U_t \psi \rangle = 1$ whenever $0 \leq t < \epsilon_n$. Since energy is bounded from below, we may apply Lemma 1 with $A = I - E_{\Delta_n}$ to conclude that $\langle U_t \psi, E_{\Delta_n} U_t \psi \rangle = 1$ for all $t \in \mathbb{R}$. That is, $E_{\Delta_n} U_t \psi = U_t \psi$ for all $t \in \mathbb{R}$. Since this holds for all $n \in \mathbb{N}$, and since (by monotonicity) $E_\Delta = \bigwedge_n E_{\Delta_n}$, it follows that $E_\Delta U_t \psi = U_t \psi$ for all $t \in \mathbb{R}$. Thus, $U_t E_\Delta U_{-t} = E_\Delta$ for all $t \in \mathbb{R}$. □

Lemma 2. *Suppose that the localization system $(\mathcal{H}, \Delta \mapsto E_\Delta, \mathbf{a} \mapsto U(\mathbf{a}))$ satisfies localizability, time-translation covariance, and no absolute velocity. Let Δ be a bounded spatial set. If $U(\mathbf{a}) E_\Delta U(\mathbf{a})^* = E_\Delta$ for all timelike translations $\mathbf{a}$ of M, then $E_\Delta = 0$.*

Proof. By no absolute velocity, there is a pair $(\mathbf{a}, \mathbf{b})$ of timelike translations such that $\Delta + (\mathbf{a} - \mathbf{b})$ is in Σ and is disjoint from Δ. By time-translation covariance, we have,

$$E_{\Delta + (\mathbf{a} - \mathbf{b})} = U(\mathbf{a}) U(\mathbf{b})^* E_\Delta U(\mathbf{b}) U(\mathbf{a})^* = E_\Delta. \tag{9}$$

Thus, localizability entails that E_Δ is orthogonal to itself, and so $E_\Delta = 0$. □

Lemma 3. *Let $\{\Delta_n : n = 0, 1, 2, \dots\}$ be a covering of Σ, and let $E = \bigvee_{n=0}^{\infty} E_{\Delta_n}$. If probability conservation and time-translation covariance hold, then $U_t E U_{-t} = E$ for all $t \in \mathbb{R}$.*

Proof. Since $\{\Delta_n + t : n \in \mathbb{N}\}$ is a covering of $\Sigma + t$, probability conservation entails that $\bigvee_n E_{\Delta_n + t} = E$. Thus,

$$U_t E U_{-t} = U_t \left[\bigvee_{n=0}^{\infty} E_{\Delta_n} \right] U_{-t} = \bigvee_{n=0}^{\infty} \left[U_t E_{\Delta_n} U_{-t} \right] \tag{10}$$

$$= \bigvee_{n=0}^{\infty} E_{\Delta_n + t} = E, \tag{11}$$

where the third equality follows from time-translation covariance. □

In order to prove the next result, we will need to invoke the following lemma from Borchers (1967).

Lemma (Borchers). *Let $U_t = e^{itH}$, where H is a self-adjoint operator with spectrum bounded from below. Let E and F be projection operators such that $EF = 0$. If there is an $\epsilon > 0$ such that*

$$[E, U_t F U_{-t}] = 0, \qquad 0 \leq t < \epsilon,$$

then $E U_t F U_{-t} = 0$ for all $t \in \mathbb{R}$.

Lemma 4. *Let $U_t = e^{itH}$, where H is a self-adjoint operator with spectrum bounded from below. Let $\{E_n : n = 0, 1, 2, \ldots\}$ be a family of projection operators such that $E_0 E_n = 0$ for all $n \geq 1$, and let $E = \bigvee_{n=0}^{\infty} E_n$. If $U_t E U_{-t} = E$ for all $t \in \mathbb{R}$, and if for each $n \geq 1$ there is an $\epsilon_n > 0$ such that*

$$[E_0, U_t E_n U_{-t}] = 0, \qquad 0 \leq t < \epsilon_n, \tag{12}$$

then $U_t E_0 U_{-t} = E_0$ for all $t \in \mathbb{R}$.

Proof. If $E_0 = 0$ then the conclusion obviously holds. Suppose then that $E_0 \neq 0$, and let ψ be a unit vector in the range of E_0. Fix $n \geq 1$. Using (12) and Borchers' lemma, it follows that $E_0 U_t E_n U_{-t} = 0$ for all $t \in \mathbb{R}$. Then,

$$\|E_n U_{-t} \psi\|^2$$

$$= \langle U_{-t}\psi, E_n U_{-t}\psi \rangle = \langle \psi, U_t E_n U_{-t}\psi \rangle \tag{13}$$

$$= \langle E_0\psi, U_t E_n U_{-t}\psi \rangle = \langle \psi, E_0 U_t E_n U_{-t}\psi \rangle = 0, \tag{14}$$

for all $t \in \mathbb{R}$. Thus, $E_n U_{-t}\psi = 0$ for all $n \geq 1$, and consequently, $[\bigvee_{n\geq 1} E_n]U_{-t}\psi = 0$. Since $E_0 = E - [\bigvee_{n\geq 1} E_n]$, and since (by assumption) $EU_{-t} = U_{-t}E$, it follows that

$$E_0 U_{-t}\psi = EU_{-t}\psi = U_{-t}E\psi = U_{-t}\psi\,, \tag{15}$$

for all $t \in \mathbb{R}$. □

Theorem 1. *Suppose that the localization system $(\mathcal{H}, \Delta \mapsto E_\Delta, t \mapsto U_t)$ satisfies localizability, probability conservation, time-translation covariance, energy bounded below, and microcausality. Then $U_t E_\Delta U_{-t} = E_\Delta$ for all Δ and all $t \in \mathbb{R}$.*

Proof. Let Δ be an open subset of Σ. If $\Delta = \Sigma$ then probability conservation and time-translation covariance entail that $E_\Delta = E_{\Delta+t} = U_t E_\Delta U_{-t}$ for all $t \in \mathbb{R}$. If $\Delta \neq \Sigma$ then, since Σ is a manifold, there is a covering $\{\Delta_n : n \in \mathbb{N}\}$ of $\Sigma\backslash\Delta$ such that the distance between Δ_n and Δ is nonzero for all n. Let $E_0 = E_\Delta$, and let $E_n = E_{\Delta_n}$ for $n \geq 1$. Then localizability entails that $E_0 E_n = 0$ when $n \geq 1$. If we let $E = \bigvee_{n=0}^{\infty} E_n$ then probability conservation entails that $U_t E U_{-t} = E$ for all $t \in \mathbb{R}$ (see Lemma 3). By time-translation covariance and microcausality, for each $n \geq 1$ there is an $\epsilon_n > 0$ such that

$$[E_0, U_t E_n U_{-t}] = 0, \qquad 0 \leq t < \epsilon_n. \tag{16}$$

Since the energy is bounded from below, Lemma 4 entails that $U_t E_0 U_{-t} = E_0$ for all $t \in \mathbb{R}$. That is, $U_t E_\Delta U_{-t} = E_\Delta$ for all $t \in \mathbb{R}$. □

Theorem 2. *Suppose that the unsharp localization system $(\mathcal{H}, \Delta \mapsto A_\Delta, \mathbf{a} \mapsto U(\mathbf{a}))$ satisfies additivity, translation covariance, energy bounded below, microcausality, and no absolute velocity. Then $A_\Delta = 0$ for all Δ.*

Proof. We prove by induction that $\|A_\Delta\| \leq (2/3)^m$, for each $m \in \mathbb{N}$, and for each bounded Δ. For this, let F_Δ denote the spectral measure for A_Δ.

(Base case: $m = 1$) Let $E_\Delta = F_\Delta(2/3, 1)$. We verify that $(\mathcal{H}, \Delta \mapsto E_\Delta, \mathbf{a} \mapsto U(\mathbf{a}))$ satisfies the conditions of Malament's theorem.

Clearly, no absolute velocity and energy bounded below hold. Moreover, since unitary transformations preserve spectral decompositions, translation covariance holds; and since spectral projections of compatible operators are also compatible, microcausality holds. To see that localizability holds, let Δ and Δ' be disjoint bounded subsets of a single hyperplane. Then microcausality entails that $[A_\Delta, A_{\Delta'}] = 0$, and therefore $E_\Delta E_{\Delta'}$ is a projection operator. Suppose for *reductio ad absurdum* that ψ is a unit vector in the range of $E_\Delta E_{\Delta'}$. By additivity, $A_{\Delta\cup\Delta'} = A_\Delta + A_{\Delta'}$, and we therefore obtain the contradiction:

$$1 \geq \langle \psi, A_{\Delta\cup\Delta'}\psi \rangle = \langle \psi, A_\Delta \psi \rangle + \langle \psi, A_{\Delta'}\psi \rangle \geq 2/3 + 2/3 . \quad (17)$$

Thus, $E_\Delta E_{\Delta'} = 0$, and Malament's theorem entails that $E_\Delta = 0$ for all Δ. Therefore, $A_\Delta = A_\Delta F_\Delta(0, 2/3)$ has spectrum lying in $[0, 2/3]$, and $\|A_\Delta\| \leq 2/3$ for all bounded Δ.

(Inductive step) Suppose that $\|A_\Delta\| \leq (2/3)^{m-1}$ for all bounded Δ. Let $E_\Delta = F_\Delta((2/3)^m, (2/3)^{m-1})$. In order to see that Malament's theorem applies to $(\mathcal{H}, \Delta \mapsto E_\Lambda, \mathbf{a} \mapsto U(\mathbf{a}))$, we need only check that localizability holds. For this, suppose that Δ and Δ' are disjoint subsets of a single hyperplane. By microcausality, $[A_\Delta, A_{\Delta'}] = 0$, and therefore $E_\Delta E_{\Delta'}$ is a projection operator. Suppose for *reductio ad absurdum* that ψ is a unit vector in the range of $E_\Delta E_{\Delta'}$. Since $\Delta \cup \Delta'$ is bounded, the induction hypothesis entails that $\|A_{\Delta\cup\Delta'}\| \leq (2/3)^{m-1}$. By additivity, $A_{\Delta\cup\Delta'} = A_\Delta + A_{\Delta'}$, and therefore we obtain the contradiction:

$$\begin{aligned} (2/3)^{m-1} &\geq \langle \psi, A_{\Delta\cup\Delta'}\psi \rangle \\ &= \langle \psi, A_\Delta \psi \rangle + \langle \psi, A_{\Delta'}\psi \rangle \geq (2/3)^m + (2/3)^m . \end{aligned} \quad (18)$$

Thus, $E_\Delta E_{\Delta'} = 0$, and Malament's theorem entails that $E_\Delta = 0$ for all Δ. Therefore, $\|A_\Delta\| \leq (2/3)^m$ for all bounded Δ. □

Theorem 3. *Suppose that the system* $(\mathcal{H}, \Delta \mapsto N_\Delta, \mathbf{a} \mapsto U(\mathbf{a}))$ *of local number operators satisfies additivity, translation covariance, energy bounded below, number conservation, microcausality, and no absolute velocity. Then,* $N_\Delta = 0$ *for all bounded* Δ.

Proof. Let N be the unique total number operator obtained from taking the sum $\sum_n N_{\Delta_n}$ where $\{\Delta_n : n \in \mathbb{N}\}$ is a disjoint covering of Σ. Note that for any $\Delta \subseteq \Sigma$, we can choose a covering containing Δ, and hence, $N = N_\Delta + A$, where A is a positive operator. By microcausality, $[N_\Delta, A] = 0$, and therefore $[N_\Delta, N] = [N_\Delta, N_\Delta + A] = 0$. Furthermore, for any vector ψ in the domain of N, $\langle \psi, N_\Delta \psi \rangle \leq \langle \psi, N\psi \rangle$.

Let E be the spectral measure for N, and let $E_n = E(0, n)$. Then, NE_n is a bounded operator with norm at most n. Since $[E_n, N_\Delta] = 0$, it follows that

$$\langle \psi, N_\Delta E_n \psi \rangle = \langle E_n \psi, N_\Delta E_n \psi \rangle \leq \langle E_n \psi, NE_n \psi \rangle \leq n\,, \tag{19}$$

for any unit vector ψ. Thus, $\|N_\Delta E_n\| \leq n$. Since $\bigcup_{n=1}^{\infty} E_n(\mathcal{H})$ is dense in $\mathcal{H}$, and since $E_n(\mathcal{H})$ is in the domain of N_Δ (for all n), it follows that if $N_\Delta E_n = 0$, for all n, then $N_\Delta = 0$. We now concentrate on proving the antecedent.

For each Δ, let $A_\Delta = (1/n) N_\Delta E_n$. We show that the structure $(\mathcal{H}, \Delta \mapsto A_\Delta, \mathbf{a} \mapsto U(\mathbf{a}))$ satisfies the conditions of Theorem 2. Clearly, energy bounded below and no absolute velocity hold. It is also straightforward to verify that additivity and microcausality hold. To check translation covariance, we compute:

$$\begin{aligned} U(\mathbf{a}) A_\Delta U(\mathbf{a})^* &= U(\mathbf{a}) N_\Delta E_n U(\mathbf{a})^* && (20) \\ &= U(\mathbf{a}) N_\Delta U(\mathbf{a})^* U(\mathbf{a}) E_n U(\mathbf{a})^* && (21) \\ &= U(\mathbf{a}) N_\Delta U(\mathbf{a})^* E_n \;=\; N_{\Delta+\mathbf{a}} E_n \;=\; A_{\Delta+\mathbf{a}}. && (22) \end{aligned}$$

The third equality follows from number conservation, and the fourth equality follows from translation covariance. Thus, $N_\Delta E_n = A_\Delta = 0$ for all Δ. Since this holds for all $n \in \mathbb{N}$, $N_\Delta = 0$ for all Δ. □

Bibliography

Barrett, J. A. (2002), On the nature of measurement records in relativistic quantum field theory, in M. Kuhlmann, H. Lyre & A. Wayne, eds, 'Ontological aspects of quantum field theory', World Scientific, River Edge, NJ, pp. 165–79.

Borchers, H.-J. (1967), 'A remark on a theorem of B. Misra', *Communications in Mathematical Physics* **4**, 315–23.

Buchholz, D. (2000), 'Algebraic quantum field theory: A status report'. E-print: math-ph/0011044.

Busch, P. (1999), 'Unsharp localization and causality in relativistic quantum theory', *Journal of Physics A* **32**, 6535–46.

Chaiken, J. M. (1968), 'Number operators for representations of the canonical commutation relations', *Communications in Mathematical Physics* **8**, 164–84.

Cushing, J. T. (1996), What measurement problem?, in R. Clifton, ed., 'Perspectives on quantum reality', Kluwer, Dordrecht, pp. 167–81.

Dickson, W. M. (1998), *Quantum Chance and Nonlocality*, Cambridge University Press, New York.

Earman, J. S. (1986), *A Primer on Determinism*, D. Reidel, Boston.

Fleming, G. & Butterfield, J. (1999), Strange positions, in J. Butterfield & C. Pagonis, eds, 'From Physics to Philosophy', Cambridge University Press, New York.

Haag, R. (1992), *Local Quantum Physics*, Springer, New York.

Halvorson, H. P. (2001), 'Reeh-Schlieder defeats Newton-Wigner: On alternative localization schemes in relativistic quantum field theory', *Philosophy of Science* **68**, 111–33.

Hegerfeldt, G. C. (1998*a*), Causality, particle localization and positivity of the energy, in A. Böhm, H.-D. Doebner & P. Kielanowski, eds, 'Irreversibility and Causality: Semigroups and Rigged Hilbert Spaces', Springer, New York, pp. 238–45.

Hegerfeldt, G. C. (1998*b*), 'Instantaneous spreading and Einstein causality in quantum theory', *Annalen der Physik* **7**, 716–25.

Holland, P. R. (1993), *The Quantum Theory of Motion*, Cambridge University Press, New York.

Malament, D. B. (1996), In defense of dogma: Why there cannot be a relativistic quantum mechanics of (localizable) particles, in R. Clifton, ed., 'Perspectives on Quantum Reality', Kluwer, Dordrecht, pp. 1–10.

Maudlin, T. (1994), *Quantum Non-Locality and Relativity*, Blackwell, Cambridge.

Redhead, M. L. G. (1995), The vacuum in relativistic quantum field theory, in D. Hull, M. Forbes & R. M. Burian, eds, 'PSA; proceedings of the biennial meeting of the Philosophy of Science Association', Vol. 2, Philosophy of Science Association, East Lansing, MI, pp. 77–87.

Requardt, M. (1982), 'Spectrum condition, analyticity, Reeh-Schlieder and cluster properties in non-relativistic Galilei-invariant quantum theory', *Journal of Physics A* **15**, 3715–23.

Schlieder, S. (1971), Zum kausalen Verhalten eines relativistischen quantenmechanischen System, in S. P. Dürr, ed., 'Quanten und Felder', Vieweg, Braunschweig, pp. 145–60.

Segal, I. E. (1964), Quantum fields and analysis in the solution manifolds of differential equations, in W. T. Martin & I. E. Segal, eds, 'Proceedings of a Conference on the Theory and Applications of Analysis in Function Space', MIT Press, Cambridge, MA, pp. 129–53.

Streater, R. F. & Wightman, A. S. (2000), *PCT, Spin and Statistics, and All That*, 3rd ed., Princeton University Press, Princeton, NJ.

Thaller, B. (1992), *The Dirac Equation*, Springer, New York.

Ticiatti, R. (1999), *Quantum Field Theory for Mathematicians*, Cambridge University Press, New York.

Chapter 9

Are Rindler quanta real? Inequivalent particle concepts in quantum field theory

with Hans Halvorson

> *Sagredo: Do we not see here another example of that all-pervading principle of complementarity which excludes the simultaneous applicability of concepts to the real objects of our world? Is it not so that, rather than being frustrated by this limitation of our conceptual grasp of the reality, we see in this unification of opposites the deepest and most satisfactory result of the dialectical process in our struggle for understanding?*
>
> — Josef Jauch, *Are Quanta Real? A Galilean Dialogue* (1973)

We are extremely grateful to John Earman and Laura Ruetsche for many stimulating discussions which provided the impetus for writing this paper, and to Aristidis Arageorgis for writing a provocative and inspiring dissertation. We would also like to thank Klaas Landsman and Rainer Verch for help with the proof of Proposition 6.

1 Introduction

Philosophical reflection on quantum field theory has tended to focus on how it revises our conception of what a particle is. For instance, though there is a self-adjoint operator in the theory representing the total number of particles of a field, the standard 'Fock space' formalism does not individuate particles from one another. Thus, Teller (1995, Chapter 2) suggests that we speak of *quanta* that can be 'aggregated', instead of (enumerable) *particles* — which implies that they can be distinguished and labeled. Moreover, because the theory *does* contain a total number of quanta observable (which, therefore, has eigenstates corresponding to different values of this number), a field state can be a nontrivial superposition of number eigenstates that fails to predict any particular number of quanta with certainty. Teller (1995, p. 105–6) counsels that we think of these superpositions as not actually containing any quanta, but only propensities to *display* various numbers of quanta when the field interacts with a 'particle detector'.

The particle concept seems so thoroughly denuded by quantum field theory that is hard to see how it could possibly underwrite the particulate nature of laboratory experience. Those for whom fields are the fundamental objects of the theory are especially aware of this explanatory burden:

> ... quantum field theory is the quantum theory of a field, not a theory of 'particles'. However, when we consider the manner in which a quantum field interacts with other systems to which it is coupled, an interpretation of the states in [Fock space] in terms of 'particles' naturally arises. It is, of course, essential that this be the case if quantum field theory is to describe observed phenomena, since 'particle-like' behavior is commonly observed. (Wald 1994, pp. 46–7)

These remarks occur in the context of Wald's discussion of yet another threat to the 'reality' of quanta.

The threat arises from the possibility of inequivalent representations of the algebra of observables of a field in terms of operators on a Hilbert space. Inequivalent representations are required in a variety of situations; for example, interacting field theories in which the scattering matrix does not exist ('Haag's theorem'), free fields whose dynamics cannot be unitarily implemented (Arageorgis *et al.* 2002), and states in quantum statistical mechanics corresponding to different temperatures (Emch 1972). The catch is that each representation carries with it a distinct notion of 'particle'. Our main goal in this chapter is to clarify the subtle relationship between inequivalent representations of a field theory and their associated particle concepts.

Most of our discussion shall apply to any case in which inequivalent representations of a field are available. However, we have a particular interest in the case of the Minkowski versus Rindler representations of a free Boson field. What makes this case intriguing is that it involves two radically different descriptions of the particle content of the field in the *very same* spacetime region. The questions we aim to answer are:

- Are the Minkowski and Rindler descriptions nevertheless, in some sense, *physically* equivalent?
- Or, are they incompatible, even theoretically *incommensurable*?
- Can they be thought of as *complementary* descriptions in the same way that the concepts of position and momentum are?
- Or, can at most one description, the 'inertial' story in terms of Minkowski quanta, be the *correct* one?

Few discussions of Minkowski versus Rindler quanta broaching these questions can be found in the philosophical literature, and what discussion there is has not been sufficiently grounded in a rigorous mathematical treatment to deliver cogent answers (as we shall see). We do not intend to survey the vast physics literature about Minkowski versus Rindler quanta, nor all physical aspects of the problem. Yet a proper appreciation of what is at stake, and of which answers to the above questions are sustainable, requires that we lay out the basics of the relevant formalism. We have strived for a self-contained treatment, in the hopes of opening up the discussion to philosophers of physics already familiar with elementary non-relativistic quantum theory. (We are inclined to agree with Torretti's recent diagnosis that most philosophers of physics tend to neglect quantum field theory because they are 'sickened by untidy math' (1999, p. 387).)

We begin in Section 2 with a general introduction to the problem of quantizing a classical field theory. This is followed by a detailed discussion of the conceptual relationship between inequivalent representations in which we reach conclusions at variance with some of the extant literature. In Section 3, we explain how the state of motion of an observer is taken into account when constructing a Fock space representation of a field, and how the Minkowski and Rindler constructions give rise to inequivalent representations. Finally, in Section 4, we examine the subtle relationship between the different particle concepts implied by these representations. In particular, we defend the idea that they supply *complementary* descriptions of the same field against the claim that they embody different, incommensurable *theories*.

A certain number of mathematical results play an important role in our exposition and in our philosophical arguments. The results are stated in the main text as propositions, and the proofs of those that cannot be found in the literature are included in an appendix.

2 Inequivalent field quantizations

In Section 2.1 we discuss the Weyl algebra, which in the case of infinitely many degrees of freedom circumscribes the basic kinematical structure of a free Boson field. After introducing in Section 2.2 some important concepts concerning representations of the Weyl algebra in terms of operators on Hilbert space, we shall be in a position to draw firm conclusions about the conceptual relation between inequivalent representations in Section 2.3.

2.1 The Weyl algebra

Consider how one constructs the quantum-mechanical analogue of a classical system with a finite number of degrees of freedom, described by a $2n$-dimensional phase space S. Each point of S is determined by a pair of vectors $\vec{a}, \vec{b} \in \mathbb{R}^n$ whose components $\{a_j\}$ and $\{b_j\}$ encode all the position and momentum components of the system

$$x(\vec{a}) = \sum_{j=1}^{n} a_j x_j, \qquad p(\vec{b}) = \sum_{j=1}^{n} b_j p_j. \tag{1}$$

To quantize the system, we impose the *canonical commutation relations* (CCRs)

$$[x(\vec{a}), x(\vec{a}')] = [p(\vec{b}), p(\vec{b}')] = 0, \qquad [x(\vec{a}), p(\vec{b})] = i(\vec{a} \cdot \vec{b})I, \tag{2}$$

and, then, seek a representation of these relations in terms of operators on a Hilbert space $\mathcal{H}$. In the standard *Schrödinger representation*, $\mathcal{H}$ is the space of square-integrable wavefunctions $L^2(\mathbb{R}^n)$, $x(\vec{a})$ becomes the operator that multiplies a wavefunction $\Psi(x_1, \ldots, x_n)$ by $\sum_{j=1}^{n} a_j x_j$, and $p(\vec{b})$ is the partial differential operator $-i\sum_{j=1}^{n} b_j \frac{\partial}{\partial x_j}$.

Note the action of $x(\vec{a})$ is not defined on an element $\Psi \in L^2(\mathbb{R}^n)$ unless $x(\vec{a})\Psi$ is again square-integrable, and $p(\vec{b})$ is not defined on Ψ unless it is suitably differentiable. This is not simply a peculiarity of the Schrödinger representation. Regardless of the Hilbert space on which they act, two self-adjoint operators whose commutator is a nonzero scalar multiple of the identity, as in (2), cannot both be everywhere defined (Kadison & Ringrose 1997, Remark 3.2.9). To avoid the technical inconvenience of dealing with domains of definition, it is standard to reformulate the representation problem in terms of unitary operators which are bounded, and hence everywhere defined.

Introducing the two n-parameter families of unitary operators

$$U(\vec{a}) := e^{ix(\vec{a})}, \; V(\vec{b}) := e^{ip(\vec{b})}, \qquad \vec{a}, \vec{b} \in \mathbb{R}^n, \tag{3}$$

it can be shown, at least formally, that the CCRs are equivalent to

$$U(\vec{a})U(\vec{a}') = U(\vec{a} + \vec{a}'), \; V(\vec{b})V(\vec{b}') = V(\vec{b} + \vec{b}'), \tag{4}$$

$$U(\vec{a})V(\vec{b}) = e^{i(\vec{a}\cdot\vec{b})}V(\vec{b})U(\vec{a}), \tag{5}$$

called the *Weyl form* of the CCRs. This equivalence holds rigorously in the Schrödinger representation, however there are 'irregular' representations in which it fails (see Segal 1967, Sec. 1; Summers 2001, Sec. 1). Thus, one reconstrues the goal as that of finding a representation of the Weyl form of the CCRs in terms of two concrete families of unitary operators $\{U(\vec{a}), V(\vec{b}) : \vec{a}, \vec{b} \in \mathbb{R}^n\}$ acting on a Hilbert space $\mathcal{H}$ that *can* be related, via (3), to canonical position and momentum operators on $\mathcal{H}$ satisfying the CCRs. We shall return to this latter 'regularity' requirement later in this section.

Though the position and momentum degrees of freedom have so far been treated on a different footing, we can simplify things further by introducing the composite *Weyl operators*

$$W(\vec{a},\vec{b}) := e^{i(\vec{a}\cdot\vec{b})/2}V(\vec{b})U(\vec{a}), \qquad \vec{a},\vec{b} \in \mathbb{R}. \tag{6}$$

Combining this definition with Eqns. (4) and (5) yields the multiplication rule

$$W(\vec{a},\vec{b})W(\vec{a}',\vec{b}') = e^{-i\sigma((\vec{a},\vec{b}),(\vec{a}',\vec{b}'))/2}W(\vec{a}+\vec{a}',\vec{b}+\vec{b}'), \tag{7}$$

where

$$\sigma((\vec{a},\vec{b}),(\vec{a}',\vec{b}')) := (\vec{a}'\cdot\vec{b}) - (\vec{a}\cdot\vec{b}'). \tag{8}$$

Observe that $\sigma(\cdot,\cdot)$ is a symplectic form (i.e., an anti-symmetric, bilinear functional) on S. (Note, also, that σ is nondegenerate; i.e., if for any $f \in S$, $\sigma(f,g) = 0$ for all $g \in S$, then $f = 0$.) We set

$$W(\vec{a},\vec{b})^* := e^{-i(\vec{a}\cdot\vec{b})/2}U(-\vec{a})V(-\vec{b}) = W(-\vec{a},-\vec{b}). \tag{9}$$

Clearly, then, any representation of the Weyl operators $W(\vec{a},\vec{b})$ on a Hilbert space $\mathcal{H}$ gives rise to a representation of the Weyl form of the CCRs, and vice versa.

Now, more generally, we allow our classical phase space S to be a vector space of arbitrary dimension; e.g., S may be an infinite-dimensional space constructed out of solutions to some relativistic wave equation. We assume S comes equipped with a (nondegenerate) symplectic form σ, and we say that a family $\{W_\pi(f) : f \in S\}$ of unitary operators acting on some Hilbert space $\mathcal{H}_\pi$ satisfies *the Weyl relations* just in case (cf. (7), (9))

$$W_\pi(f)W_\pi(g) = e^{-i\sigma(f,g)/2}W_\pi(f+g), \qquad f, g \in S, \tag{10}$$

$$W_\pi(f)^* = W_\pi(-f), \qquad f \in S. \tag{11}$$

We may go on to form arbitrary linear combinations of the Weyl operators, and thus obtain (at least some of) the self-adjoint operators that will serve as observables of the system.

Let $\mathcal{F}$ be the complex linear span of the set of Weyl operators $\{W_\pi(f) : f \in S\}$ acting on $\mathcal{H}_\pi$. (It follows from (10) that $\mathcal{F}$ is closed under taking operator products.) We say that a bounded operator A on $\mathcal{H}_\pi$ may be *uniformly* approximated by operators in $\mathcal{F}$ just in case for every $\epsilon > 0$, there is an operator $\tilde{A} \in \mathcal{F}$ such that

$$\|(A - \tilde{A})x\| < \epsilon, \text{ for all unit vectors } x \in \mathcal{H}_\pi. \tag{12}$$

If we let $\mathcal{W}_\pi$ denote the set of all bounded operators on $\mathcal{H}_\pi$ that can be uniformly approximated by elements in $\mathcal{F}$, then $\mathcal{W}_\pi$ is the *C^*-algebra* generated by the Weyl operators $\{W_\pi(f) : f \in S\}$. In particular, $\mathcal{W}_\pi$ is a C^*-subalgebra of the algebra $\mathfrak{B}(\mathcal{H}_\pi)$ of *all* bounded operators on $\mathcal{H}_\pi$, which is itself uniformly closed and closed under taking adjoints $A \mapsto A^*$.

Suppose now that $\{W_\pi(f) : f \in S\}$ and $\{W_\phi(f) : f \in S\}$ are two systems of Weyl operators representing the same classical system but acting, respectively, on two different Hilbert spaces $\mathcal{H}_\pi$ and $\mathcal{H}_\phi$. Let $\mathcal{W}_\pi, \mathcal{W}_\phi$ denote the corresponding C^*-algebras. A bijective mapping $\alpha : \mathcal{W}_\pi \mapsto \mathcal{W}_\phi$ is called a *$*$-isomorphism* just in case α is linear, multiplicative, and commutes with the adjoint operation. We then have the following uniqueness result for the C^*-algebra generated by Weyl operators (see Bratteli & Robinson 1996, Thm. 5.2.8).

Proposition 1. *There is a $*$-isomorphism α from $\mathcal{W}_\pi$ onto $\mathcal{W}_\phi$ such that $\alpha(W_\pi(f)) = W_\phi(f)$ for all $f \in S$.*

This Proposition establishes that the C^*-algebra constructed from any representation of the Weyl relations is, in fact, a unique object, independent of the representation in which we chose to construct it. We shall denote this abstract algebra, called the *Weyl algebra over* (S, σ), by $\mathcal{W}[S, \sigma]$ (and, when no confusion can result, simply say 'Weyl algebra' and write $\mathcal{W}$ for $\mathcal{W}[S, \sigma]$). So our problem boils down to choosing a *representation* $(\pi, \mathcal{H}_\pi)$ of $\mathcal{W}[S, \sigma]$ given by a mapping $\pi : \mathcal{W}[S, \sigma] \mapsto \mathfrak{B}(\mathcal{H}_\pi)$ preserving all algebraic relations. Note, also, that since the image $\pi(\mathcal{W}[S, \sigma])$ will always be an isomorphic copy of $\mathcal{W}[S, \sigma]$, π will always be one-to-one, and hence provide a *faithful* representation of $\mathcal{W}[S, \sigma]$.

With the representation-independent character of $\mathcal{W}[S, \sigma]$, why should we care any longer to choose a representation? After all, there is no technical obstacle to proceeding abstractly. We can take the self-adjoint elements of $\mathcal{W}$ to be the quantum-mechanical observables of our system. A linear functional ω on $\mathcal{W}$ is called a *state* just in case ω is positive (i.e., $\omega(A^*A) \geq 0$) and normalized (i.e., $\omega(I) = 1$). As usual, a state ω is taken to be *pure* (and *mixed* otherwise) just in case it is not a nontrivial convex combination of other states of $\mathcal{W}$. The dynamics of the system can be represented by a one-parameter group α_t of automorphisms of $\mathcal{W}$ (i.e., each α_t is just a map of $\mathcal{W}$ onto itself that preserves all algebraic relations). Hence, if we have some initial state ω_0, the final state will be given by $\omega_t = \omega_0 \circ \alpha_t$. We can even supply definitions for the probability in the state ω_t that a self-adjoint element $A \in \mathcal{W}$ takes a value lying in some Borel subset of its spectrum (Wald 1994, pp. 79–80), and for transition probabilities between, and superpositions of, pure states of $\mathcal{W}$ (Roberts & Roepstorff 1969). At no stage, it seems, need we ever introduce a Hilbert space as an essential element of the formalism. In fact, Haag & Kastler (1964, p. 852) and Robinson (1966, p. 488) maintain that the choice of a representation is largely a matter of analytical convenience without physical implications.

Nonetheless, the abstract Weyl algebra does not contain unbounded operators, many of which are naturally taken as corresponding to important physical quantities. For instance, the total energy of the system, the canonically conjugate position and momentum observables — which in field theory play the role of the local field observables — and the total number of particles are all represented by unbounded operators. Also, we shall see later that not even any *bounded* function of the total number of particles (apart from zero and the identity) lies in the Weyl algebra. Surprisingly, Irving Segal (one of the founders of the mathematically rigorous approach to quantum field theory) has written that this:

> ... has the simple if quite rough and somewhat oversimplified interpretation that the total number of 'bare' particles is devoid of physical meaning (Segal 1963, p. 56; see also Segal 1959, p. 12).

We shall return to this issue of physical meaning shortly. First, let us see how a representation can be used to expand the observables of a system beyond the abstract Weyl algebra.

Let $\mathcal{F}$ be a family of bounded operators acting on a representation space $\mathcal{H}_\pi$. We say that a bounded operator A on $\mathcal{H}_\pi$ can be *weakly* approximated by elements of $\mathcal{F}$ just in case for any vector $x \in \mathcal{H}$, and any $\epsilon > 0$, there is some $\tilde{A} \in \mathcal{F}$ such that

$$\left|\langle x, Ax\rangle - \langle x, \tilde{A}x\rangle\right| < \epsilon. \tag{13}$$

(Note the important quantifier change between the definitions of uniform and weak approximation, and that weak approximation has no abstract representation-independent counterpart.) Consider the family $\pi(\mathcal{W})^-$ of bounded operators that can be weakly approximated by elements of $\pi(\mathcal{W})$, i.e., $\pi(\mathcal{W})^-$ is the weak closure of $\pi(\mathcal{W})$. By von Neumann's double commutant theorem, $\pi(\mathcal{W})^- = \pi(\mathcal{W})''$, where the prime operation on a family of operators (here applied twice) denotes the set of all bounded operators on $\mathcal{H}_\pi$ commuting with each member of that family. $\pi(\mathcal{W})''$ is called the *von Neumann algebra* generated by $\pi(\mathcal{W})$. Clearly $\pi(\mathcal{W}) \subseteq \pi(\mathcal{W})''$, however we can hardly expect that $\pi(\mathcal{W}) = \pi(\mathcal{W})''$ when $\mathcal{H}_\pi$ is infinite-dimensional (which it *must* be, since there is no finite-dimensional representation of the Weyl algebra for even a single degree of freedom). Nor should we generally expect that $\pi(\mathcal{W})'' = \mathfrak{B}(\mathcal{H}_\pi)$, though this does hold in 'irreducible' representations, as we explain in the next subsection.

We may now expand our observables to include all self-adjoint operators in $\pi(\mathcal{W})''$. And, although $\pi(\mathcal{W})''$ still contains only bounded operators, it is easy to associate (potentially physically significant) unbounded observables with this algebra as well. We say that a (possibly unbounded) self-adjoint operator A on $\mathcal{H}_\pi$ is *affiliated* with $\pi(\mathcal{W})''$ just in case all A's spectral projections lie in $\pi(\mathcal{W})''$. Of course, we could have adopted the same definition for self-adjoint operators 'affiliated to' $\pi(\mathcal{W})$ itself, but C^*-algebras do not generally contain nontrivial projections (or, if they do, will not generally contain even the spectral projections of their self-adjoint members).

As an example, suppose we now demand to have a (so-called) *regular* representation π, in which the mappings $\mathbb{R} \ni t \mapsto \pi(W(tf))$, for all $f \in S$, are all weakly continuous. Then Stone's theorem will guarantee the existence of unbounded self-adjoint operators $\{\Phi(f) : f \in S\}$ on $\mathcal{H}_\pi$ satisfying $\pi(W(tf)) = e^{i\Phi(f)t}$, and it can be shown that all these operators are affiliated to $\pi(\mathcal{W})''$ (Kadison & Ringrose 1997, Ex. 5.7.53(ii)). In this way, we can recover as observables our original canonically conjugate positions and momenta (cf. Eqn. (3)), which the Weyl relations ensure will satisfy the original unbounded form of the CCRs.

It is important to recognize, however, that by enlarging the set of observables to include those affiliated to $\pi(\mathcal{W})''$, we have now left ourselves open to arbitrariness. In contrast to Proposition 1, we now have

Proposition 2. *There are (even regular) representations π, ϕ of $\mathcal{W}[S, \sigma]$ for which there* is no *$*$-isomorphism α from $\pi(\mathcal{W})''$ onto $\phi(\mathcal{W})''$ such that $\alpha(\pi(W(f))) = \phi(W(f))$ for all $f \in S$.*

This occurs when the representations are 'disjoint', which we discuss in the next subsection.

Proposition 2 is what motivates Segal to argue that observables affiliated to the weak closure $\pi(\mathcal{W})''$ in a representation of the Weyl algebra are 'somewhat unphysical' and 'have only analytical significance' (1963, pp. 11–4, 134).[1] Segal is explicit that by 'physical'

[1] Actually, Segal consistently finds it convenient to work with a strictly larger algebra than our (minimal) Weyl algebra, sometimes called the *mode finite* or *tame* Weyl algebra. However, both Proposition 1 (see Baez *et al*. 1992, Thm. 5.1) and

he means 'empirically measurable in principle' (1963, p. 11). We should not be confused by the fact that he often calls observables that fail this test 'conceptual' (suggesting they are more than mere analytical crutches). For in Baez *et al.* (1992, p. 145), Segal gives as an example the bounded self-adjoint operator $\cos p + (1 + x^2)^{-1}$ on $L^2(\mathbb{R})$ 'for which no known "Gedanken experiment" will actually directly determine the spectrum, and so [it] represents an observable in a purely conceptual sense'. Thus, the most obvious reading of Segal's position is that he subscribes to an operationalist view about the physical significance of theoretical quantities. Indeed, since good reasons *can* be given for the impossibility of exact ('sharp') measurements of observables in the von Neumann algebra generated by a C^*-algebra (see Wald 1994, Halvorson 2001*a*), operationalism explains Segal's dismissal of the physical (as opposed to analytical) significance of observables not in the Weyl algebra *per se*. (And it is worth recalling that Bridgman himself was similarly unphased by having to relegate much of the mathematical structure of a physical theory to 'a ghostly domain with no physical relevance' (1936, p. 116).)

Of course, insofar as operationalism is philosophically defensible at all, it does not compel assent. And, in this instance, Segal's operationalism has not dissuaded others from taking the more liberal view advocated by Wald:

> ... one should not view [the Weyl algebra] as encompassing *all* observables of the theory; rather, one should view [it] as encompassing a 'minimal' collection of observables, which is sufficiently large to enable the theory to be formulated. One may later wish to enlarge [the algebra] and/or further restrict the notion of 'state' in order to accommodate the existence of additional observables. (Wald 1994, p. 75)

The conservative and liberal views entail quite different commitments about the physical equivalence of representations — or so we shall argue.

Proposition 2 continue to hold for the tame Weyl algebra (also cf. Segal 1967, pp. 128–9).

2.2 Equivalence and disjointness of representations

It is essential that precise mathematical definitions of equivalence be clearly distinguished from the, often dubious, arguments that have been offered for their conceptual significance. We confine this section to discussing the definitions.

Since our ultimate goal is to discuss the Minkowski and Rindler quantizations of the Weyl algebra, we only need to consider the case where one of the two representations at issue, say π, is 'irreducible' and the other, ϕ, is 'factorial'. A representation π of $\mathcal{W}$ is called *irreducible* just in case no nontrivial subspace of the Hilbert space $\mathcal{H}_\pi$ is invariant under the action of all operators in $\pi(\mathcal{W})$. It is not difficult to see that this is equivalent to $\pi(\mathcal{W})'' = \mathfrak{B}(\mathcal{H}_\pi)$ (using the fact that an invariant subspace will exist just in case the projection onto it commutes with all operators in $\pi(\mathcal{W})$). A representation ϕ of $\mathcal{W}$ is called *factorial* whenever the von Neumann algebra $\phi(\mathcal{W})''$ is a *factor*, i.e., it has trivial center (the only operators in $\phi(\mathcal{W})''$ that commute with all other operators in $\phi(\mathcal{W})''$ are multiples of the identity). Since $\mathfrak{B}(\mathcal{H}_\pi)$ is a factor, it is clear that π's irreducibility entails its factoriality. Thus, the Schrödinger representation of the Weyl algebra is both irreducible and factorial.

The strongest form of equivalence between representations is unitary equivalence: ϕ and π are said to be *unitarily equivalent* just in case there is a unitary operator U mapping $\mathcal{H}_\phi$ isometrically onto $\mathcal{H}_\pi$, and such that

$$U\phi(A)U^{-1} = \pi(A) \qquad \forall A \in \mathcal{W}. \tag{14}$$

There are two other weaker definitions of equivalence.

Given a family π_i of irreducible representations of the Weyl algebra on Hilbert spaces $\mathcal{H}_i$, we can construct another (reducible) representation ϕ of the Weyl algebra on the direct sum Hilbert space $\sum_i \oplus \mathcal{H}_i$, by setting

$$\phi(A) = \sum_i \oplus \, \pi_i(A), \qquad A \in \mathcal{W}. \tag{15}$$

If each representation $(\pi_i, \mathcal{H}_i)$ is unitarily equivalent to some fixed representation $(\pi, \mathcal{H})$, we say that $\phi = \sum \oplus \pi_i$ is a *multiple* of the representation π. Furthermore, we say that two representations of the

Weyl algebra, ϕ (factorial) and π (irreducible), are *quasi-equivalent* just in case ϕ is a multiple of π. It should be obvious from this characterization that quasi-equivalence weakens unitary equivalence. Another way to see this is to use the fact (Kadison & Ringrose 1997, Def. 10.3.1, Cor. 10.3.4) that quasi-equivalence of ϕ and π is equivalent to the existence of a $*$-isomorphism α from $\phi(\mathcal{W})''$ onto $\pi(\mathcal{W})''$ such that $\alpha(\phi(A)) = \pi(A)$ for all $A \in \mathcal{W}$. Unitary equivalence is then just the special case where the $*$-isomorphism α can be implemented by a unitary operator.

If ϕ is not even quasi-equivalent to π, then we say that ϕ and π are *disjoint* representations of $\mathcal{W}$.[2] Note, then, that if both π and ϕ are irreducible, they are either unitarily equivalent or disjoint.

We can now state the following pivotal result (von Neumann 1931).

Stone-von Neumann Uniqueness Theorem. *When S is finite-dimensional, every regular representation of $\mathcal{W}[S, \sigma]$ is quasiequivalent to the Schrödinger representation.*

This theorem is usually interpreted as saying that there is a unique quantum theory corresponding to a classical theory with finitely-many degrees of freedom. The theorem *fails* in field theory — where S is infinite-dimensional — opening the door to disjoint representations and Proposition 2.

There is another way to think of the relations between representations, in terms of states. Recall the abstract definition of a state of a C^*-algebra, as simply a positive normalized linear functional on the algebra. Since, in any representation π, $\pi(\mathcal{W})$ is just a faithful copy of $\mathcal{W}$, π induces a one-to-one correspondence between the abstract states of $\mathcal{W}$ and the abstract states of $\pi(\mathcal{W})$. Note now that *some* of the abstract states on $\pi(\mathcal{W})$ are the garden-variety density

[2]In general, disjointness is not defined as the negation of quasi-equivalence, but by the more cumbersome formulation: Two representations π, ϕ are disjoint just in case π has no 'subrepresentation' quasi-equivalent to ϕ, and ϕ has no subrepresentation quasi-equivalent to π. Since we are only interested, however, in the special case where π is irreducible (and hence has no nontrivial subrepresentations) and ϕ is 'factorial' (and hence is quasi-equivalent to each of its subrepresentations), the cumbersome formulation reduces to our definition.

operator states that we are familiar with from elementary quantum mechanics. In particular, define ω_D on $\pi(\mathcal{W})$ by picking a density operator D on $\mathcal{H}_\pi$ and setting

$$\omega_D(A) := \mathrm{Tr}(DA), \qquad A \in \pi(\mathcal{W}). \tag{16}$$

In general, however, there will be abstract states of $\pi(\mathcal{W})$ that are not given by density operators via Eqn. (16).[3] We say then that an abstract state ω of $\pi(\mathcal{W})$ is *normal* just in case it is given (via Eqn. (16)) by some density operator D on $\mathcal{H}_\pi$. We let $\mathfrak{F}(\pi)$ denote the subset of the abstract state space of $\mathcal{W}$ consisting of those states that correspond to normal states in the representation π, and we call $\mathfrak{F}(\pi)$ the *folium* of the representation π. That is, $\omega \in \mathfrak{F}(\pi)$ just in case there is a density operator D on $\mathcal{H}_\pi$ such that

$$\omega(A) = \mathrm{Tr}(D\pi(A)), \qquad A \in \mathcal{W}. \tag{17}$$

We then have the following equivalences (Kadison & Ringrose 1997, Prop. 10.3.13):

$$\begin{aligned} \pi \text{ and } \phi \text{ are quasi-equivalent} &\iff \mathfrak{F}(\pi) = \mathfrak{F}(\phi), \\ \pi \text{ and } \phi \text{ are disjoint} &\iff \mathfrak{F}(\pi) \cap \mathfrak{F}(\phi) = \emptyset. \end{aligned}$$

In other words, π and ϕ are quasi-equivalent just in case they share the same normal states. And π and ϕ are disjoint just in case they have *no* normal states in common.

In fact, if π is disjoint from ϕ, then all normal states in the representation π are 'orthogonal' to all normal states in the representation ϕ. We may think of this situation intuitively as follows. Define a third representation ψ of $\mathcal{W}$ on $\mathcal{H}_\pi \oplus \mathcal{H}_\phi$ by setting

$$\psi(A) = \pi(A) \oplus \phi(A), \qquad A \in \mathcal{W}. \tag{18}$$

[3] Gleason's theorem does not rule out these states because it is not part of the definition of an abstract state that it be countably additive over mutually orthogonal projections. Indeed, such additivity does not even make sense abstractly, because an infinite sum of orthogonal projections can never converge uniformly, only weakly (in a representation).

Then, every normal state of the representation π is orthogonal to every normal state of the representation ϕ.[4] This makes sense of the oft-repeated phrase (see, e.g., Gerlach 1989) that 'The Rindler vacuum is orthogonal to all states in the Minkowski vacuum representation'.

While not every abstract state of $\mathcal{W}$ will be in the folium of a given representation, there is always *some* representation of $\mathcal{W}$ in which the state *is* normal, as a consequence of the following (Kadison & Ringrose 1997, Thms. 4.5.2 and 10.2.3).

Gelfand-Naimark-Segal Theorem. *Any abstract state ω of a C^*-algebra $\mathcal{A}$ gives rise to a unique (up to unitary equivalence) representation $(\pi_\omega, \mathcal{H}_\omega)$ of $\mathcal{A}$ and vector $\Omega_\omega \in \mathcal{H}_\omega$ such that*

$$\omega(A) = \langle \Omega_\omega, \pi_\omega(A)\Omega_\omega \rangle, \qquad A \in \mathcal{A}, \tag{19}$$

and such that the set $\{\pi_\omega(A)\Omega_\omega : A \in \mathcal{A}\}$ is dense in $\mathcal{H}_\omega$. Moreover, π_ω is irreducible just in case ω is pure.

The triple $(\pi_\omega, \mathcal{H}_\omega, \Omega_\omega)$ is called the *GNS representation* of $\mathcal{A}$ induced by the state ω, and Ω_ω is called a *cyclic* vector for the representation. We shall see in the next main section how the Minkowski and Rindler vacuums induce disjoint GNS representations of the Weyl algebra.

There is a third notion of equivalence of representations, still weaker than quasi-equivalence. Let π be a representation of $\mathcal{W}$, and let $\mathfrak{F}(\pi)$ be the folium of π. We say that an abstract state ω of $\mathcal{W}$ can be *weak* approximated* by states in $\mathfrak{F}(\pi)$ just in case for each $\epsilon > 0$, and for each finite collection $\{A_i : i = 1, \ldots, n\}$ of operators in $\mathcal{W}$, there is a state $\omega' \in \mathfrak{F}(\pi)$ such that

$$\left|\omega(A_i) - \omega'(A_i)\right| < \epsilon, \qquad i = 1, \ldots, n. \tag{20}$$

Two representations π, ϕ are then said to be *weakly equivalent* just in case all states in $\mathfrak{F}(\pi)$ may be weak* approximated by states in $\mathfrak{F}(\phi)$ and vice versa. We then have the following fundamental result (Fell 1960).

[4] This intuitive picture may be justified by making use of the 'universal representation' of $\mathcal{W}$ (Kadison & Ringrose 1997, Thm. 10.3.5).

Fell's Theorem. *Let π be a faithful representation of a C^*-algebra $\mathcal{A}$. Then, every abstract state of $\mathcal{A}$ may be weak* approximated by states in $\mathfrak{F}(\pi)$.*

In particular, then, it follows that *all* representations of $\mathcal{W}$ are weakly equivalent.

In summary, we have the following implications for any two representations π, ϕ:

Unitarily equivalent $\Longrightarrow$ Quasiequivalent $\Longrightarrow$ Weakly equivalent.

If π and ϕ are both irreducible, then the first arrow is reversible.

2.3 Physical equivalence of representations

Do disjoint representations yield *physically* inequivalent theories? It depends on what one takes to be the physical content of a theory, and what one means by 'equivalent theories' — subjects about which philosophers of science have had plenty to say.

Recall that Reichenbach (1938) deemed two theories 'the same' just in case they are empirically equivalent, i.e., they are confirmed equally under all possible evidence. Obviously this criterion, were we to adopt it here, would beg the question against those who (while agreeing that, strictly speaking, only self-adjoint elements of the Weyl algebra can actually be measured) attribute physical significance to 'global' quantities only definable in a represenxtation, like the total number of particles.

A stronger notion of equivalence, formulated by Glymour (1971) (who proposed it only as a *necessary* condition), is that two theories are equivalent only if they are 'intertranslatable'. This is often cashed out in logical terms as the possibility of defining the primitives of one theory in terms of those of the other so that the theorems of the first appear as logical consequences of those of the second, and vice versa. Prima facie, this criterion is ill-suited to the present context, because the different 'theories' are not presented to us as syntactic structures or formalized logical systems, but rather two competing algebras of observables whose states represent physical predictions. In addition, intertranslatability *per se* has nothing to say about what portions of the mathematical formalism of the two physical theories being compared ought to be intertranslatable, and what should be regarded as 'surplus mathematical structure' not required to be part of the translation.

Nevertheless, we believe the intertranslatability thesis can be naturally expressed in the present context and rendered neutral as between the conservative and liberal approaches to physical observables discussed earlier. Think of the Weyl operators $\{\phi(W(f)) : f \in S\}$ and $\{\pi(W(f)) : f \in S\}$ as the primitives of our two 'theories', in analogy with the way the natural numbers can be regarded as the primitives of a 'theory' of real numbers. Just as we may define rational numbers as ratios of natural numbers, and then construct real numbers as the limits of Cauchy sequences of rationals, we construct the Weyl algebras $\phi(\mathcal{W})$ and $\pi(\mathcal{W})$ by taking linear combinations of the Weyl operators and then closing in the uniform topology. We then close in the weak topology of the two representations to obtain the von Neumann algebras $\phi(\mathcal{W})''$ and $\pi(\mathcal{W})''$. Whether the observables affiliated with this second closure have physical significance is up for grabs, as is whether we should be conservative and take only normal states in the given representation to be physical, or be more liberal and admit a broader class of algebraic states. The analogue of the 'theorems' of the theory are then statements about the expectation values dictated by the physical states for the self-adjoint elements in the physically relevant algebra of the theory.

We therefore propose the following formal rendering of Glymour's inter-translatability thesis adapted to the present context. Representations ϕ and π are *physically equivalent* only if there exists a bijective mapping α from the physical observables of the representation ϕ to the physical observables of the representation π, and another bijective mapping β from the physical states of the representation ϕ to the physical states of the representation π, such that

$$\alpha(\phi(W(f))) = \pi(W(f)), \ \forall f \in S, \tag{21}$$

('primitives')

$$\beta(\omega)(\alpha(A)) = \omega(A), \ \forall \text{ states } \omega, \ \forall \text{ observables } A. \tag{22}$$

('theorems')

Of course, the notion of equivalence we obtain depends on how we construe the phrases 'physical observables of a representation π' and 'physical states of a representation π'. According to a conservative rendering of observables, only the self-adjoint elements of the Weyl algebra $\pi(\mathcal{W})$ are genuine physical observables of the representation π. (More generally, an unbounded self-adjoint operator on $\mathcal{H}_\pi$ is a physical observable only if all of its bounded functions lie in $\pi(\mathcal{W})$.) On the other hand, a liberal rendering of observables considers all self-adjoint operators in the weak closure $\pi(\mathcal{W})^-$ of $\pi(\mathcal{W})$ as genuine physical observables. (More generally, those unbounded self-adjoint operators whose bounded functions lie in $\pi(\mathcal{W})^-$, i.e., all such operators affiliated with $\pi(\mathcal{W})^-$, should be considered genuine physical observables.) A conservative with respect to states claims that only the density operator states (i.e., normal states) of the algebra $\pi(\mathcal{W})$ are genuine physical states. On the other hand, a liberal with respect to states claims that all algebraic states of $\pi(\mathcal{W})$ should be thought of as genuine physical states. We thereby obtain *four distinct* necessary conditions for physical equivalence, according to whether one is conservative or liberal about observables, and conservative or liberal about states.[5]

Arageorgis (1995, p. 302) and Arageorgis *et al.* (2002) also take the correct notion of physical equivalence in this context to be intertranslatability. On the basis of informal discussions (with rather less supporting argument than one would have liked), they claim that physical equivalence of representations requires that they be unitarily equivalent. (They do not discuss quasi-equivalence.) We disagree with this conclusion, but there is still substantial overlap between us. For instance, with our precise necessary condition for

[5]The distinction between the conservative and liberal positions about observables could be further ramified by taking into account the distinction — which is suppressed throughout this chapter — between local and global observables. In particular, if all (and only) locally measurable observables have genuine physical status, then physical equivalence of π and ϕ would require a bijection α between *local* observables in $\pi(\mathcal{W})^-$ and *local* observables in $\phi(\mathcal{W})^-$. Similarly, the distinction between the conservative and liberal positions about states could be further ramified by taking into account the distinction between normal states and 'locally normal' states.

physical equivalence above, we may establish the following elementary result.

Proposition 3. *Under the conservative approach to states, ϕ (factorial) and π (irreducible) are physically equivalent representations of $\mathcal{W}$ only if they are quasi-equivalent.*

With somewhat more work, the following result may also be established.[6]

Proposition 4. *Under the liberal approach to observables, ϕ (factorial) and π (irreducible) are physically equivalent representations of $\mathcal{W}$ only if they are quasi-equivalent.*

The above results leave only the position of the 'conservative about observables/liberal about states' undecided. However, we claim, *pace* Arageorgis *et al.*, that a proponent of this position can satisfy conditions (21),(22) *without* committing himself to quasi-equivalence of the representations. Since he is conservative about observables, Proposition 1 already guarantees the existence of a bijective mapping α — in fact, a *-isomorphism from the whole of $\phi(\mathcal{W})$ to the whole of $\pi(\mathcal{W})$ — satisfying (21). And if he is liberal about states, the state mapping β need not map any normal state of $\phi(\mathcal{W})$ into a normal state of $\pi(\mathcal{W})$, bypassing the argument for Proposition 3. Indeed, since the liberal takes *all* algebraic states of $\phi(\mathcal{W})$ and $\pi(\mathcal{W})$ to be physically significant, for any algebraic state ω of $\phi(\mathcal{W})$, the bijective mapping β that sends ω to the state $\omega \circ \alpha^{-1}$ on $\pi(\mathcal{W})$ trivially satisfies condition (22) even when ϕ and π are disjoint.

Though we have argued that Segal was conservative about observables, we are not claiming he was liberal about states. In fact, Segal consistently maintained that only the 'regular states' of the Weyl algebra have physical relevance (1961, p. 7; 1967, pp. 120, 132). A state ω of $\mathcal{W}[S,\sigma]$ is called *regular* just in case the map $f \mapsto \omega(W(f))$ is continuous on all finite-dimensional subspaces of

[6]Our proof in the appendix makes rigorous Arageorgis' brief (and insufficient) reference to Wigner's symmetry representation theorem in his (1995, p. 302, footnote).

S; or, equivalently, just in case the GNS representation of $\mathcal{W}[S,\sigma]$ determined by ω is regular (Segal 1967, p. 134). However, note that, unlike normality of a state, regularity is representation-*independent*. Taking the set of all regular states of the Weyl algebra to be physical is therefore still liberal enough to permit satisfaction of condition (22). For the mapping β of the previous paragraph trivially preserves regularity, insofar as both ω and $\omega \circ \alpha^{-1}$ induce the same abstract regular state of $\mathcal{W}$.

Our verdict, then, is that Segal, for one, is not committed to saying only quasi-equivalent representations can be physically equivalent. And this explains why he sees fit to *define* physical equivalence of representations in such a way that Proposition 1 secures the physical equivalence of all representations (see Segal 1961, Defn. 1(c)). (Indeed, Segal regards Proposition 1 as the appropriate generalization of the Stone-von Neumann uniqueness theorem to infinite-dimensional S!) One might still ask what the point of passing to a concrete Hilbert space representation of $\mathcal{W}$ is if one is going to allow as physically possible regular states not in the folium of the chosen representation. The point, we take it, is that if we are interested in drawing out the predictions of some particular regular state, such as the Minkowski vacuum or the Rindler vacuum, then passing to a particular representation will put at our disposal all the standard analytical techniques of Hilbert space quantum mechanics to facilitate calculations in that particular state.[7]

Haag & Kastler (1964, p. 852) and Robinson (1966, p. 488) have argued that *by itself* the *weak* equivalence of all representations of the Weyl algebra entails their physical equivalence.[8] Their argument starts from the fact that, by measuring the expectations of a fi-

[7]In support of not limiting the physical states of the Weyl algebra to any one representation's folium, one can also cite the cases of non-unitarily implementable dynamics discussed by Arageorgis *et al.* (2002) in which dynamical evolution occurs between regular states that induce disjoint GNS representations. In such cases, it would hardly be coherent to maintain that regular states *dynamically accessible to one another* are not physically co-possible.

[8]Indeed, the term 'physical equivalence' is often used synonymously with weak equivalence; for example, by Emch (1972, p. 108), who, however, issues the warning that 'we should be seriously wary of semantic extrapolations' from this usage. Indeed!

nite number of observables $\{A_i\}$ in the Weyl algebra, each to a finite degree of accuracy ϵ, we can only determine the state of the system to within a weak* neighborhood. But by Fell's density theorem, states from the folium of *any* representation lie in this neighborhood. So for all practical purposes, we can never determine which representation is the physically 'correct' one and they all, in some (as yet, unarticulated!) sense, carry the same physical content. And as a corollary, choosing a representation is simply a matter of convention.

Clearly the necessary condition for physical equivalence we have proposed constitutes a very different notion of equivalence than weak equivalence, so we are not disposed to agree with this argument. Evidently it presupposes that only the observables in the Weyl algebra itself are physically significant, which we have granted *could* be grounded in operationalism. However, there is an additional layer of operationalism that the argument must presuppose: skepticism about the physical meaning of postulating an *absolutely precise* state for the system. If we follow this skepticism to its logical conclusion, we should instead think of physical states of the Weyl algebra as represented by weak* neighborhoods of algebraic states. What it would then mean to falsify a state, so understood, is that some finite number of expectation values measured to within finite accuracy are found to be incompatible with all the algebraic states in some proposed weak* neighborhood. Unfortunately, no particular 'state' in this sense can ever be fully empirically adequate, for any hypothesized state (= weak* neighborhood) will be subject to constant revision as the accuracy and number of our experiments increase. We agree with Summers (2001) that this would do irreparable damage to the predictive power of the theory — damage that can only be avoided by maintaining that there is a correct algebraic state.

We do not, however, agree with Summers' (2001) presumption (tacitly endorsed by Arageorgis *et al.* (2002)) that we not only need the correct algebraic state, but '... the correct state *in the correct representation*' (p. 13; italics ours). This added remark of Summers' is directed against the conventionalist corollary to Fell's theorem. Yet we see nothing in the point about predictive power that privileges any particular representation, not even the GNS representation of the predicted state. We might well have good reason to deliberately choose a representation in which the *precise* algebraic state predicted is not normal. (For example, Kay (1985) does exactly this, by 'constructing' the Minkowski vacuum as a thermal state in the Rindler quantization.) The role Fell's theorem plays is then, at best, methodological. All it guarantees is that when we calculate with density operators in our chosen representation, we can always get a reasonably good indication of the predictions of *whatever* precise algebraic state we have postulated for the system.

So much for the conservative stance on observables. An interpreter of quantum field theory is not likely to find it attractive, if only because none of the observables that have any chance of underwriting the particle concept lie in the Weyl algebra. But suppose, as interpreters, we adopt the liberal approach to observables. Does the physical inequivalence of disjoint representations entail their incompatibility, or even incommensurability? By this, we do not mean to conjure up Kuhnian thoughts about incommensurable 'paradigms', whose proponents share no methods to resolve their disputes. Rather, we are pointing to the (more Feyerabendian?) possibility of an unanalyzable shift in meaning between disjoint representations as a consequence of the fact that the concepts (observables and/or states) of one representation are not wholly definable or translatable in terms of those of the other.

One might think of neutralizing this threat by viewing disjoint representations as sub-theories or models of a more general theory built upon the Weyl algebra. Consider the analogy of *two different* classical systems, modeled, say, by phase spaces of different dimension. Though not physically equivalent, these models hardly define incommensurable theories insofar as they share the characteristic kinematical and dynamical features that warrant the term 'classical'. Surely the same could be said of disjoint representations of the Weyl algebra?

There is, however, a crucial disanalogy that needs to be taken into account. In the case of the Minkowski and Rindler representations, physicists freely switch between them to describe the quantum state of the *very same* 'system' — in this case, the quantum field in a fixed region of spacetime (see, e.g., Unruh & Wald 1984; Wald 1994, Sec. 5.1). And, as we shall see later, the weak closures of these representations provide physically inequivalent descriptions of the particle content in the region. So it is tempting to view this switching back and forth between disjoint representations as conceptually incoherent (Arageorgis 1995, p. 268), and to see the particle concepts associated to the representations as not just different, but outright incommensurable (Arageorgis *et al.* 2002).

We shall argue that this view, tempting as it is, goes too far. For suppose we *do* take the view that the observables affiliated to the von Neumann algebras generated by two disjoint representations ϕ and π simply represent different physical aspects of the same physical system. If we are also liberal about states (not restricting ourselves to any one representation's folium), then it is natural to ask what implications a state ω of our system, that happens to be in the folium of ϕ, has for the observables in $\pi(\mathcal{W})''$. In many cases, it is possible to extract a definite answer.

In particular, any abstract state ω of $\mathcal{W}$ gives rise to a state on $\pi(\mathcal{W})$, which may be extended to a state on the weak closure $\pi(\mathcal{W})''$ (Kadison & Ringrose 1997, Thm. 4.3.13). The only catch is that unless $\omega \in \mathfrak{F}(\pi)$, this extension will not be unique. For, only normal states of $\pi(\mathcal{W})$ possess sufficiently nice continuity properties to ensure that their values on $\pi(\mathcal{W})$ uniquely fix their values on the weak closure $\pi(\mathcal{W})''$ (see Kadison & Ringrose 1997, Thm. 7.1.12). *However*, it may happen that all extensions of ω agree on the expectation value they assign to a *particular observable* affiliated to $\pi(\mathcal{W})''$. This is the strategy we shall use to make sense of assertions such as 'The Minkowski vacuum in a (Rindler) spacetime wedge is full of Rindler quanta' (cf., e.g., DeWitt 1979*a*). The very fact that such assertions can be made sense of *at all* takes the steam out of claims that disjoint representations are necessarily incommensurable. Indeed, we shall ultimately argue that this shows disjoint representations should not be treated as *competing* 'theories' in the first place. Rather, they are better viewed as supplying physically different, 'complementary' perspectives on the same quantum system from within a broader theoretical framework that does not privilege a particular representation.

3 Constructing representations

We now explain how to construct 'Fock representations' of the CCRs. In Sections 3.1 and 3.2 we show how this construction depends on one's choice of preferred timelike motion in Minkowski spacetime. In Section 3.3, we show that alternative choices of preferred timelike motion can result in unitarily inequivalent — indeed, disjoint — representations.

3.1 First quantization ('Splitting the frequencies')

The first step in the quantization scheme consists in turning the classical phase space (S, σ) into a quantum-mechanical 'one particle space' — i.e., a Hilbert space. *The non-uniqueness of the quantization scheme comes in at this very first step.*

Depending on our choice of preferred timelike motion, we will have a one-parameter group T_t of linear mappings from S onto S representing the evolution of the classical system in time. The flow $t \mapsto T_t$ should also preserve the symplectic form. A bijective real-linear mapping $T : S \mapsto S$ is called a *symplectomorphism* just in case T preserves the symplectic form; i.e., $\sigma(Tf, Tg) = \sigma(f, g)$ for all $f, g \in S$.

We say that J is a *complex structure* for (S, σ) just in case

1. J is a symplectomorphism,
2. $J^2 = -I$,
3. $\sigma(f, Jf) > 0, \qquad 0 \neq f \in S.$

Relative to a complex structure J, we may extend the scalar multiplication on S to complex numbers; viz., take multiplication by

$a + ib$ as given by $(a + ib)f := af + bJf \in S$. We may also define an inner product $(\cdot, \cdot)_J$ on the resulting complex vector space by setting

$$(f, g)_J := \sigma(f, Jg) + i\sigma(f, g), \qquad f, g \in S. \tag{23}$$

We let $\mathcal{S}_J$ denote the Hilbert space that results when we equip (S, σ) with the extended scalar multiplication and inner product $(\cdot, \cdot)_J$.

A symplectomorphism T is (by assumption) a real-linear operator on S. However, it does not automatically follow that T is a *complex*-linear operator on $\mathcal{S}_J$, since $T(if) = i(Tf)$ may fail. If, however, T commutes with J, then T will be a complex-linear operator on $\mathcal{S}_J$, and it is easy to see that $(Tf, Tg)_J = (f, g)_J$ for all $f, g \in \mathcal{S}_J$, so T would in fact be unitary. Accordingly, we say that a group T_t of symplectomorphisms on (S, σ) is *unitarizable* relative to J just in case $[J, T_t] = 0$ for all $t \in \mathbb{R}$.

If T_t is unitarizable and $t \mapsto T_t$ is weakly continuous, so that we have $T_t = e^{itH}$ (by Stone's theorem), we say that T_t has *positive energy* just in case H is a positive operator. In general, we say that $(\mathcal{H}, U_t)$ is a *quantum one particle system* just in case $\mathcal{H}$ is a Hilbert space and U_t is a weakly continuous one-parameter unitary group on $\mathcal{H}$ with positive energy. Kay (1979) proved:

Proposition 5. *Let T_t be a one-parameter group of symplectomorphisms of (S, σ). If there is a complex structure J on (S, σ) such that $(\mathcal{S}_J, T_t)$ is a quantum one particle system, then J is unique.*

Physically, the time translation group T_t determines a natural decomposition (or 'splitting') of the solutions of the relativistic wave equation we are quantizing into those that oscillate with purely positive and with purely negative frequency with respect to the motion. This has the effect of uniquely fixing a choice of J, and the Hilbert space $\mathcal{S}_J$ then provides a representation of the positive frequency solutions alone.[9]

We shall see in the next section how the representation space of a 'Fock' representation of the Weyl algebra is constructed directly from the Hilbert space $\mathcal{S}_J$. Thus, as we claimed, the non-uniqueness of the resulting representation stems entirely from the

[9]For more physical details, see Fulling (1972, Secs. VIII.3,4) and Wald (1994, pp. 41–2, 63, 111).

arbitrary choice of the time translation group T_t in Minkowski spacetime and the complex structure J on S it determines.

3.2 Second quantization (Fock space)

Once we have used some time translation group T_t to fix the Hilbert space $\mathcal{S}_J$, the 'second quantization' procedure yields a unique representation $(\pi, \mathcal{H}_\pi)$ of the Weyl algebra $\mathcal{W}[S, \sigma]$.

Let $\mathcal{H}^n$ denote the n-fold symmetric tensor product of $\mathcal{S}_J$ with itself. That is, using $\mathcal{S}_J^n$ to denote $\mathcal{S}_J \otimes \cdots \otimes \mathcal{S}_J$ (n times), $\mathcal{H}^n = P_+(\mathcal{S}_J^n)$ where P_+ is the projection onto the symmetric subspace. Then we define a Hilbert space

$$\mathcal{F}(\mathcal{S}_J) := \mathbb{C} \oplus \mathcal{H}^1 \oplus \mathcal{H}^2 \oplus \mathcal{H}^3 \oplus \cdots, \tag{24}$$

called the *bosonic Fock space over* $\mathcal{S}_J$. Let

$$\Omega := 1 \oplus 0 \oplus 0 \oplus \cdots, \tag{25}$$

denote the privileged 'Fock vacuum' state in $\mathcal{F}(\mathcal{S}_J)$.

Now, we define creation and annihilation operators on $\mathcal{F}(\mathcal{S}_J)$ in the usual way. For any fixed $f \in S$, we first consider the unique bounded linear extensions of the mappings $a_n^*(f) : \mathcal{S}_J^{n-1} \to \mathcal{S}_J^n$ and $a_n(f) : \mathcal{S}_J^n \to \mathcal{S}_J^{n-1}$ defined by the following actions on product vectors

$$a_n^*(f)(f_1 \otimes \cdots \otimes f_{n-1}) = f \otimes f_1 \otimes \cdots \otimes f_{n-1}, \tag{26}$$

$$a_n(f)(f_1 \otimes \cdots \otimes f_n) = (f, f_1)_J \, f_2 \otimes \cdots \otimes f_n. \tag{27}$$

We then define the *unbounded* creation and annihilation operators on $\mathcal{F}(\mathcal{S}_J)$ by

$$a^*(f) := a_1^*(f) \oplus \sqrt{2}P_+ a_2^*(f) \oplus \sqrt{3}P_+ a_3^*(f) \oplus \cdots, \tag{28}$$

$$a(f) := 0 \oplus a_1(f) \oplus \sqrt{2}a_2(f) \oplus \sqrt{3}a_3(f) \oplus \cdots. \tag{29}$$

(Note that the mapping $f \mapsto a^*(f)$ is linear while $f \mapsto a(f)$ is *anti-linear*.)

As the definitions and notation suggest, $a^*(f)$ and $a(f)$ are each other's adjoint, $a^*(f)$ is the creation operator for a particle with wavefunction f, and $a(f)$ the corresponding annihilation operator. The unbounded self-adjoint operator $N(f) = a^*(f)a(f)$ represents the number of particles in the field with wavefunction f (unbounded, because we are describing bosons to which no exclusion principle applies). Summing $N(f)$ over any J-orthonormal basis of wavefunctions in $\mathcal{S}_J$, we obtain the *total* number operator N on $\mathcal{F}(\mathcal{S}_J)$, which has the form

$$N = 0 \oplus 1 \oplus 2 \oplus 3 \oplus \cdots . \tag{30}$$

Next, we define the self-adjoint 'field operators"

$$\Phi(f) := 2^{-1/2}(a^*(f) + a(f)), \qquad f \in S. \tag{31}$$

(In heuristic discussions of free quantum field theory, these are normally encountered as 'operator-valued solutions' $\Phi(x)$ to a relativistic field equation at some fixed time. However, if we want to associate a properly defined self-adjoint field operator with the spatial point x, we must consider a neighborhood of x, and an operator of form $\Phi(f)$, where the 'test-function' $f \in S$ has support in the neighborhood.[10]) Defining the unitary operators

$$\pi(W(tf)) := \exp(it\Phi(f)), \qquad t \in \mathbb{R},\ f \in S, \tag{32}$$

it can then be verified (though it is not trivial) that the $\pi(W(f))$ satisfy the Weyl form of the CCRs. In fact, the mapping $W(f) \mapsto \pi(W(f))$ gives an irreducible regular representation π of $\mathcal{W}$ on $\mathcal{F}(\mathcal{S}_J)$.

We also have

$$\langle \Omega, \pi(W(f))\Omega \rangle = e^{-(f,f)_J/4}, \qquad f \in S. \tag{33}$$

(We shall always distinguish the inner product of $\mathcal{F}(\mathcal{S}_J)$ from that of $\mathcal{S}_J$ by using angle brackets.) The vacuum vector $\Omega \in \mathcal{F}(\mathcal{S}_J)$ defines

[10]The picture of a quantum field as an operator-valued *field* — or, as Teller (1995, Chap. 5) aptly puts it, a field of 'determinables' — unfortunately, has no mathematically rigorous foundation.

an abstract regular state ω_J of $\mathcal{W}$ via $\omega_J(A) := \langle \Omega, \pi(A)\Omega \rangle$ for all $A \in \mathcal{W}$. Since the action of $\pi(\mathcal{W})$ on $\mathcal{F}(\mathcal{S}_J)$ is irreducible, $\{\pi(A)\Omega : A \in \mathcal{W}\}$ is dense in $\mathcal{F}(\mathcal{S}_J)$ (else its closure would be a nontrivial subspace invariant under all operators in $\pi(\mathcal{W})$). Thus, the Fock representation of $\mathcal{W}$ on $\mathcal{F}(\mathcal{S}_J)$ is unitarily equivalent to the GNS representation of $\mathcal{W}$ determined by the pure state ω_J.

In sum, a complex structure J on (S, σ) gives rise to an abstract vacuum state ω_J on $\mathcal{W}[S, \sigma]$ whose GNS representation $(\pi_{\omega_J}, \mathcal{H}_{\omega_J}, \Omega_{\omega_J})$ is just the standard Fock vacuum representation $(\pi, \mathcal{F}(\mathcal{S}_J), \Omega)$. Note also that inverting Eqn. (31) yields

$$\begin{aligned} a^*(f) &= 2^{-1/2}(\Phi(f) - i\Phi(if)), \\ a(f) &= 2^{-1/2}(\Phi(f) + i\Phi(if)), \end{aligned} \tag{34}$$

for all $f \in S$. Thus, we could just as well have arrived at the Fock representation of $\mathcal{W}$ 'abstractly' by *starting* with the pure regular state ω_J on $\mathcal{W}[S, \sigma]$ as our proposed vacuum, exploiting its regularity to guarantee the existence of field operators $\{\Phi(f) : f \in S\}$ acting on $\mathcal{H}_{\omega_J}$, and then using Eqns. (34) to *define* $a^*(f)$ and $a(f)$ (and, from thence, the number operators $N(f)$ and N).

There is a natural way to construct operators on $\mathcal{F}(\mathcal{S}_J)$ out of operators on the one-particle space $\mathcal{S}_J$, using the *second quantization map* Γ and its 'derivative' $d\Gamma$. Unlike the representation map π, the operators on $\mathcal{F}(\mathcal{S}_J)$ in the range of Γ and $d\Gamma$ do not 'come from' $\mathcal{W}[S, \sigma]$, but rather $\mathfrak{B}(\mathcal{S}_J)$. Since the latter depends on how S was complexified, we cannot expect second quantized observables to be representation-independent.

To define $d\Gamma$, first let H be a self-adjoint (possibly unbounded) operator on $\mathcal{S}_J$. We define H_n on $\mathcal{H}^n$ by setting $H_0 = 0$ and

$$H_n(P_+(f_1 \otimes \cdots \otimes f_n)) = P_+ \left(\sum_{i=1}^{n} f_1 \otimes f_2 \otimes \cdots \otimes Hf_i \otimes \cdots \otimes f_n \right), \tag{35}$$

for all f_i in the domain of H, and then extending by continuity. It then follows that $\oplus_{n \geq 0} H_n$ is an 'essentially self-adjoint' operator on

$\mathcal{F}(\mathcal{S}_J)$ (see Bratteli & Robinson 1996, p. 8). We let

$$d\Gamma(H) := \overline{\bigoplus_{n\geq 0} H_n}, \tag{36}$$

denote the resulting (closed) self-adjoint operator. The simplest example occurs when we take $H = I$, in which case it is easy to see that $d\Gamma(H) = N$. However, the total number operator N is not affiliated with the Weyl algebra.[11]

Proposition 6. *When S is infinite-dimensional, $\pi(\mathcal{W}[S,\sigma])$ contains no nontrivial bounded functions of the total number operator on $\mathcal{F}(\mathcal{S}_J)$.*

In particular, $\pi(\mathcal{W})$ does not contain any of the spectral projections of N. Thus, while the *conservative* about observables is free to refer to the abstract state ω_J of $\mathcal{W}$ as a 'vacuum' state, he cannot use that language to underwrite the claim that ω_J is a state of 'no particles'!

To define Γ, let U be a unitary operator on $\mathcal{S}_J$. Then $U_n = P_+(U \otimes \cdots \otimes U)$ is a unitary operator on $\mathcal{H}^n$. We define the unitary operator $\Gamma(U)$ on $\mathcal{F}(\mathcal{S}_J)$ by

$$\Gamma(U) := \bigoplus_{n\geq 0} U_n. \tag{37}$$

If $U_t = e^{itH}$ is a weakly continuous unitary group on $\mathcal{S}_J$, then $\Gamma(U_t)$ is a weakly continuous group on $\mathcal{F}(\mathcal{S}_J)$, and we have

$$\Gamma(U_t) = e^{itd\Gamma(H)}. \tag{38}$$

In particular, the one-particle evolution $T_t = e^{itH}$ that was used to fix J 'lifts' to a field evolution given by $\Gamma(T_t)$, where $d\Gamma(H)$ represents the energy of the field and has the vacuum Ω as a ground state.

It can be shown that the representation and second quantization maps interact as follows:

$$\pi(W(Uf)) = \Gamma(U)^*\pi(W(f))\Gamma(U), \qquad f \in S, \tag{39}$$

[11] Our proof in the appendix reconstructs the argument briefly sketched in Segal (1959, p. 12).

for any unitary operator U on $\mathcal{S}_J$. Taking the phase transformation $U = e^{it}I$, it follows that

$$\pi(W(e^{it}f)) = e^{-itN}\pi(W(f))e^{itN}, \qquad f \in S,\ t \in \mathbb{R}. \tag{40}$$

Using Eqn. (33), it also follows that

$$\langle \Gamma(U)\Omega, \pi(W(f))\Gamma(U)\Omega \rangle = \langle \Omega, \pi(W(Uf))\Omega \rangle = \langle \Omega, \pi(W(f))\Omega \rangle. \tag{41}$$

Since the states induced by the vectors Ω and $\Gamma(U)\Omega$ are both normal in π and agree on $\pi(\mathcal{W})$, they determine the same state of $\pi(\mathcal{W})'' = \mathfrak{B}(\mathcal{F}(\mathcal{S}_J))$. Thus Ω must be an eigenvector of $\Gamma(U)$ for any unitary operator U on $\mathcal{S}_J$. In particular, the vacuum is invariant under the group $\Gamma(T_t)$, and is therefore time-translation invariant.

3.3 Disjointness of the Minkowski and Rindler representations

We omit the details of the construction of the classical phase space (S, σ), since they are largely irrelevant to our concerns. The only information we need is that the space S may be taken (roughly) to be solutions to some relativistic wave equation, such as the Klein-Gordon equation. More particularly, S may be taken to consist of pairs of smooth, compactly supported functions on $\mathbb{R}^3$: one function specifies the values of the field at each point in space at some initial time (say $t = 0$), and the other function is the time-derivative of the field (evaluated at $t = 0$). If we then choose a 'timelike flow' in Minkowski spacetime, we will get a corresponding flow in the solution space S; and, in particular, this flow will be given by a one-parameter group T_t of symplectomorphisms on (S, σ).

First, consider the group T_t of symplectomorphisms of (S, σ) induced by the standard inertial timelike flow. (See Figure 9.1, which suppresses two spatial dimensions. Note that it is irrelevant which *inertial* frame's flow we pick, since they all determine the same representation of $\mathcal{W}[S, \sigma]$ up to unitary equivalence; see Wald 1994, p. 106.) It is well known that there is a complex structure M on (S, σ) such that $(\mathcal{S}_M, T_t)$ is a quantum one-particle system (see Kay 1985; Horuzhy 1988, Chap. 4). We call the associated pure regular state

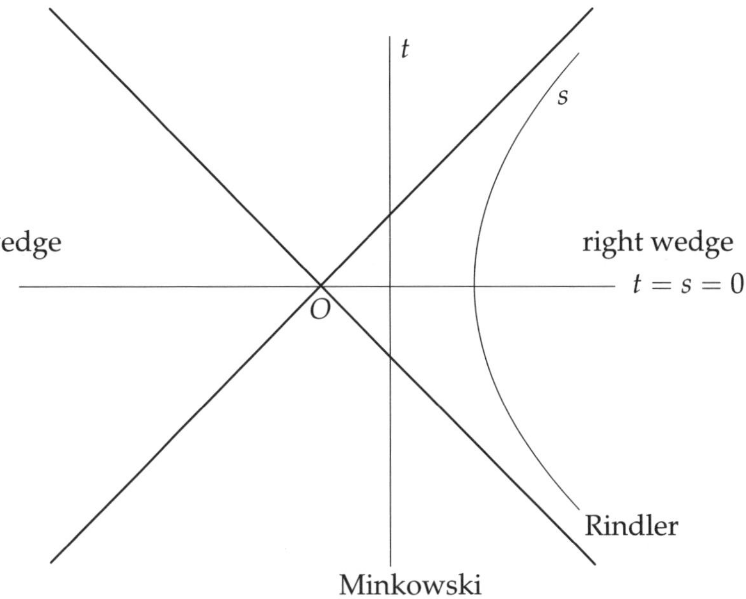

Figure 9.1: Minkowski and Rindler motions

ω_M of $\mathcal{W}[S, \sigma]$ the *Minkowski vacuum state*. As we have seen, it gives rise via the GNS construction to a unique Fock vacuum representation π_{ω_M} on the Hilbert space $\mathcal{H}_{\omega_M} = \mathcal{F}(\mathcal{S}_M)$.

Next, consider the group of Lorentz boosts about a given center point O in spacetime. This also gives rise to a one-parameter group T_s of symplectomorphisms of (S, σ) (cf. Figure 9.1). Let $S(\triangleleft)$ be the subspace of S consisting of Cauchy data with support in the right Rindler wedge ($x_1 > 0$); i.e., at $s = 0$, both the field and its first derivative vanish when $x_1 \leq 0$. Let $\mathcal{W}_\triangleleft := \mathcal{A}[S(\triangleleft), \sigma]$ be the Weyl algebra over the symplectic space $(S(\triangleleft), \sigma)$. Then, T_s leaves $S(\triangleleft)$ invariant, and hence gives rise to a one-parameter group of symplectomorphisms of $(S(\triangleleft), \sigma)$. Kay (1985) has shown rigorously that there is indeed a complex structure R on $(S(\triangleleft), \sigma)$ such that $(\mathcal{S}(\triangleleft)_R, T_s)$ is a quantum one-particle system. We call the resulting state $\omega_R^\triangleleft$ of $\mathcal{W}_\triangleleft$ the *(right) Rindler vacuum state*. It gives rise to a unique GNS-Fock representation $\pi_{\omega_R^\triangleleft}$ of $\mathcal{A}_\triangleleft$ on $\mathcal{H}_{\omega_R^\triangleleft} = \mathcal{F}(\mathcal{S}(\triangleleft)_R)$ and, hence, a quantum field theory for the spacetime consisting of the right wedge *alone*.

The Minkowski vacuum state ω_M of $\mathcal{W}$ also determines, by restriction, a state $\omega_M^\triangleleft$ of $\mathcal{W}_\triangleleft$ (i.e., $\omega_M^\triangleleft := \omega_M|_{\mathcal{W}_\triangleleft}$). Thus, we may apply the GNS construction to obtain the Minkowski representation $(\pi_{\omega_M^\triangleleft}, \mathcal{H}_{\omega_M^\triangleleft})$ of $\mathcal{W}_\triangleleft$. It can be shown (using the 'Reeh-Schlieder theorem' — see Chap. 7) that $\omega_M^\triangleleft$ is a highly mixed state (unlike $\omega_R^\triangleleft$). Therefore, $\pi_{\omega_M^\triangleleft}$ is reducible.

To obtain a concrete picture of this representation, note that (again, as a consequence of the 'Reeh-Schlieder theorem') Ω_{ω_M} is a cyclic vector for the subalgebra $\pi_{\omega_M}(\mathcal{W}_\triangleleft)$ acting on the 'global' Fock space $\mathcal{F}(\mathcal{S}_M)$. Thus, by the uniqueness of the GNS representation $(\pi_{\omega_M^\triangleleft}, \mathcal{H}_{\omega_M^\triangleleft})$, it is unitarily equivalent to the representation $(\pi_{\omega_M}|_{\mathcal{W}_\triangleleft}, \mathcal{F}(\mathcal{S}_M))$. It can be shown that $\pi_{\omega_M}(\mathcal{W}_\triangleleft)''$ is a factor (Horuzhy, 1988, Thm. 3.3.4). Thus, while reducible, $\pi_{\omega_M^\triangleleft}$ is still factorial.

Under the liberal approach to observables, the representations $\pi_{\omega_M^\triangleleft}$ (factorial) and $\pi_{\omega_R^\triangleleft}$ (irreducible) provide physically inequivalent descriptions of the physics in the right wedge.[12]

Proposition 7. *The Minkowski and Rindler representations of $\mathcal{W}_\triangleleft$ are disjoint.*

Now let $\triangleright$ denote the left Rindler wedge, and define the subspace $S(\triangleright)$ of S as $S(\triangleleft)$ was defined above. (Of course, by symmetry, Proposition 7 holds for $\mathcal{W}_\triangleright$ as well.) Let $\mathcal{W}_{\bowtie} := \mathcal{W}[S(\triangleright) \oplus S(\triangleleft), \sigma]$ denote the Weyl algebra over the symplectic space $(S(\triangleright) \oplus S(\triangleleft), \sigma)$. Then $\mathcal{W}_{\bowtie} = \mathcal{W}_\triangleright \otimes \mathcal{W}_\triangleleft$, and $\omega_M^{\bowtie} := \omega_M|_{\mathcal{W}_{\bowtie}}$ is *pure* (Kay 1985, Defn., Thm. 1.3(iii)).[13] The GNS representation $\omega_M^{\bowtie}$ induces is therefore

[12]If only locally measurable observables have genuine physical significance (see note 5), then the Minkowski and Rindler representations *are* physically equivalent. Indeed, since both $\omega_M^\triangleleft$ and $\omega_R^\triangleleft$ are 'of Hadamard form', it follows that $\pi_{\omega_M^\triangleleft}$ and $\pi_{\omega_R^\triangleleft}$ are 'locally quasi-equivalent' (Verch 2000; cf. Verch 1994, Theorem 3.9). That is, for each algebra $\mathcal{A}(O)$ of local observables in $\mathcal{W}_\triangleleft$, $\pi_{\omega_M^\triangleleft}|_{\mathcal{A}(O)}$ is quasi-equivalent to $\pi_{\omega_R^\triangleleft}|_{\mathcal{A}(O)}$. Clearly, this fact only strengthens our case against the claim that the Minkowski and Rindler representations correspond to incommensurable theories of the quantum field.

[13]The restriction of ω_M to $\mathcal{W}_{\bowtie}$ is a pure 'quasifree' state. Thus, there is a complex

irreducible, and (again invoking the uniqueness of the GNS representation) it is equivalent to $(\pi_{\omega_M}|_{\mathcal{W}_{\bowtie}}, \mathcal{F}(\mathcal{S}_M))$ (since $\Omega_{\omega_M} \in \mathcal{F}(\mathcal{S}_M)$ is a cyclic vector for the subalgebra $\pi_{\omega_M}(\mathcal{W}_{\bowtie})$ as well).

The tensor product of the pure left and right Rindler vacua $\omega_R^{\bowtie} := \omega_R^{\triangleright} \otimes \omega_R^{\triangleleft}$ is of course also a pure state of $\mathcal{W}_{\bowtie}$.[14] It will induce a GNS representation of the latter on the Hilbert space $\mathcal{H}_{\omega_R^{\bowtie}}$ given by $\mathcal{F}(\mathcal{S}_R) \equiv \mathcal{F}(\mathcal{S}(\triangleright)_R) \otimes \mathcal{F}(\mathcal{S}(\triangleleft)_R)$. It is not difficult to show that $\omega_R^{\bowtie}$ and $\omega_M^{\bowtie}$, both now irreducible, are also disjoint.[15]

Proposition 8. *The Minkowski and Rindler representations of $\mathcal{W}_{\bowtie}$ are disjoint.*

In our final main section we shall discuss the conceptually problematic implications that the *M-vacuum* states $\omega_M^{\bowtie}$ and $\omega_M^{\triangleleft}$ have for the *presence* of R-quanta in the double and right wedge spacetime regions. However, we note here an important difference between Rindler and Minkowski observers.

The total number of R-quanta, according to a Rindler observer confined to the left (resp., right) wedge, is represented by the number operator $N_{\triangleright}$ (resp., $N_{\triangleleft}$) on $\mathcal{F}(\mathcal{S}(\triangleright)_R)$ (resp., $\mathcal{F}(\mathcal{S}(\triangleleft)_R)$). However, because of the spacelike separation of the wedges, no single Rindler observer has access, even in principle, to the expectation value of the 'overall' total Rindler number operator $N_R = N_{\triangleright} \otimes I + I \otimes N_{\triangleleft}$ acting on $\mathcal{F}(\mathcal{S}(\triangleright)_R) \otimes \mathcal{F}(\mathcal{S}(\triangleleft)_R)$.

The reverse is true for a Minkowski observer. While she has access, at least in principle, to the total number of M-quanta op-

structure M' on $S(\triangleright) \oplus S(\triangleleft)$ such that

$$\omega_M(W(f)) = \exp(-\sigma(f, M'f)/4) = \exp(-\sigma(f, Mf)/4), \tag{42}$$

for all $f \in S(\triangleright) \oplus S(\triangleleft)$ (Petz 1990, Prop. 3.9). It is not difficult to see then that $M|_{S(\triangleright)\oplus S(\triangleleft)} = M'$ and therefore that M leaves $S(\triangleright) \oplus S(\triangleleft)$ invariant. Hereafter, we will use M to denote the complex structure on S as well as its restriction to $S(\triangleright) \oplus S(\triangleleft)$.

[14] More precisely, $\omega_R^{\triangleleft}$ arises from a complex structure $R_{\triangleleft}$ on $S(\triangleleft)$, $\omega_R^{\triangleright}$ arises from a complex structure $R_{\triangleright}$ on $S(\triangleright)$, and $\omega_R^{\bowtie}$ arises from the complex structure $R_{\triangleright} \oplus R_{\triangleleft}$ of $S(\triangleright) \oplus S(\triangleleft)$. When no confusion can result, we will use R to denote the complex structure on $S(\triangleright) \oplus S(\triangleleft)$ and its restriction to $S(\triangleleft)$.

[15] We give proofs of Propositions 7 and 8 in the appendix. For another proof, employing quite different methods, see the appendix of Beyer (1991).

erator N_M acting on $\mathcal{F}(\mathcal{S}_M)$, N_M is a purely global observable that does not split into the sum of two separate number operators associated with the left and right wedges (as a general consequence of the 'Reeh-Schlieder theorem' — see Redhead 1995). In fact, since the Minkowski complex structure M is an 'anti-local' operator (Segal & Goodman 1965), it fails to leave either of the subspaces $S(\triangleright)$ or $S(\triangleleft)$ invariant, and it follows that no M-quanta number operator is affiliated with $\pi_{\omega_M^{\triangleleft}}(\mathcal{W}_{\triangleleft})''$.[16] Thus, even a liberal about observables must say that a Minkowski observer with access only to the right wedge does not have the capability of counting M-quanta.

So, while it might be sensible to ask for the probability in state $\omega_M^{\triangleleft}$ that a Rindler observer detects particles in the right wedge, it is *not* sensible to ask, conversely, for the probability in state $\omega_R^{\triangleleft}$ that a Minkowski observer will detect particles in the right wedge. Note also that since N_M is a purely global observable (i.e., there is no sense to be made of 'the number of Minkowski quanta in a bounded spatial or spacetime region'), what a Minkowski observer might *locally* detect with a 'particle detector' (over an extended, but finite, interval of time) can at best give an approximate indication of the global Minkowski particle content of the field.

4 Minkowski versus Rindler quanta

We have seen that a Rindler observer will construct 'his quantum field theory' of the right wedge spacetime region differently from a Minkowski observer. He will use the complex structure R picked out uniquely by the boost group about O, and build up a representation of $\mathcal{W}_{\triangleleft}$ on the Fock space $\mathcal{F}(\mathcal{S}(\triangleleft)_R)$. However, suppose that the state of $\mathcal{W}_{\triangleleft}$ is the state $\omega_M^{\triangleleft}$ of *no* particles (globally!) according to a Minkowski observer. What, if anything, will our Rindler observer say about the particle content in the right wedge? And does *this* question even make sense?

We shall argue that this question does make sense, notwithstanding the disjointness of the Minkowski and Rindler represen-

[16]See Halvorson (2001*b*) for further details and a critical analysis of different approaches to the problem of particle localization in quantum field theory.

tations. And the answer is surprising: Not only does a Rindler observer have a nonzero chance of detecting the presence of R-quanta, but if a Rindler observer were to measure the *total* number of R-quanta in the right wedge, he would always find (as we show in Section 4.2) that the probability of an *infinite* total number is *one*!

We begin in section 4.1 by discussing the paradox of observer-dependence of particles to which such results lead. In particular, we shall criticize Teller's (1995, 1996) resolution of this paradox. Later, in Section 4.3, we shall also criticize the arguments of Arageorgis (1995) and Arageorgis *et al.* (2002) for the incommensurability of inequivalent particle concepts, and argue, instead, for their complementarity (in *support* of Teller).

4.1 The paradox of the observer-dependence of particles

Not surprisingly, physicists initially found a Rindler observer's ability to detect particles in the Minkowski vacuum paradoxical (see Rüger 1989, p. 571; Teller 1995, p. 110). After all, particles are the sorts of things that are either there or not there, so how could their presence depend on an observer's state of motion?

One way to resist this paradox is to reject from the outset the physicality of the Rindler representation, thereby withholding bona fide particle status from Rindler quanta. For instance, one could be bothered by the fact that the Rindler representation cannot be globally defined over the whole of Minkowski spacetime, or that the one-particle Rindler Hamiltonian lacks a mass gap, allowing an arbitrarily large number of R-quanta to have a fixed finite amount of energy ('infrared divergence'). Arageorgis (1995, Chap. 6) gives a thorough discussion of these and other 'pathologies' of the Rindler representation.[17] In consequence, he argues that the phenomenology associated with a Rindler observer's 'particle detections' in the Minkowski vacuum ought to be explained entirely in terms of observables affiliated to the Minkowski representation (such as garden-variety Minkowski vacuum fluctuations of the local field observables).

[17] See also, more recently, Belinskiĭ (1997), Fedotov *et al.* (1999), Nikolić (2000).

This is not the usual response to the paradox of observer-dependence. Rüger (1989) has characterized the majority of physicists' responses in terms of the *field approach* and the *detector approach*. Proponents of the field approach emphasize the need to forfeit particle talk at the fundamental level, and to focus the discussion on measurement of local field quantities. Those of the detector approach emphasize the need to relativize particle talk to the behavior of concrete detectors following specified world-lines. Despite their differing emphases, and the technical difficulties in unifying these programs (well documented by Arageorgis 1995), neither eschews the Rindler representation as unphysical, presumably because of its deep connections with quantum statistical mechanics and blackhole thermodynamics (Sciama *et al.* 1981). Moreover, pathological or not, it remains of philosophical interest to examine the consequences of taking the Rindler representation seriously — just as the possibility of time travel in general relativity admitted by certain 'pathological' solutions to Einstein's field equations is of interest. And it is remarkable that there should be *any* region of Minkowski spacetime that admits two physically inequivalent quantum field descriptions.

Teller (1995, 1996) has recently offered his own resolution of the paradox. We reproduce below the relevant portions of his discussion in Teller (1995, p. 111). However, note that he does not distinguish between left and right Rindler observers, $|0; M\rangle$ refers, in our notation, to the Minkowski vacuum vector $\Omega_{\omega_M} \in \mathcal{F}(\mathcal{S}_M)$, and $|1,0,0,\ldots\rangle_M$ (resp., $|1,0,0,\ldots\rangle_R$) is a one-particle state $0 \oplus f \oplus 0 \oplus 0 \oplus \cdots \in \mathcal{F}(\mathcal{S}_M)$ (resp., $\in \mathcal{F}(\mathcal{S}_R)$).

> ... Rindler raising and lowering operators are expressible as superpositions of the Minkowski raising and lowering operators, and states with a definite number of Minkowski quanta are superpositions of states with different numbers of Rindler quanta. In particular, $|0; M\rangle$ is a superposition of Rindler quanta states, including states for arbitrarily large numbers of Rindler quanta. In other words, $|0; M\rangle$ has an exact value of zero for the Minkowski number operator, and is simultaneously

> highly indefinite for the Rindler number operator.
>
> ... In $|0; M\rangle$ there is *no* definite number of Rindler quanta. There is only a propensity for detection of one or another number of Rindler quanta by an accelerating detector. A state in which a quantity has no exact value is one in which no values for that quantity are definitely, and so actually, exemplified. Thus in $|0; M\rangle$ no Rindler quanta actually occur, so the status of $|0; M\rangle$ as a state completely devoid of quanta is not impugned.
>
> To be sure, this interpretive state of affairs is surprising. To spell it out one step further, in $|1, 0, 0, \ldots\rangle_M$ there is one actual Minkowski quantum, no actual Rindler quanta, and all sorts of propensities for manifestation of Rindler quanta, among other things. In $|1, 0, 0, \ldots\rangle_R$ the same comment applies with the role of Minkowski and Rindler reversed. It turns out that there are various kinds of quanta, and a state in which one kind of quanta actually occurs is a state in which there are only propensities for complementary kinds of quanta. Surprising, but perfectly consistent and coherent.

Teller's point is that R-quanta only exist (so to speak) potentially in the M-vacuum, not actually. Thus it is still an invariant observer-independent fact that there are no *actual quanta* in the field, and the paradox evaporates. Similarly for Minkowski states of one or more particles as seen by Rindler observers. There is the same definite number of *actual quanta* for all observers. Thus, since actual particles are the 'real stuff', the real stuff *is* invariant!

Notice, however, that there is something self-defeating in Teller's final concession, urged by advocates of the field and detector approaches, that different kinds of quanta need to be distinguished. For if we do draw the distinction sharply, it is no longer clear why even the actual presence of R-quanta in the M-vacuum should bother us. Teller seems to want to have it both ways: while there are different kinds of quanta, there is still only one kind of *actual* quanta, and it better be invariant.

Does this invariance really hold? In one sense, yes. Disjointness does not prevent us from building Rindler creation and annihilation operators on the Minkowski representation space $\mathcal{F}(\mathcal{S}_M)$. We simply need to define Rindler analogues, $a_R^*(f)$ and $a_R(f)$, of the Minkowski creation and annihilation operators via Eqns. (34) with $\Phi(Rf)$ in place of $\Phi(if)$ $(= \Phi(Mf))$ (noting that $f \mapsto a_R(f)$ will now be anti-linear with respect to the *Rindler* conjugation R). It is then easy to see, using (31), that

$$a_R(f) = 2^{-1}[a_M^*((I + MR)f) + a_M((I - MR)f)]. \tag{43}$$

This linear combination would be trivial if $R = \pm M$. However, we know $R \neq M$, and $R = -M$ is ruled out because it is inconsistent with both complex structures being positive definite. Consequently, $\Omega_{\omega_M^{\bowtie}}$ must be a nontrivial superposition of eigenstates of the Rindler number operator $N_R(f) := a_R^*(f)a_R(f)$; for an easy calculation, using (43), reveals that

$$N_R(f)\Omega_{\omega_M^{\bowtie}} = 2^{-2}[\Omega_{\omega_M^{\bowtie}} + a_M^*((I - MR)f)a_M^*((I + MR)f)\Omega_{\omega_M^{\bowtie}}], \tag{44}$$

which (the presence of the nonzero second term guarantees) is not simply a multiple of $\Omega_{\omega_M^{\bowtie}}$. Thus, Teller would be correct to conclude that the Minkowski vacuum implies dispersion in the number operator $N_R(f)$. And the same conclusion would follow if, instead, we considered the Minkowski creation and annihilation operators as acting on the Rindler representation space $\mathcal{F}(\mathcal{S}_R)$. Since only finitely many degrees of freedom are involved, this is guaranteed by the Stone-von Neumann theorem.

However, therein lies the rub. $N_R(f)$ merely represents the number of R-quanta with a specified wavefunction f. What about the *total* number of R-quanta in the M-vacuum (which involves *all* degrees of freedom)? If Teller cannot assure us that this too has dispersion, his case for the invariance of 'actual quanta' is left in tatters. In his discussion, Teller fails to distinguish $N_R(f)$ from the total number operator N_R, but the distinction is crucial. It is a well known consequence of the disjointness of $\pi_{\omega_R^{\bowtie}}$ and $\pi_{\omega_M^{\bowtie}}$ that neither representation's total number operator is definable on the Hilbert space of the other (Bratteli & Robinson 1996, Thm. 5.2.14). Therefore, it is literally *nonsense* to speak of $\Omega_{\omega_M^{\bowtie}}$ as a superposition of eigenstates of N_R![18] If $x_n, x_m \in \mathcal{F}(\mathcal{S}_R)$ are eigenstates of N_R with eigenvalues n, m respectively, then $x_n + x_m$ again lies in $\mathcal{F}(\mathcal{S}_R)$, and so is 'orthogonal' to all eigenstates of the Minkowski number operator N_M acting on $\mathcal{F}(\mathcal{S}_M)$. And, indeed, taking infinite sums of Rindler number eigenstates will again leave us in the folium of the Rindler representation. As Arageorgis (1995, p. 303) has also noted: 'The Minkowski vacuum state is not a superposition of Rindler quanta states, despite "appearances"'.[19]

Yet this point, by itself, does not tell us that Teller's discussion cannot be salvaged. Recall that a state ρ is *dispersion-free* on a (bounded) observable X just in case $\rho(X^2) = \rho(X)^2$. Suppose, now, that Y is a possibly *un*bounded observable that is definable in some representation π of $\mathcal{W}$. We can then rightly say that an algebraic state ρ of $\mathcal{W}$ *predicts dispersion in* Y just in case, for *every* extension $\hat{\rho}$ of ρ to $\pi(\mathcal{W})''$, $\hat{\rho}$ is *not* dispersion-free on all bounded functions of

[18] In their review of Teller's (1995) book, Huggett & Weingard (1996) question whether Teller's 'quanta interpretation' of quantum field theory can be implemented in the context of inequivalent representations. However, when they discuss Teller's resolution of the observer-dependence paradox, in terms of mere *propensities to display R*-quanta in the *M*-vacuum, they write 'This seems all well and good' (1996, p. 309)! Their only criticism is the obvious one: legitimizing such propensity talk ultimately requires a solution to the measurement problem. Teller's response to their review is equally unsatisfactory. Though he pays lip-service to the possibility of inequivalent representations (Teller 1998, pp. 156–7), he fails to notice how inequivalence undercuts his discussion of the paradox.

[19] Arageorgis presumes Teller's discussion is based upon the appearance of the following purely formal (i.e., nonnormalizable) expression for $\Omega_{\omega_M^{\bowtie}}$ as a superposition in $\mathcal{F}(\mathcal{S}_R) \equiv \mathcal{F}(\mathcal{S}(\triangleright)_R) \otimes \mathcal{F}(\mathcal{S}(\triangleleft)_R)$ over left ('I') and right ('II') Rindler modes (Wald 1994, Eqn. 5.1.27):

$$\prod_i \left\{ \sum_{n=0}^{\infty} \exp(-n\pi\omega_i/a)\, |n_{iI}\rangle \otimes |n_{iII}\rangle \right\} . \tag{45}$$

However, it bears mentioning that, as this expression suggests: (a) the restriction of $\omega_M^{\bowtie}$ to either $\mathcal{W}_{\triangleright}$ or $\mathcal{W}_{\triangleleft}$ is indeed mixed; (b) $\omega_M^{\bowtie}$ can be shown rigorously to be an entangled state of $\mathcal{W}_{\triangleright} \otimes \mathcal{W}_{\triangleleft}$ (Chapter 7 of this volume); and (c) the thermal properties of the 'reduced density matrix' for either wedge obtained from this formal expression can be derived rigorously (Kay 1985). In addition, see Propositions 9 and 10 below!

Y. We then have the following result.

Proposition 9. *If J_1, J_2 are distinct complex structures on (S, σ), then ω_{J_1} (resp., ω_{J_2}) predicts dispersion in N_{J_2} (resp., N_{J_1}).*

As a consequence, the Minkowski vacuum $\omega_M^{\bowtie}$ indeed predicts dispersion in the Rindler total number operator N_R (and in both $N_{\triangleright} \otimes I$ and $I \otimes N_{\triangleleft}$, invoking the symmetry between the wedges).

Teller also writes of the Minkowski vacuum as being a superposition of eigenstates of the Rindler number operator with *arbitrary large* eigenvalues. Eschewing the language of superposition, the idea that there is no finite number of R-quanta to which the M-vacuum assigns probability one can also be rendered sensible. The relevant result was first proved by Fulling (1972, Appendix F) (see also Fulling 1989, p. 145):

Fulling's "Theorem". *Two Fock vacuum representations $(\pi, \mathcal{F}(\mathcal{H}), \Omega)$ and $(\pi', \mathcal{F}(\mathcal{H}'), \Omega')$ of $\mathcal{W}$ are unitarily equivalent if and only if $\langle \Omega, N'\Omega \rangle < \infty$ (or, equivalently, $\langle \Omega', N\Omega' \rangle < \infty$).*

As stated, this 'theorem' also fails to make sense, because it is only in the case where the representations are *already* equivalent that the primed total number operator is definable on the unprimed representation space and an expression like '$\langle \Omega, N'\Omega \rangle$' is well defined. (We say more about why this is so in the next section.) However, there *is* a way to understand the expression '$\langle \Omega, N'\Omega \rangle < \infty$' (resp., '$\langle \Omega, N'\Omega \rangle = \infty$') in a rigorous, non-question-begging way. We can take it to be the claim that all extensions $\hat{\rho}$ of the abstract unprimed vacuum state of $\mathcal{W}$ to $\mathfrak{B}(\mathcal{F}(\mathcal{H}'))$ assign (resp., do *not* assign) N' a finite value; i.e., for any such extension, $\sum_{n'=1}^{\infty} \hat{\rho}(P_{n'})n'$ converges (resp., does not converge), where $\{P_{n'}\}$ are the spectral projections of N'. With this understanding, the following rigorization of Fulling's 'theorem' can then be proved.

Proposition 10. *A pair of Fock representations $\pi_{\omega_{J_1}}, \pi_{\omega_{J_2}}$ are unitarily equivalent if and only if ω_{J_1} assigns N_{J_2} a finite value (equivalently, ω_{J_2} assigns N_{J_1} a finite value).*

It follows that $\omega_M^{\bowtie}$ cannot assign probability one to any finite number of R-quanta (and vice versa, with $R \leftrightarrow M$).

Unfortunately, neither Proposition 9 nor 10 is sufficient to rescue Teller's 'actual quanta' invariance argument, for these propositions give no further information about the shape of the probability distribution that $\omega_M^{\bowtie}$ prescribes for N_R's eigenvalues. In particular, both propositions are compatible with there being a probability of *one* that *at least* $n > 0$ R-quanta obtain in the M-vacuum, for any $n \in \mathbb{N}$. If that were the case, Teller would then be forced to withdraw and concede that at least *some*, and perhaps many, Rindler quanta *actually* occur in a state with no actual Minkowski quanta. In the next section, we shall show that this — Teller's worst nightmare — is in fact the case.

4.2 Minkowski probabilities for Rindler number operators

We now defend the claim that a Rindler observer will say that there are actually *infinitely many* quanta while the field is in the Minkowski vacuum state (or, indeed, in any other state of the Minkowski folium).[20] This result applies more generally to any pair of disjoint regular representations, at least one of which is the GNS representation of an abstract Fock vacuum state. We shall specialize back down to the Minkowski/Rindler case later on.

Let ρ be a regular state of $\mathcal{W}$ inducing the GNS representation $(\pi_\rho, \mathcal{H}_\rho)$, and let ω_J be the abstract vacuum state determined by a complex structure J on (S, σ). The case we are interested in is, of course, when π_ρ, π_{ω_J} are disjoint. We first want to show how to define representation-independent probabilities in the state ρ for any J-quanta number operator that 'counts' the number of quanta with wavefunctions in a fixed *finite*-dimensional subspace $F \subseteq \mathcal{S}_J$. (Parts of our exposition below follow Bratteli & Robinson (1996, pp.

[20] In fact, this was first proved, in effect, by Chaiken (1967). However his lengthy analysis focused on comparing Fock with non-Fock (so-called 'strange') representations of the Weyl algebra, and the implications of his result for disjoint Fock representations based on inequivalent one-particle structures seem not to have been carried down into the textbook tradition of the subject. (The closest result we have found is Bratteli & Robinson (1996, Thm. 5.2.14), which we are able to employ as a lemma to recover Chaiken's result for disjoint Fock representations — see Section 5.)

26–30), which may be consulted for further details.)

We know that, for any $f \in S$, there exists a self-adjoint operator $\Phi_\rho(f)$ on $\mathcal{H}_\rho$ such that

$$\pi_\rho(W(tf)) = \exp\left(it\Phi_\rho(f)\right), \qquad t \in \mathbb{R}. \tag{46}$$

We can also define unbounded annihilation and creation operators on $\mathcal{H}_\rho$ for J-quanta by

$$\begin{aligned} a_\rho(f) &= 2^{-1/2}(\Phi_\rho(f) + i\Phi_\rho(Jf)), \\ a^*_\rho(f) &= 2^{-1/2}(\Phi_\rho(f) - i\Phi_\rho(Jf)). \end{aligned} \tag{47}$$

Earlier, we denoted these operators by $a_J(f)$ and $a^*_J(f)$. However, we now want to emphasize the representation space upon which they act; and only the single complex structure J shall concern us in our general discussion, so there is no possibility of confusion with others.

Next, define a 'quadratic form' $n_\rho(F) : \mathcal{H}_\rho \mapsto \mathbb{R}^+$. The domain of $n_\rho(F)$ is

$$D(n_\rho(F)) := \bigcap_{f \in F} D(a_\rho(f)), \tag{48}$$

where $D(a_\rho(f))$ is the domain of $a_\rho(f)$. Now let $\{f_k : k = 1, \ldots, m\}$ be some J-orthonormal basis for F, and define

$$[n_\rho(F)](\psi) := \sum_{k=1}^{m} \|a_\rho(f_k)\psi\|^2, \tag{49}$$

for any $\psi \in D(n_\rho(F))$. It can be shown that the sum in (49) is independent of the chosen orthonormal basis for F, and that $D(n_\rho(F))$ lies dense in $\mathcal{H}_\rho$. Given any densely defined, positive, closed quadratic form t on $\mathcal{H}_\rho$, there exists a unique positive self-adjoint operator T on $\mathcal{H}_\rho$ such that $D(t) = D(T^{1/2})$ and

$$t(\psi) = \langle T^{1/2}\psi, T^{1/2}\psi \rangle, \qquad \psi \in D(t). \tag{50}$$

We let $N_\rho(F)$ denote the finite-subspace J-quanta number operator on $\mathcal{H}_\rho$ arising from the quadratic form $n_\rho(F)$.

We seek a representation-*independent* value for 'Prob$^{\rho}(N(F) \in \Delta)$', where $\Delta \subseteq \mathbb{N}$. So let τ be *any* regular state of $\mathcal{W}$, and let $N_\tau(F)$ be the corresponding number operator on $\mathcal{H}_\tau$. Let $\mathcal{W}_F$ be the Weyl algebra over $(F, \sigma|_F)$, and let $E_\tau(F)$ denote the spectral measure for $N_\tau(F)$ acting on $\mathcal{H}_\tau$. Then, $[E_\tau(F)](\Delta)$ (the spectral projection representing the proposition '$N_\tau(F) \in \Delta$') is in the weak closure of $\pi_\tau(\mathcal{W}_F)$, by the Stone-von Neumann uniqueness theorem. In particular, there is a net $\{A_i\} \subseteq \mathcal{W}_F$ such that $\pi_\tau(A_i)$ converges weakly to $[E_\tau(F)](\Delta)$. Now, the Stone-von Neumann uniqueness theorem also entails that there is a density operator D_ρ on $\mathcal{H}_\tau$ such that

$$\rho(A) = \mathrm{Tr}(D_\rho \pi_\tau(A)), \qquad A \in \mathcal{W}_F. \tag{51}$$

We therefore define

$$\begin{aligned} \mathrm{Prob}^{\rho}(N(F) \in \Delta) &:= \lim_i \rho(A_i) && (52)\\ &= \lim_i \mathrm{Tr}(D_\rho \pi_\tau(A_i)) && (53)\\ &= \mathrm{Tr}(D_\rho [E_\tau(F)](\Delta)). && (54) \end{aligned}$$

The final equality displays that this definition is independent of the chosen approximating net $\{\pi_\tau(A_i)\}$, and the penultimate equality displays that this definition is independent of the (regular) representation π_τ. In particular, since we may take $\tau = \rho$, it follows that

$$\mathrm{Prob}^{\rho}(N(F) \in \Delta) \quad = \quad \langle \Omega_\rho, [E_\rho(F)](\Delta)\Omega_\rho \rangle, \tag{55}$$

exactly as expected.

We can also define a positive, closed quadratic form on $\mathcal{H}_\rho$ corresponding to the *total* J-quanta number operator by:

$$\begin{aligned} n_\rho(\psi) &= \sup_{F \in \mathbb{F}} [n_\rho(F)](\psi), && (56)\\ D(n_\rho) &= \left\{ \psi \in \mathcal{H}_\rho : \psi \in \bigcap_{f \in S} D(a_\rho(f)),\ n_\rho(\psi) < \infty \right\}, && (57) \end{aligned}$$

where $\mathbb{F}$ denotes the collection of all finite-dimensional subspaces of $\mathcal{S}_J$. If $D(n_\rho)$ is dense in $\mathcal{H}_\rho$, then it makes sense to say that the total

J-quanta number operator N_ρ exists on the Hilbert space $\mathcal{H}_\rho$. In general, however, $D(n_\rho)$ will not be dense, and may contain only the 0 vector. Accordingly, we cannot use a direct analogue to Eqn. (54) to define the probability, in the state ρ, that there are, say, n or fewer J-quanta.

However, we can still proceed as follows. Fix $n \in \mathbb{N}$, and suppose $F \subseteq F'$ with both $F, F' \in \mathbb{F}$. Since any state with n or fewer J-quanta with wavefunctions in F' cannot have *more* than n J-quanta with wavefunctions in the (smaller) subspace F,

$$\text{Prob}^\rho(N(F) \in [0,n]) \ \geq \ \text{Prob}^\rho(N(F') \in [0,n]). \tag{58}$$

Thus, whatever value we obtain for 'Prob$^\rho(N \in [0,n])$', it should satisfy the inequality

$$\text{Prob}^\rho(N(F) \in [0,n]) \ \geq \ \text{Prob}^\rho(N \in [0,n]), \tag{59}$$

for any finite-dimensional subspace $F \subseteq \mathcal{S}_J$. However, the following result holds.

Proposition 11. *If ρ is a regular state of $\mathcal{W}$ disjoint from the Fock state ω_J, then* $\inf_{F\in\mathbb{F}} \left\{ \text{Prob}^\rho(N(F) \in [0,n]) \right\} = 0$ *for every $n \in \mathbb{N}$.*

Thus ρ must assign every finite number of J-quanta probability zero; i.e., ρ predicts an infinite number of J-quanta with probability one!

Let us tighten this up some more. Suppose that we are in any regular representation $(\pi_\omega, \mathcal{H}_\omega)$ in which the total J-quanta number operator N_ω exists and is affiliated to $\pi_\omega(\mathcal{W})''$. (For example, we may take the Fock representation where $\omega = \omega_J$.) Let E_ω denote the spectral measure of N_ω on $\mathcal{H}_\omega$. Considering ρ as a state of $\pi_\omega(\mathcal{W})$, it is then reasonable to define

$$\text{Prob}^\rho(N \in [0,n]) := \widehat{\rho}(E_\omega([0,n])), \tag{60}$$

where $\widehat{\rho}$ is any extension of ρ to $\pi_\omega(\mathcal{W})''$, provided the right-hand side takes the same value for all extensions. (And, of course, it will when $\rho \in \mathfrak{F}(\pi_\omega)$, where (60) reduces to the standard definition.) Now clearly

$$[E_\omega(F)]([0,n]) \geq E_\omega([0,n]),\ F \in \mathbb{F}. \tag{61}$$

('If there are at most n J-quanta in total, then there are at most n J-quanta whose wavefunctions lie in any finite-dimensional subspace of $\mathcal{S}_J$'.) Since states preserve order relations between projections, every extension $\widehat{\rho}$ must therefore satisfy

$$\mathrm{Prob}^\rho(N(F) \in [0,n]) = \widehat{\rho}([E_\omega(F)]([0,n])) \geq \widehat{\rho}(E_\omega([0,n])). \tag{62}$$

Thus, if ρ is disjoint from ω, Proposition 11 entails that $\mathrm{Prob}^\rho(N \in [0,n]) = 0$ for all finite n.[21]

As an immediate consequence of this and the disjointness of the Minkowski and Rindler representations, we have (reverting back to our earlier number operator notation):

$$\mathrm{Prob}^{\omega_M^{\bowtie}}(N_R \in [0,n]) = 0 = \mathrm{Prob}^{\omega_R^{\bowtie}}(N_M \in [0,n]),\ \text{for all } n \in \mathbb{N}, \tag{63}$$

$$\mathrm{Prob}^{\omega_M^{\triangleright}}(N_\triangleright \in [0,n]) = 0 = \mathrm{Prob}^{\omega_M^{\triangleleft}}(N_\triangleleft \in [0,n]),\ \text{for all } n \in \mathbb{N}. \tag{64}$$

The same probabilities obtain when the Minkowski vacuum is replaced with any other state normal in the Minkowski representation.[22] So it could not be farther from the truth to say that there is merely the potential for Rindler quanta in the Minkowski vacuum, or any other eigenstate of N_M.

One must be careful, however, with an informal statement like 'The M-vacuum contains infinitely many R-quanta with probability 1'. Since Rindler wedges are unbounded, there is nothing unphysical, or otherwise metaphysically incoherent, about thinking of

[21]Notice that such a prediction could never be made by a state in the folium of π_ω, since density operator states are countably additive (see note 3).

[22]This underscores the utter bankruptcy, from the standpoint of the liberal about observables, in taking the weak equivalence of the Minkowski and Rindler representations to be sufficient for their physical equivalence. Yes, every Rindler state of the Weyl algebra is a weak* limit of Minkowski states. But the former all predict a finite number of Rindler quanta with probability one, while the latter all predict an *infinite* number with probability one! (Wald (1994, pp. 82–3) makes the exact same point with respect to states that do and do not satisfy the 'Hadamard' property.)

wedges as containing an infinite number of Rindler quanta. But we must not equate this with the quite different *empirical* claim 'A Rindler observer's particle detector has the sure-fire disposition to register the value "∞"'. There is no such value! Rather, the empirical content of equations (63) and (64) is simply that an idealized 'two-state' measuring apparatus designed to register whether there are $> n$ Rindler quanta in the Minkowski vacuum will always return the answer 'Yes'. This is a perfectly sensible physical disposition for a measuring device to have. Of course, we are not pretending to have in hand a specification of the physical details of such a device. Indeed, when physicists model particle detectors, these are usually assumed to couple to specific 'modes' of the field, represented by finite-subspace, not total, number operators (cf., e.g., Wald 1994, Sec. 3.3). But this is really beside the point, since Teller advertises his resolution of the paradox as a way to *avoid* a 'retreat to instrumentalism' about the particle concept (1995, p. 110).

On Teller's behalf, one might object that there are still no grounds for saying any R-quanta obtain in the M-vacuum, since for any particular number n of R-quanta you care to name, equations (63) and (64) entail that n is *not* the number of R-quanta in the M-vacuum. But remember that the same is true for $n = 0$, and that, therefore, $n \geq 1$ R-quanta has probability 1! A further tack might be to deny that probability 0 for $n = 0$, or any other n, entails impossibility or non-actuality of that number of R-quanta. This would be similar to a common move made in response to the lottery paradox, in the hypothetical case where there are an infinite number of ticket holders. Since *someone* has to win, each ticket holder must still have the *potential* to win, even though his or her probability of winning is zero. The difficulty with this response is that in the Rindler case, we have no independent reason to think that some particular finite number of R-quanta *has* to be detected at all. Moreover, if we were to go soft on taking probability 0 to be sufficient for 'not actual', we should equally deny that probability 1 is sufficient for 'actual', and by Teller's lights the paradox would go away at a stroke (because there could never be actual Rindler *or* Minkowski quanta in *any* field state).

We conclude that Teller's resolution of the paradox of observer-dependence of particles fails. And so be it, since it was ill-motivated in the first place. We already indicated in the previous subsection that it should be enough of a resolution to recognize that there are different kinds of quanta. We believe the physicists of the field and detector approaches are correct to bite the bullet hard on this, even though it means abandoning naïve realism about particles (though not, of course, about detection events). We turn, next, to arguing that a coherent story can still be told about the relationship between the different kinds of particle talk used by different observers.

4.3 Incommensurable or complementary?

At the beginning of this paper, we reproduced a passage from Jauch's amusing Galilean dialogue on the question 'Are Quanta Real?'. In that passage, Sagredo is glorying in the prospect that complementarity may be applicable even in classical physics; and, more generally, to solving the philosophical problem of the specificity of individual events versus the generality of scientific description. It is well known that Bohr himself sought to extend the idea of complementarity to all different walks of life, beyond its originally intended application in quantum theory. And even within the confines of quantum theory, it is often the case that when the going gets tough, tough quantum theorists cloak themselves in the mystical profundity of complementarity, sometimes just to get philosophers off their backs.

So it seems with the following notorious comments of a well known advocate of the detector approach that have received a predictably cool reception from philosophers:

> Bohr taught us that quantum mechanics is an algorithm for computing the results of measurements. Any discussion about what is a 'real, physical vacuum', must therefore be related to the behavior of real, physical measuring devices, in this case particle-number detectors. Armed with such heuristic devices, we may then assert the following. There are quantum states and there

> are particle detectors. Quantum field theory enables us to predict probabilistically how a particular detector will respond to that state. That is all. That is all there can ever be in physics, because physics is about the observations and measurements that we can make in the world. We can't talk meaningfully about whether such-and-such a state contains particles except in the context of a specified particle detector measurement. To claim (as some authors occasionally do!) that when a detector responds (registers particles) in somebody's cherished vacuum state that the particles concerned are 'fictitious' or 'quasi-particles', or that the detector is being 'misled' or 'distorted', is an empty statement. (Davies 1984, 69)

We shall argue that, cleansed of Davies' purely operationalist reading of Bohr, complementarity *does*, after all, shed light on the relation between inequivalent particle concepts in quantum field theory.

Rüger (1989) balks at this idea. He writes:

> The 'real problem' — how to understand how there might be particles for one observer, but none at all for another observer in a different state of motion — is not readily solved by an appeal to Copenhagenism .... Though quantum mechanics can tell us that the *properties* of micro-objects (like momentum or energy) depend in a sense on observers measuring them, the standard interpretation of the theory still does not tell us that whether there is a micro-object or not depends on observers. At least the common form of this interpre-xtation is not of immediate help here. (Rüger 1989, 575–6)

Well, let us consider the 'common form' of the Copenhagen interpretation. Whatever one's preferred embellishment of the interpretation, it must at least imply that observables represented by non-commuting 'complementary' self-adjoint operators cannot have simultaneously determinate values in all states. Since field quantizations are built upon an abstract noncommutative algebra, the Weyl algebra, complementarity retains its application to quantum field theory. In particular, in any *single* Fock space representation — setting aside inequivalent representations for the moment — there will be a total number operator and nontrivial superpositions of its eigenstates. In these superpositions, which are eigenstates of observables failing to commute with the number operator, it is therefore perfectly in line with complementarity that we say they contain no actual particles in any substantive sense.[23] In addition, there will be different number operators on Fock space that count the number of quanta with wavefunctions lying in different subspaces of the one-particle space, and they will only commute if the corresponding subspaces are compatible. So even before we consider inequivalent particle concepts, we must already accept that there are different *complementary* 'kinds' of quanta, according to what their wavefunctions are.

Does complementarity extend to the particle concepts associated with inequivalent Fock representations? *Contra* Rüger (1989), we claim that it does. We saw earlier that one can build finite-subspace J-quanta number operators in *any* regular representation of $\mathcal{W}[S,\sigma]$, provided only that J defines a proper complex structure on S that leaves it invariant. In particular, using the canonical commutation relation $[\Phi(f),\Phi(g)] = i\sigma(f,g)I$, a tedious but elementary calculation reveals that, for any $f,g \in S$,

$$\begin{aligned}[N_{J_1}(f),N_{J_2}(g)] = i/2\{&\sigma(f,g)[\Phi(f),\Phi(g)]_+\\ &+\sigma(f,J_2g)[\Phi(f),\Phi(J_2g)]_+\\ &+\sigma(J_1f,g)[\Phi(J_1f),\Phi(g)]_+\\ &+\sigma(J_1f,J_2g)[\Phi(J_1f),\Phi(J_2g)]_+\},\end{aligned} \tag{65}$$

in any regular representation.[24] Thus, there are well-defined and, in general, *nontrivial* commutation relations between finite-subspace

[23] As Rüger notes earlier (1989, p. 571), in ordinary non-field-theoretic quantum theory, complementarity only undermined a naïve substance-properties ontology. However, this was only because there was no 'number of quanta' observable in the theory!

[24] As a check on expression (65), note that it is invariant under the one-particle

number operators, even when the associated particle concepts are inequivalent. We also saw in Eqn. (44) that when $J_2 \neq J_1$, no $N_{J_2}(f)$, for any $f \in \mathcal{S}_{J_2}$, will leave the zero-particle subspace of N_{J_1} invariant. Since it is a necessary condition that this nondegenerate eigenspace be left invariant by any self-adjoint operator commuting with N_{J_1}, it follows that $[N_{J_2}(f), N_{J_1}] \neq 0$ for all $f \in \mathcal{S}_{J_2}$. Thus finite-subspace number operators for one kind of quanta are complementary to the total number operators of inequivalent kinds of quanta.

Of course, we cannot give the same argument for complementarity between the *total* number operators N_{J_1} and N_{J_2} pertaining to inequivalent kinds of quanta, because, as we know, they cannot even be defined as operators on the same Hilbert space. However, we disagree with Arageorgis (1995, pp. 303–4) that this means Teller's 'complementarity talk' in relation to the Minkowski and Rindler total number operators is wholly inapplicable. We have two reasons for the disagreement.

First, since it is a necessary condition that a (possibly unbounded) self-adjoint observable Y on $\mathcal{H}_{\omega_{J_1}}$ commuting with N_{J_1} have $\Omega_{\omega_{J_1}}$ as an eigenvector, it is also necessary that the abstract vacuum state ω_{J_1} be dispersion-free on Y. But this latter condition is purely algebraic and makes sense even when Y does *not* act on $\mathcal{H}_{\omega_{J_1}}$. Moreover, as Proposition 9 shows, this condition fails when Y is taken to be the total number operator of any Fock representation inequivalent to $\pi_{\omega_{J_1}}$. So it is entirely natural to treat Proposition 9 as a vindication of the idea that inequivalent pairs of total number operators are complementary.

Secondly, we have seen that any state in the folium of a representation associated with one kind of quanta assigns probability zero to any finite number of an inequivalent kind of quanta. This has a direct analogue in the most famous instance of complementarity: that which obtains between the concepts of position and momentum.

Consider the unbounded position and momentum operators,

---

space phase transformations $f \to (\cos t + J_1 \sin t)f$ and $g \to (\cos t + J_2 \sin t)g$, and when $J_1 = J_2 = J$, reduces to zero just in case the rays generated by f and g are compatible subspaces of $\mathcal{S}_J$.

x and p ($= -i\frac{\partial}{\partial x}$), acting on $L^2(\mathbb{R})$. Let E_x and E_p be their spectral measures. We say that a state ρ of $\mathfrak{B}(L^2(\mathbb{R}))$ assigns x a *finite* dispersion-free value just in case ρ is dispersion-free on x and there is a $\lambda \in \mathbb{R}$ such that $\rho(E_x((a,b))) = 1$ if and only if $\lambda \in (a,b)$. (Similarly, for p.) Then the following is a direct consequence of the canonical commutation relation $[x,p] = iI$ (see Halvorson & Clifton 1999, Prop. 3.7).

Proposition 12. *If ρ is a state of $\mathfrak{B}(L^2(\mathbb{R}))$ that assigns x (resp., p) a finite dispersion-free value, then $\rho(E_p((a,b))) = 0$ (resp., $\rho(E_x((a,b))) = 0$) for any $a, b \in \mathbb{R}$.*

This result makes rigorous the fact, suggested by Fourier analysis, that if either of x or p has a sharp finite value in any state, the other is 'maximally indeterminate'. But the same goes for pairs of inequivalent number operators (N_{J_1}, N_{J_2}): if a regular state ρ assigns N_{J_1} a finite dispersion-free value, then $\rho \in \mathfrak{F}(\pi_{\omega_{J_1}})$ which, in turn, entails that ρ assigns probability zero to any finite set of eigenvalues for N_{J_2}. Thus, (N_{J_1}, N_{J_2}) are, in a natural sense, *maximally* complementary, despite the fact that they have no well-defined commutator.

One might object that our analogy is only skin deep; after all, x and p still act on the *same* Hilbert space, $L^2(\mathbb{R})$! So let us deepen the analogy. Let $\mathcal{W}$ be the Weyl algebra for one degree of freedom, and let $U(a) \equiv W(a,0)$ and $V(b) \equiv W(0,b)$ be the unitary operators corresponding, respectively, to position and momentum. Now, if we think of position as analogous to the Minkowski number operator and momentum as analogous to the Rindler number operator, the standard Schrödinger representation is not the analogue of the Minkowski vacuum representation — since the Minkowski vacuum representation is constructed so as to have eigenvectors for N_M, whereas the Schrödinger representation obviously does not have eigenvectors for x. Thus, to find a representation analogous to the Minkowski vacuum representation, first choose a state ρ of $\mathcal{W}$ that is dispersion-free on all elements $\{U(a) : a \in \mathbb{R}\}$. In particular, we may choose ρ such that $\rho(U(a)) = e^{ia\lambda}$ for all $a \in \mathbb{R}$. If we then let $(\pi_\rho, \mathcal{H}_\rho, \Omega_\rho)$ denote the GNS representation of $\mathcal{W}$ induced by ρ, it follows that we may construct an unbounded position operator x on $\mathcal{H}_\rho$ which has Ω_ρ as an eigenvector with eigenvalue λ. But, lo and behold, it is not possible to define a momentum operator p on the Hilbert space $\mathcal{H}_\rho$.

Indeed, since ρ is dispersion-free on $U(a)$, it is multiplicative for the product of $U(a)$ with any other element of $\mathcal{W}$ (Kadison & Ringrose 1997, Ex. 4.6.16). In particular,

$$\rho(U(a))\rho(V(b)) = e^{iab}\rho(V(b))\rho(U(a)), \qquad a, b \in \mathbb{R}. \tag{66}$$

Since $\rho(U(a)) = e^{ia\lambda} \neq 0$, this implies

$$\rho(V(b)) = e^{iab}\rho(V(b)), \qquad a, b \in \mathbb{R}. \tag{67}$$

However, when $a \neq 0$, (67) cannot hold for all $b \neq 0$ unless $\rho(V(b)) = 0$. Thus,

$$\langle \Omega_\rho, \pi_\rho(V(b))\Omega_\rho \rangle = 0, \qquad \forall b \neq 0. \tag{68}$$

On the other hand,

$$\langle \Omega_\rho, \pi_\rho(V(0))\Omega_\rho \rangle = \langle \Omega_\rho, I\Omega_\rho \rangle = 1. \tag{69}$$

Thus, $\pi_\rho(V(b))$ is not weakly continuous in b, and there can be no self-adjoint operator p on $\mathcal{H}_\rho$ such that $\pi_\rho(V(b)) = e^{ibp}$. On the other hand, since $\mathbb{R} \ni a \mapsto \rho(U(a)) = e^{ia\lambda}$ is continuous, and hence π_ρ is regular with respect to the subgroup of unitary operators $\{U(a) : a \in \mathbb{R}\}$, there *is* a position operator on $\mathcal{H}_\rho$.

Similarly, if ω is a state of $\mathcal{W}$ that is dispersion-free on the momentum unitary operators $\{V(b) : b \in \mathbb{R}\}$, then it is not possible to define a position operator on the Hilbert space $\mathcal{H}_\omega$. Moreover, the GNS representations π_ρ and π_ω are disjoint — precisely as in the case of the GNS representations induced by the Minkowski and Rindler vacuum states. Indeed, suppose for *reductio* that there is a unitary operator T from $\mathcal{H}_\omega$ to $\mathcal{H}_\rho$ such that $T^{-1}\pi_\rho(A)T = \pi_\omega(A)$ for all $A \in \mathcal{W}$. Then, it would follow that $\pi_\omega(U(a)) = T^{-1}\pi_\rho(U(a))T$ is weakly continuous in a, in contradiction to the fact that x cannot be defined on $\mathcal{H}_\omega$.

So we maintain that there are compelling formal reasons for thinking of Minkowski and Rindler quanta as complementary.

What's more, when a Minkowski observer sets out to detect particles, her state of motion determines that her detector will be sensitive to the presence of Minkowski quanta. Similarly for a Rindler observer and his detector. This is borne out by the analysis of Unruh and Wald (1984) in which they show how his detector will *itself* 'define' (in a 'nonstandard' way) what solutions of the relativistic wave equation are counted as having positive frequency, via the way the detector couples to the field. So we may think of the choice of an observer to follow an inertial or Rindler trajectory through spacetime as analogous to the choice between measuring the position or momentum of a particle. Each choice requires a distinct kind of coupling to the system, and both measurements cannot be executed on the field simultaneously and with arbitrarily high precision.[25] Moreover, execution of one type of measurement precludes meaningful discourse about the values of the observable that the observer did not choose to measure. All this is the essence of 'Copenhagenism'.

And it should *not* be equated with operationalism! The goal of the detector approach to the paradox of observer-dependence was to achieve clarity on the problem by reverting back to operational definitions of the word 'particle' with respect to the concrete behavior of particular kinds of detectors (cf., e.g., DeWitt 1979*b*, p. 692). But, as with early days of special relativity and quantum theory, operationalism can serve its purpose and then be jettisoned. Rindler

[25]Why can't *both* a Minkowski and a Rindler observer set off in different spacetime directions and *simultaneously* measure their respective (finite-subspace or total) number operators? Would it not, then, be a violation of relativistic causality when the Minkowski observer's measurement disturbs the statistics of the Rindler observer's measurement outcomes? No. We must remember that the Minkowski particle concept is global, so our Minkowski observer cannot make a precise measurement of any of her number operators unless it is executed throughout the whole of spacetime, which would necessarily destroy her spacelike separation from the Rindler observer. On the other hand, if she is content with only an approximate measurement of one of her number operators in a bounded spacetime region, it is well known that simultaneous, nondisturbing 'unsharp' measurements of incompatible observables *are* possible. For an analysis of the case of simultaneous measurements of unsharp position and momentum, see Busch *et al.* (1995).

quanta get their status as such not because they are, *by definition*, the sort of thing that accelerated detectors detect. This gets things backwards. Rindler detectors display Rindler quanta in the Minkowski vacuum *because* they couple to *Rindler* observables of the field that are distinct from, and indeed complementary to, Minkowski observables.

Arageorgis (1995) himself, together with his collaborators (Arageorgis *et al.* 2002), prefer to characterize inequivalent particle concepts, not as complementary, but *incommensurable*. At first glance, this looks like a trivial semantic dispute between us. For instance Glymour, in a recent introductory text on the philosophy of science, summarizes complementarity using the language of incommensurability:

> Changing the experiments we conduct is like changing conceptual schemes or paradigms: we experience a different world. Just as no world of experience combines different conceptual schemes, no reality we can experience (even indirectly through our experiments) combines precise position and precise momentum.
> (Glymour 1992, p. 128)

However, philosophers of science usually think of incommensurability as a relation between theories *in toto*, not different parts of the same physical theory. Arageorgis *et al.* maintain that inequivalent quantizations define incommensurable *theories*.

Arageorgis (1995) makes the claim that 'the degrees of freedom of the field in the Rindler model *simply cannot be described* in terms of the ground state and the elementary excitations of the degrees of freedom of the field in the Minkowski model' (1995, p. 268; our italics). Yet so much of our earlier discussion proves the contrary. Disjoint representations *are* commensurable, via the abstract Weyl algebra they share. The result is that the ground state of one Fock representation makes definite, if sometimes counterintuitive, predictions for the 'differently complexified' degrees of freedom of other Fock representations.

Arageorgis *et al.* (2002) offer an *argument* for incommensurability — based on Fulling's 'theorem'. They begin by discussing the case where the primed and unprimed representations are unitarily equivalent. (Notice that they speak of two different 'theorists', rather than two different observers.)

> ... while *different*, these particle concepts can nevertheless be deemed to be *commensurable*. The two theorists are just labeling the particle states in different ways, since each defines particles of a given type by mixing the creation and annihilation operators of the other theorist. Insofar as the primed and unprimed theorists disagree, they disagree over which of two inter-translatable descriptions of the same physical situation to use.
>
> The gulf of disagreement between two theorists using unitarily inequivalent Fock space representations is much deeper. If in this case the primed-particle theorist can speak sensibly of the unprimed-particle theorist's vacuum at all, he will say that its primed-particle content is infinite (or more properly, undefined), and the unprimed-theorist will say the same of the unprimed-particle content of the primed vacuum. Such disagreement is profound enough that we deem the particle concepts affiliated with unitarily inequivalent Fock representations *incommensurable*. (Arageorgis *et al.* 2002, p. 26)

The logic of this argument is curious. In order to make Fulling's 'theorem' do the work for incommensurability that Arageorgis *et al.* want it to, one must first have in hand a rigorous version of the theorem (otherwise their argument would be built on sand). But any rigorous version, like our Proposition 10, has to presuppose that there is sense to be made of using a vector state from one Fock representation to generate a prediction for the expectation value of the total number operator in another inequivalent representation. Thus, one cannot even *entertain* the philosophical implications of Fulling's result if one has not first granted a certain level of commensurability between inequivalent representations.

Moreover, while it may be tempting to *define* what one means by 'incommensurable representations' in terms of Fulling's characterization of inequivalent representations, it is difficult to see the exact motivation for such a definition. Even vector states *in the folium* of the unprimed 'theorist's' Fock representation can fail to assign his total number operator a finite expectation value (just consider any vector not in the operator's domain). Yet it would be alarmist to claim that, were the field in such a state, the unprimed 'theorist' would lose his conceptual grasp on, or his ability to talk about, his *own* unprimed kind of quanta! So long as a state prescribes a well-defined probability measure over the spectral projections of the unprimed 'theorist's' total number operator — and all states in his *and* the folium of any primed 'theorist's' representation *will* — we fail to see the difficulty.

5 Conclusion

Let us return to answer the questions we raised in our introduction.

We have argued that a conservative operationalist about physical observables is not committed to the physical inequivalence of disjoint representations, so long as he has no attachment to states in a particular folium being the only physical ones. On the other hand, a liberal about physical observables, no matter what his view on states, *must* say that disjoint representations yield physically inequivalent descriptions of a field. However, we steadfastly resisted the idea that this means an interpreter of quantum field theory must say disjoint representations are incommensurable, or even different, *theories*.

Distinguishing 'potential' from 'actual' quanta won't do to resolve the paradox of observer-dependence. Rather, the paradox forces us to thoroughly abandon the idea that Minkowski and Rindler observers moving through the same field are both trying to detect the presence of particles *simpliciter*. Their motions cause their detectors to couple to *different* incompatible particle observables of the field, making their perspectives on the field necessarily complementary. Furthermore, taking this complementarity seriously means saying that neither the Minkowski nor Rindler perspective yields the uniquely 'correct' story about the particle content of the field, and that *both* are necessary to provide a complete picture.

So, 'Are Rindler Quanta Real?' This is a loaded question that can be understood in two different ways.

First, we could be asking 'Are *Any* Quanta Real?' without regard to inequivalent notions of quanta. Certainly particle detection events, modulo a resolution of the measurement problem, are real. But it should be obvious by now that detection events do not generally license naïve talk of individuatable, localizable, particles that come in determinate numbers in the *absence* of being detected.

A fuller response would be that quantum field theory is 'fundamentally' a theory of a field, not particles. This is a reasonable response given that: (i) the field operators $\{\Phi(f) : f \in S\}$ exist in every regular representation; (ii) they can be used to construct creation, annihilation, and number operators; and (iii) their expectation values evolve in significant respects like the values of the counterpart classical field, modulo nonlocal Bell-type correlations. This 'field approach' response might seem to leave the ontology of the theory somewhat opaque. The field operators, being subject to the canonical commutation relations, do not all commute; so we cannot speak sensibly of them all simultaneously having determinate values! However, the right way to think of the field approach, compatible with complementarity, is to see it as viewing a quantum field as a collection of correlated 'objective propensities' to display values of the field operators in more or less localized regions of spacetime, relative to various measurement contexts. This view makes room for the reality of quanta, but only as a kind of epiphenomenon of the field associated with certain functions of the field operators.

Second, we could be specifically interested in knowing whether it is sensible to say that *Rindler*, as opposed to just Minkowski, quanta are real. An uninteresting answer would be 'No' — on the grounds that quantum field theory on flat spacetime is not a serious candidate for describing our actual universe, or that the Rindler representation is too 'pathological'. But, as philosophers, we are content to leave to the physicists the task of deciding the question 'Are

Rindler Quanta *Empirically Verified*?'. All we have tried to determine (to echo words of van Fraassen) is how the world *could possibly be* if both the Rindler and Minkowski representations were 'true'. We have argued that the antecedent of this counterfactual makes perfect sense, and that it forces us to view Rindler and Minkowski quanta as complementary. Thus, Rindler and Minkowski would be equally amenable to achieving 'reality status' provided the appropriate measurement context were in place. As Wald has put it:

> Rindler particles are 'real' to accelerating observers! This shows that different notions of 'particle' are useful for different purposes. (1994, p. 116)

Appendix: Proofs of selected theorems

Proposition 3. *Under the conservative approach to states, ϕ (factorial) and π (irreducible) are physically equivalent representations of $\mathcal{W}$ only if they are quasi-equivalent.*

Proof. Let ω be a normal state of $\phi(\mathcal{W})$. Then, by hypothesis, $\beta(\omega)$ is a normal state of $\pi(\mathcal{W})$. Define a state ρ on $\mathcal{W}$ by

$$\rho(A) = \omega(\phi(A)), \qquad A \in \mathcal{W}. \tag{70}$$

Since ω is normal, $\rho \in \mathfrak{F}(\phi)$. Define a state ρ' on $\mathcal{W}$ by

$$\rho'(A) = \beta(\omega)(\pi(A)), \qquad A \in \mathcal{W}. \tag{71}$$

Since $\beta(\omega)$ is normal, $\rho' \in \mathfrak{F}(\pi)$. Now, conditions (21) and (22) entail that

$$\omega(\phi(A)) = \beta(\omega)(\alpha(\phi(A))) = \beta(\omega)(\pi(A)), \tag{72}$$

for any $A = W(f) \in \mathcal{W}$, and thus $\rho(W(f)) = \rho'(W(f))$ for any $f \in S$. However, a state of the Weyl algebra is uniquely determined (via linearity and uniform continuity) by its action on the generators $\{W(f) : f \in S\}$. Thus, $\rho = \rho'$ and since $\rho \in \mathfrak{F}(\phi) \cap \mathfrak{F}(\pi)$, it follows that ϕ and π are quasi-equivalent. $\square$

Proposition 4. *Under the liberal approach to observables, ϕ (factorial) and π (irreducible) are physically equivalent representations of $\mathcal{W}$ only if they are quasi-equivalent.*

Proof. By hypothesis, the bijective mapping α must map the self-adjoint part of $\phi(\mathcal{W})''$ onto that of $\pi(\mathcal{W})''$. Extend α to *all* of $\phi(\mathcal{W})''$ by defining

$$\alpha(X) := \alpha(\mathrm{Re}(X)) + i\alpha(\mathrm{Im}(X)), \qquad X \in \phi(\mathcal{W})''. \tag{73}$$

Clearly, then, α preserves adjoints.

Recall that a family of states S_0 on a C^*-algebra is called *full* just in case S_0 is convex, and for any $A \in \mathcal{A}$, $\rho(A) \geq 0$ for all $\rho \in S_0$ only if $A \geq 0$. By hypothesis, there is a bijective mapping β from the 'physical' states of $\phi(\mathcal{W})''$ onto the 'physical' states of $\pi(\mathcal{W})''$. According to both the conservative and liberal construals of physical states, the set of physical states includes normal states. Since the normal states are full, the domain and range of β contain full sets of states of the respective C^*-algebras.

By condition (22) and the fact that the domain and range of β are full sets of states, α arises from a *symmetry* between the C^*-algebras $\phi(\mathcal{W})''$ and $\pi(\mathcal{W})''$ in the sense of Roberts & Roepstorff (1969, Sec. 3).[26] Their Propositions 3.1 and 6.3 then apply to guarantee that α must be linear and preserve Jordan structure (i.e., anti-commutator brackets). Thus α is a Jordan $*$-isomorphism.

Now both $\phi(\mathcal{W})''$ and $\pi(\mathcal{W})'' = \mathfrak{B}(\mathcal{H}_\pi)$ are von Neumann algebras, and the latter has a trivial commutant. Thus Exercise 10.5.26 of Kadison & Ringrose (1997) applies, and α is either a $*$-isomorphism or a $*$-anti-isomorphism, that reverses the order of products. However, such reversal is ruled out, otherwise we would have, using the Weyl relations (10),

$$\alpha(\phi(W(f))\phi(W(g))) = e^{-i\sigma(f,g)/2}\alpha(\phi(W(f+g))), \tag{74}$$

$$\Rightarrow \quad \alpha(\phi(W(g)))\alpha(\phi(W(f))) = e^{-i\sigma(f,g)/2}\alpha(\phi(W(f+g))), \tag{75}$$

$$\Rightarrow \quad \pi(W(g))\pi(W(f)) = e^{-i\sigma(f,g)/2}\pi(W(f+g)), \tag{76}$$

$$\Rightarrow \quad e^{i\sigma(f,g)/2}\pi(W(f+g)) = e^{-i\sigma(f,g)/2}\pi(W(f+g)), \tag{77}$$

[26] Actually, they consider only symmetries of a C^*-algebra onto *itself*, but their results remain valid for our case.

for all $f, g \in S$. This entails that the value of σ on any pair of vectors is always a multiple of 2π which, since σ is bilinear, cannot happen unless $\sigma = 0$ identically (and hence $S = \{0\}$). It follows that α is in fact a $*$-isomorphism. And, by condition (21), α must map $\phi(A)$ to $\pi(A)$ for all $A \in \mathcal{W}$. Thus ϕ is quasi-equivalent to π. □

Proposition 6. *When S is infinite-dimensional, $\pi(\mathcal{W}[S,\sigma])$ contains no nontrivial bounded functions of the total number operator on $\mathcal{F}(\mathcal{S}_J)$.*

Proof. For clarity, we suppress reference to the representation map π. Suppose that $F : \mathbb{N} \mapsto \mathbb{C}$ is a bounded function. We show that if $F(N) \in \mathcal{W}$, then $F(n) = F(n+1)$ for all $n \in \mathbb{N}$.

The Weyl operators on $\mathcal{F}(\mathcal{S}_J)$ satisfy the commutation relation (Bratteli & Robinson 1996, Prop. 5.2.4(1,2)):

$$W(g)\Phi(f)W(g)^* = \Phi(f) - \sigma(g,f)I. \tag{78}$$

Using the definition of the inner product $(\cdot,\cdot)_J$ (Eqn. (23)) and the equation $a^*(f) = 2^{-1/2}(\Phi(f) - i\Phi(if))$, we find

$$W(g)a^*(f)W(g)^* = a^*(f) + 2^{-1/2}i(g,f)_J I, \tag{79}$$

and from this, $[W(g), a^*(f)] = 2^{-1/2}i(g,f)_J W(g)$. Now let $\psi \in \mathcal{F}(\mathcal{S}_J)$ be in the domain of $a^*(f)$. Then a straightforward calculation shows that

$$\begin{aligned}\langle a^*(f)\psi, W(g)a^*(f)\psi\rangle = {} & 2^{-1/2}i(g,f)_J\langle a^*(f)\psi, W(g)\psi\rangle \\ & + \langle a(f)a^*(f)\psi, W(g)\psi\rangle.\end{aligned} \tag{80}$$

Let $\{f_k\}$ be an infinite orthonormal basis for $\mathcal{S}_J$, and let $\psi \in \mathcal{F}(\mathcal{S}_J)$ be the vector whose n-th component is $P_+(f_1 \otimes \cdots \otimes f_n)$ and whose other components are zero. Now, for any $k > n$, we have $a(f_k)a^*(f_k)\psi = (n+1)\psi$. Thus, Eqn. (80) gives

$$\begin{aligned}\langle a^*(f_k)\psi, W(g)a^*(f_k)\psi\rangle = {} & 2^{-1/2}i(g,f_k)_J\langle a^*(f_k)\psi, W(g)\psi\rangle \\ & + (n+1)\langle \psi, W(g)\psi\rangle.\end{aligned} \tag{81}$$

Hence,

$$\lim_{k\to\infty} \langle a^*(f_k)\psi, W(g)a^*(f_k)\psi\rangle = (n+1)\langle \psi, W(g)\psi\rangle. \tag{82}$$

Since $\mathcal{W}$ is generated by the $W(g)$, Eqn. (82) holds when $W(g)$ is replaced with any element in $\mathcal{W}$. On the other hand, ψ is an eigenvector with eigenvalue n for N while $a^*(f_k)\psi$ is an eigenvector with eigenvalue $n+1$ for N. Thus, $\langle\psi, F(N)\psi\rangle = F(n)\|\psi\|^2$ while

$$\begin{aligned} \langle a^*(f_k)\psi, F(N)a^*(f_k)\psi\rangle &= F(n+1)\,\|a^*(f_k)\psi\|^2 && (83) \\ &= (n+1)F(n+1)\,\|\psi\|^2, && (84) \end{aligned}$$

for all $k > n$. Thus, the assumption that $F(N)$ is in $\mathcal{W}$ (and hence satisfies (82)) entails that $F(n+1) = F(n)$. □

Proposition 7. *The Minkowski and Rindler representations of $\mathcal{W}_\triangleleft$ are disjoint.*

Proof. By Theorem 3.3.4 of Horuzhy (1988), $\pi_{\omega_M^\triangleleft}(\mathcal{W}_\triangleleft)''$ is a 'type III' von Neumann algebra which, in particular, contains no atomic projections. Since $\pi_{\omega_R^\triangleleft}$ is irreducible and $\pi_{\omega_M^\triangleleft}$ factorial, either $\pi_{\omega_R^\triangleleft}$ and $\pi_{\omega_M^\triangleleft}$ are disjoint, or they are quasi-equivalent. However, since $\pi_{\omega_R^\triangleleft}(\mathcal{W}_\triangleleft)'' = \mathfrak{B}(\mathcal{F}(\mathcal{S}(\triangleleft)_R))$, the weak closure of the Rindler representation clearly contains atomic projections. Moreover, $*$-isomorphisms preserve the ordering of projection operators. Thus there can be no $*$-isomorphism of $\pi_{\omega_M^\triangleleft}(\mathcal{W}_\triangleleft)''$ onto $\pi_{\omega_R^\triangleleft}(\mathcal{W}_\triangleleft)''$, and the Minkowski and Rindler representations of $\mathcal{W}_\triangleleft$ are disjoint. □

Proposition 8. *The Minkowski and Rindler representations of $\mathcal{W}_\bowtie$ are disjoint.*

Proof. Again, we use the fact that $\pi_{\omega_M^\bowtie}(\mathcal{W}_\triangleleft)''$ ($\equiv \pi_{\omega_M^\triangleleft}(\mathcal{W}_\triangleleft)''$) does not contain atomic projections, whereas $\pi_{\omega_R^\bowtie}(\mathcal{W}_\triangleleft)''$ ($\equiv \pi_{\omega_R^\triangleleft}(\mathcal{W}_\triangleleft)''$) does. Suppose, for *reductio ad absurdum*, that $\pi_{\omega_R^\bowtie}$ and $\pi_{\omega_M^\bowtie}$ are quasi-equivalent. Since the states $\omega_M^\bowtie$ and $\omega_R^\bowtie$ are pure, the representations $\pi_{\omega_M^\bowtie}$ and $\pi_{\omega_R^\bowtie}$ are irreducible and therefore unitarily equivalent. Thus, there is a *weakly continuous* $*$-isomorphism α from $\pi_{\omega_M^\bowtie}(\mathcal{W}_\bowtie)''$ onto $\pi_{\omega_R^\bowtie}(\mathcal{W}_\bowtie)''$ such that $\alpha(\pi_{\omega_M^\bowtie}(A)) = \pi_{\omega_R^\bowtie}(A)$ for each $A \in \mathcal{W}_\bowtie$. In particular, α maps $\pi_{\omega_M^\bowtie}(\mathcal{W}_\triangleleft)$ onto $\pi_{\omega_R^\bowtie}(\mathcal{W}_\triangleleft)$; and, since α is weakly continuous, it maps $\pi_{\omega_M^\bowtie}(\mathcal{W}_\triangleleft)''$ onto $\pi_{\omega_R^\bowtie}(\mathcal{W}_\triangleleft)''$. Consequently, $\pi_{\omega_M^\bowtie}(\mathcal{W}_\triangleleft)''$ contains an atomic projection, in contradiction with the fact that $\pi_{\omega_M^\bowtie}(\mathcal{W}_\triangleleft)''$ is a type III von Neumann algebra. □

Proposition 9. *If J_1, J_2 are distinct complex structures on (S, σ), then ω_{J_1} (resp., ω_{J_2}) predicts dispersion in N_{J_2} (resp., N_{J_1}).*

Proof. We shall prove the contrapositive. Suppose, then, that there is some extension $\hat{\omega}_{J_1}$ of ω_{J_1} to $\mathfrak{B}(\mathcal{F}(\mathcal{S}_{J_2}))$ that is dispersion-free on all bounded functions of N_{J_2}. Then $\hat{\omega}_{J_1}$ is multiplicative for the product of the bounded operator $e^{\pm itN_{J_2}}$ with any other element of $\mathfrak{B}(\mathcal{F}(\mathcal{S}_{J_2}))$ (Kadison & Ringrose 1997, Ex. 4.6.16). Hence, by Eqn. (40),

$$\omega_{J_1}(W(\cos t + \sin t J_2 f)) = \hat{\omega}_{J_1}\left(e^{-itN_{J_2}}\pi_{\omega_{J_2}}(W(f))e^{itN_{J_2}}\right) \tag{85}$$

$$= \hat{\omega}_{J_1}(e^{-itN_{J_2}})\,\omega_{J_1}(W(f))\,\hat{\omega}_{J_1}(e^{itN_{J_2}}) \tag{86}$$

$$= \omega_{J_1}(W(f)), \tag{87}$$

for all $f \in S$ and $t \in \mathbb{R}$. In particular, we may set $t = \pi/2$, and it follows that $\omega_{J_1}(W(J_2 f)) = \omega_{J_1}(W(f))$ for all $f \in S$. Since e^{-x} is a one-to-one function of $x \in \mathbb{R}$, it follows from (33) that

$$(f, f)_{J_1} = (J_2 f, J_2 f)_{J_1}, \qquad f \in S, \tag{88}$$

and J_2 is a real-linear isometry of the Hilbert space $\mathcal{S}_{J_1}$. We next show that J_2 is in fact a unitary operator on $\mathcal{S}_{J_1}$.

Since J_2 is a symplectomorphism, $\mathrm{Im}(J_2 f, J_2 g)_{J_1} = \mathrm{Im}(f, g)_{J_1}$ for any two elements $f, g \in S$. We also have

$$|f + g|^2_{J_1} = |f|^2_{J_1} + |g|^2_{J_1} + 2\mathrm{Re}(f, g)_{J_1}, \tag{89}$$

$$|J_2 f + J_2 g|^2_{J_1} = |J_2 f|^2_{J_1} + |J_2 g|^2_{J_1} + 2\mathrm{Re}(J_2 f, J_2 g)_{J_1} \tag{90}$$

$$= |f|^2_{J_1} + |g|^2_{J_1} + 2\mathrm{Re}(J_2 f, J_2 g)_{J_1}, \tag{91}$$

using the fact that J_2 is isometric. But $J_2(f + g) = J_2 f + J_2 g$, since J_2 is real-linear. Thus,

$$|J_2 f + J_2 g|^2_{J_1} = |J_2(f + g)|^2_{J_1} = |f + g|^2_{J_1}, \tag{92}$$

using again the fact that J_2 is isometric. Cancellation with Eqns. (89) and (91) then gives $\mathrm{Re}(f, g)_{J_1} = \mathrm{Re}(J_2 f, J_2 g)_{J_1}$. Thus, J_2 preserves the inner product between any two vectors in $\mathcal{S}_{J_1}$. All that remains to show is that J_2 is complex-linear. So let $f \in \mathcal{S}_{J_1}$. Then,

$$\begin{aligned}(J_2(if), J_2 g)_{J_1} &= (if, g)_{J_1} = -i(f,g)_{J_1} \\ &= -i(J_2 f, J_2 g)_{J_1} = (iJ_2 f, J_2 g)_{J_1}, \end{aligned} \tag{93}$$

for all $g \in \mathcal{H}$. Since J_2 is onto, it follows that $(J_2(if), g)_{J_1} = (iJ_2 f, g)_{J_1}$ for all $g \in \mathcal{H}$ and therefore $J_2(if) = iJ_2 f$.

Finally, since J_2 is unitary and $J_2^2 = -I$, it follows that $J_2 = \pm iI = \pm J_1$. However, if $J_2 = -J_1$, then

$$-\sigma(f, J_1 f) = \sigma(f, J_2 f) \geq 0, \qquad f \in S, \tag{94}$$

since J_2 is a complex structure. Since J_1 is also a complex structure, it follows that $\sigma(f, J_1 f) = 0$ for all $f \in S$ and $S = \{0\}$. Therefore, $J_2 = J_1$. □

Proposition 10. *A pair of Fock representations $\pi_{\omega_{J_1}}, \pi_{\omega_{J_2}}$ are unitarily equivalent if and only if ω_{J_1} assigns N_{J_2} a finite value (equivalently, ω_{J_2} assigns N_{J_1} a finite value).*

Proof. S may be thought of as a *real* Hilbert space relative to either of the inner products μ_1, μ_2 defined by

$$\mu_{1,2}(\cdot,\cdot) := \mathrm{Re}(\cdot,\cdot)_{J_{1,2}} = \sigma(\cdot, J_{1,2}\cdot). \tag{95}$$

We shall use Theorem 2 of (van Daele & Verbeure 1971): $\pi_{\omega_{J_1}}, \pi_{\omega_{J_2}}$ are unitarily equivalent if and only if the positive operator $-[J_1, J_2]_+ - 2I$ on S is trace-class relative to μ_2. (Since unitary equivalence is symmetric, the same 'if and only if' holds with $1 \leftrightarrow 2$.)

As we know, we can build any number operator $N_{J_2}(f)$ ($f \in S$) on $\mathcal{H}_{\omega_{J_1}}$ by using the complex structure J_2 in Eqns. (34). In terms of field operators, the result is

$$N_{J_2}(f) = 2^{-1}(\Phi(f)^2 + \Phi(J_2 f)^2 + i[\Phi(f), \Phi(J_2 f)]). \tag{96}$$

Observe that $N_{J_2}(J_2 f) = N_{J_2}(f)$, which had better be the case, since $N_{J_2}(f)$ represents the number of J_2-quanta with wavefunction in the *subspace* of $\mathcal{S}_{J_2}$ generated by f. The expectation value of an arbitrary 'two-point function' in the J_1-vacuum state is given by

$$\langle \Omega_{\omega_{J_1}}, \phi(f_1)\phi(f_2)\Omega_{\omega_{J_1}} \rangle \tag{97}$$

$$= (-i)^2 \frac{\partial^2}{\partial t_1 \partial t_2} \omega_{J_1}(W(t_1 f_1)W(t_2 f_2))|_{t_1=t_2=0} \tag{98}$$

$$= -\frac{\partial^2}{\partial t_1 \partial t_2} \exp\Bigg(-\frac{1}{2} t_1 t_2 (f_1, f_2)_{J_1} - \frac{1}{4} t_1^2 (f_1, f_1)_{J_1} - \frac{1}{4} t_2^2 (f_2, f_2)_{J_1} \Bigg) |_{t_1=t_2=0} \tag{99}$$

$$= \frac{1}{2}(f_1, f_2)_{J_1}, \tag{100}$$

invoking (32) in the first equality, and the Weyl relations (10) together with Eqns. (23), (33) to obtain the second. Plugging Eqn. (100) back into (96) and using (95) eventually yields

$$\langle \Omega_{\omega_{J_1}}, N_{J_2}(f)\Omega_{\omega_{J_1}} \rangle = 2^{-2}\mu_2(f, (-[J_1, J_2]_+ - 2I)f). \tag{101}$$

Next, recall that on the Hilbert space $\mathcal{H}_{\omega_{J_2}}$, $N_{J_2} = \sum_{k=1}^{\infty} N_{J_2}(f_k)$, where $\{f_k\} \subseteq \mathcal{S}_{J_2}$ is any orthonormal basis. Let $\widehat{\omega}_{J_1}$ be any extension of ω_{J_1} to $\mathfrak{B}(\mathcal{H}_{\omega_{J_2}})$. The calculation that resulted in expression (101) was done in $\mathcal{H}_{\omega_{J_1}}$, however, only finitely many-degrees of freedom were involved. Thus the Stone-von Neumann uniqueness theorem ensures that (101) gives the value of each individual $\widehat{\omega}_{J_1}(N_{J_2}(f_k))$. Since for any finite m, $\sum_{k=1}^{m} N_{J_2}(f_k) \leq N_{J_2}$ as positive operators, we must also have

$$\sum_{k=1}^{m} \widehat{\omega}_{J_1}(N_{J_2}(f_k)) = \widehat{\omega}_{J_1}\left(\sum_{k=1}^{m} N_{J_2}(f_k)\right) \leq \widehat{\omega}_{J_1}(N_{J_2}). \tag{102}$$

Thus, $\widehat{\omega}_{J_1}(N_{J_2})$ will be defined only if the sum

$$\sum_{k=1}^{\infty} \widehat{\omega}_{J_1}(N_{J_2}(f_k)) = \sum_{k=1}^{\infty} \widehat{\omega}_{J_1}(N_{J_2}(J_2 f_k)) \tag{103}$$

converges. Using (101), this is, in turn, equivalent to

$$\sum_{k=1}^{\infty} \mu_2(f_k, (-[J_1, J_2]_+ - 2I)f_k) + \sum_{k=1}^{\infty} \mu_2(J_2 f_k, (-[J_1, J_2]_+ - 2I)J_2 f_k) < \infty. \tag{104}$$

However, it is easy to see that $\{f_k\}$ is a J_2-orthonormal basis just in case $\{f_k, J_2 f_k\}$ forms an orthonormal basis in S relative to the inner product μ_2. Thus, Eqn. (104) is none other than the statement that the operator $-[J_1, J_2]_+ - 2I$ on S is trace-class relative to μ_2, which is equivalent to the unitary equivalence of $\pi_{\omega_{J_1}}, \pi_{\omega_{J_2}}$. (The same argument, of course, applies with $1 \leftrightarrow 2$ throughout.) $\square$

Proposition 11. *If ρ is a regular state of $\mathcal{W}$ disjoint from the Fock state ω_J, then $\inf_{F \in \mathbb{F}} \left\{ \text{Prob}^{\rho}(N(F) \in [0,n]) \right\} = 0$ for every $n \in \mathbb{N}$.*

Proof. Suppose that ω_J and ρ are disjoint; i.e., $\mathfrak{F}(\omega_J) \cap \mathfrak{F}(\rho) = \emptyset$. First, we show that $D(n_\rho) = \{0\}$, where n_ρ is the quadratic form on $\mathcal{H}_\rho$ which, if densely defined, would correspond to the total J-quanta number operator.

Suppose for *reductio ad absurdum* that $D(n_\rho)$ contains some unit vector ψ. Let ω be the state of $\mathcal{W}$ defined by

$$\omega(A) = \langle \psi, \pi_\rho(A)\psi \rangle, \qquad A \in \mathcal{W}. \tag{105}$$

Since $\omega \in \mathfrak{F}(\rho)$, it follows that ω is a regular state of $\mathcal{W}$ (since ρ itself is regular), and that $\omega \notin \mathfrak{F}(\omega_J)$. Let P be the projection onto the closed subspace in $\mathcal{H}_\rho$ generated by the set $\pi_\rho(\mathcal{W})\psi$. If we let $P\pi_\rho$ denote the subrepresentation of π_ρ on $P\mathcal{H}_\rho$, then $(P\pi_\rho, P\mathcal{H}_\rho)$ is a representation of $\mathcal{W}$ with cyclic vector ψ. By the uniqueness of the GNS representation, it follows that $(P\pi_\rho, P\mathcal{H}_\rho)$ is unitarily equivalent to $(\pi_\omega, \mathcal{H}_\omega)$. In particular, since Ω_ω is the image in $\mathcal{H}_\omega$ of $\psi \in P\mathcal{H}_\rho$, $D(n_\omega)$ contains a vector cyclic for $\pi_\omega(\mathcal{W})$ in $\mathcal{H}_\omega$. However, by Theorem 5.2.14 of Bratteli & Robinson (1996), this implies that $\omega \in \mathfrak{F}(\omega_J)$ — a contradiction. Therefore, $D(n_\rho) = \{0\}$.

Now suppose, again for *reductio ad absurdum*, that

$$\inf_{F \in \mathbb{F}} \left\{ \text{Prob}^{\rho}(N(F) \in [0,n]) \right\} \neq 0. \tag{106}$$

Let $E_F := [E_\rho(F)]([0,n])$ and let $E := \bigwedge_{F \in \mathbb{F}} E_F$. Since the family $\{E_F\}$ of projections is downward directed (i.e., $F \subseteq F'$ implies $E_F \geq E_{F'}$), we have

$$0 \neq \inf_{F \in \mathbb{F}} \{\langle \Omega_\rho, E_F \Omega_\rho \rangle\} = \langle \Omega_\rho, E\Omega_\rho \rangle = \|E\Omega_\rho\|^2. \tag{107}$$

Now since $E_F E\Omega_\rho = E\Omega_\rho$, it follows that

$$[n_\rho(F)](E\Omega_\rho) \leq n, \tag{108}$$

for all $F \in \mathbb{F}$. Thus, $E\Omega_\rho \in D(n_\rho)$ and $D(n_\rho) \neq \{0\}$, in contradiction with the conclusion of the previous paragraph. □

Bibliography

Arageorgis, A. (1995), Fields, Particles, and Curvature: Foundations and Philosophical Aspects of Quantum Field Theory on Curved Spacetime, PhD thesis, University of Pittsburgh.

Arageorgis, A., Earman, J. & Ruetsche, L. (2002), 'Weyling the time away: the non-unitary implementability of quantum field dynamics on curved spacetime', *Studies in History and Philosophy of Modern Physics* **33**, 151–84.

Baez, J., Segal, I. E. & Zhou, Z. (1992), *Introduction to Algebraic and Constructive Quantum Field Theory*, Princeton University Press, Princeton.

Belinskiĭ, V. (1997), 'Does the Unruh effect exist?', *Journal of European Theoretical Physics Letters* **65**, 902–8.

Beyer, H. (1991), 'Remarks on Fulling's quantization', *Classical and Quantum Gravity* **8**, 1091–112.

Bratteli, O. & Robinson, D. (1996), *Operator Algebras and Quantum Statistical Mechanics*, Vol. 2, Springer, New York.

Bridgman, P. (1936), *The Nature of Physical Theory*, Dover, New York.

Busch, P., Grabowski, M. & Lahti, P. (1995), *Operational Quantum Physics*, Springer, New York.

Chaiken, J. M. (1967), 'Finite-particle representations and states of the canonical commutation relations', *Annals of Physics* **42**, 23–80.

Davies, P. C. W. (1984), Particles do not exist, in S. Christensen, ed., 'Quantum Theory of Gravity', Adam-Hilger, Bristol.

DeWitt, B. (1979*a*), 'Quantum field theory on curved spacetime', *Physics Reports* **19**, 295–357.

DeWitt, B. (1979*b*), Quantum gravity: the new synthesis, in S. Hawking *et al.*, eds, 'General Relativity', Cambridge University Press, Cambridge, pp. 680–745.

Emch, G. G. (1972), *Algebraic Methods in Statistical Mechanics and Quantum Field Theory*, Wiley, New York.

Fedotov, A., Mur, V., Narozhny, N., Belinskiĭ, V. & Karnakov, B. (1999), 'Quantum field aspect of the Unruh problem', *Physics Letters A* **254**, 126–32.

Fell, J. (1960), 'The dual spaces of C^*-algebras', *Transactions of the American Mathematical Society* **94**, 365–403.

Fulling, S. (1972), Scalar quantum field theory in a closed universe of constant curvature, PhD thesis, Princeton University.

Fulling, S. (1989), *Aspects of quantum field theory in curved space-time*, Cambridge University Press, New York.

Gerlach, U. (1989), 'Quantum states of a field partitioned by an accelerated frame', *Physical Review D* **40**, 1037–1047.

Glymour, C. (1971), Theoretical realism and theoretical equivalence, in R. Buck & R. Cohen, eds, 'Boston Studies in Philosophy of Science, VIII', Reidel, Dordrecht, pp. 275–88.

Glymour, C. (1992), Realism and the nature of theories, in M. Salmon *et al.*, eds, 'Introduction to the Philosophy of Science', Prentice-Hall, Upper Saddle River, NJ, pp. 104–31.

Haag, R. & Kastler, D. (1964), 'An algebraic approach to quantum field theory', *Journal of Mathematical Physics* **5**, 848–61.

Halvorson, H. P. (2001*a*), 'On the nature of continuous physical quantities in classical and quantum mechanics', *Journal of Philosophical Logic* **30**, 27–50.

Halvorson, H. P. (2001*b*), 'Reeh-Schlieder defeats Newton-Wigner: On alternative localization schemes in relativistic quantum field theory', *Philosophy of Science* **68**, 111–33.

Halvorson, H. P. & Clifton, R. K. (1999), 'Maximal beable subalgebras of quantum mechanical observables', *International Journal of Theoretical Physics* **38**, 2441–84.

Horuzhy, S. S. (1988), *Introduction to Algebraic Quantum Field Theory*, Kluwer, Dordrecht.

Huggett, N. & Weingard, R. (1996), 'Critical review: Paul Teller's "Interpretive introduction to quantum field theory"', *Philosophy of Science* **63**, 302–14.

Kadison, R. & Ringrose, J. (1997), *Fundamentals of the Theory of Operator Algebras*, American Mathematical Society, Providence, RI.

Kay, B. S. (1979), 'A uniqueness result in the Segal-Weinless approach to linear Bose fields', *Journal of Mathematical Physics* **20**, 1712–3.

Kay, B. S. (1985), 'The double-wedge algebra for quantum fields on Schwarzschild and Minkowski spacetimes', *Communications in Mathematical Physics* **100**, 57–81.

Nikolić, H. (2000), 'Can the Unruh-DeWitt detector extract energy from the vacuum?'. E-print: hep-th/0005240.

Petz, D. (1990), *An Invitation to the Algebra of the Canonical Commutation Relations*, Leuven University Press, Belgium.

Redhead, M. L. G. (1995), The vacuum in relativistic quantum field theory, in D. Hull, M. Forbes & R. M. Burian, eds, 'PSA; proceedings of the biennial meeting of the Philosophy of Science Association', Vol. 2, Philosophy of Science Association, East Lansing, MI, pp. 77–87.

Reichenbach, H. (1938), *Experience and Prediction: An Analysis of the Foundations and Structure of Knowledge*, University of Chicago Press, Chicago.

Roberts, J. E. & Roepstorff, G. (1969), 'Some basic concepts of algebraic quantum theory', *Communications in Mathematical Physics* **11**, 321–38.

Robinson, D. (1966), Algebraic aspects of relativistic quantum field theory, in M. Chretien & S. Deser, eds, 'Axiomatic Field Theory', Vol. 1, Gordon and Breach, New York, pp. 389–516.

Rüger, A. (1989), 'Complementarity meets general relativity: A study in ontological commitments and theory unification', *Synthese* **79**, 559–80.

Sciama, D., Candelas, P. & Deutsch, D. (1981), 'Quantum field theory, horizons, and thermodynamics', *Advances in Physics* **30**, 327–66.

Segal, I. E. (1959), 'Foundations of the theory of dynamical systems of infinitely many degrees of freedom, I', *Matematisk-Fysike Meddelelser. Kongelige Dansk Videnskabernes Selskab* **31**, 1–39.

Segal, I. E. (1961), 'Foundations of the theory of dynamical systems of infinitely many degrees of freedom, II', *Canadian Journal of Mathematics* **13**, 1–18.

Segal, I. E. (1963), *Mathematical Problems of Relativistic Physics*, American Mathematical Society, Providence, RI.

Segal, I. E. (1967), Representations of the canonical commutation relations, in F. Lurçat, ed., 'Cargèse Lectures in Theoretical Physics', Gordon and Breach, New York, pp. 107–70.

Segal, I. E. & Goodman, R. W. (1965), 'Anti-locality of certain Lorentz-invariant operators', *Journal of Mathematics and Mechanics* **14**, 629–38.

Summers, S. J. (2001), On the Stone-von Neumann uniqueness theorem and its ramifications, in M. Rédei & M. Stölzner, eds, 'John von Neumann and the Foundations of Quantum Mechanics', Kluwer, Norwell, MA, pp. 135–52.

Teller, P. (1995), *An Interpretive Introduction to Quantum Field Theory*, Princeton University Press, Princeton.

Teller, P. (1996), Wave and particle concepts in quantum field theory, in R. K. Clifton, ed., 'Perspectives on Quantum Reality', Kluwer, Dordrecht, pp. 143–54.

Teller, P. (1998), 'On Huggett and Weingard's review of "An interpretive introduction to quantum field theory": Continuing the discussion', *Philosophy of Science* **65**, 151–61.

Torretti, R. (1999), *The Philosophy of Physics*, Cambridge University Press, New York.

Unruh, W. & Wald, R. M. (1984), 'What happens when an accelerating observer detects a Rindler particle?', *Physical Review D* **29**, 1047–56.

van Daele, A. & Verbeure, A. (1971), 'Unitary equivalence of Fock representations of the Weyl algebra', *Communications in Mathematical Physics* **20**, 268–78.

Verch, R. (1994), 'Local definiteness, primarity and quasiequivalence of quasifree Hadamard quantum states in curved spacetime', *Communications in Mathematical Physics* **160**, 507–36.

Verch, R. (2000). Personal communication.

von Neumann, J. (1931), 'Die Eindeutigkeit der Schrödingerschen Operatoren', *Mathematische Annalen* **104**, 570–8.

Wald, R. M. (1994), *Quantum Field Theory in Curved Spacetime and Black Hole Thermodynamics*, University of Chicago Press, Chicago.

Part IV

New Light on Complementarity, Hidden Variables, and Entanglement

Chapter 10

Nonlocal correlations are generic in infinite-dimensional bipartite systems

with Hans Halvorson and Adrian Kent

The observables of a bipartite quantum system are represented by the set of all self-adjoint operators on the tensor product of two Hilbert spaces $\mathcal{H}_1 \otimes \mathcal{H}_2$, whose dimensions we shall denote by d_1 and d_2, taking $d_1 \leq d_2$ without loss of generality. It is well known that when $d_1 \geq 2$ the states of the system can be nonseparable, and it is this possibility that much of the new technology associated with quantum information and computation theory relies upon. Prompted by concerns about whether the very noisy mixed states exploited by certain models of NMR quantum computing are truly nonseparable (Braunstein *et al.* 1999, Linden & Popescu 2001), detailed investigations have shown that, whenever $d_2 < \infty$, there is always an open neighborhood of separable states surrounding the maximally mixed state $(d_1 d_2)^{-1} I \otimes I$ (Życzkowski *et al.* 1998, Caves

A. K. thanks the Royal Society for financial support.

& Schack 2000, Życzkowski 1999).

Complementing these results, two of us (Clifton & Halvorson 2000) have recently shown that if $d_2 = \infty$, the set of nonseparable states is dense, and, therefore, there can be no open neighborhood of separable states in that case. It was then conjectured (Clifton & Halvorson 2000) that the same density result ought to hold for states which violate some Bell inequality, at least in the case $d_1 = d_2 = \infty$. This does not follow immediately from the main theorem in Clifton & Halvorson (2000), since the nonseparability of a mixed state (in contrast to the pure case (Gisin 1991, Popescu & Rohrlich 1992)) is not known to imply that it violates a Bell inequality or that its correlations cannot be reproduced by a local hidden variables model. No counterexample is known either; however, Werner (1989) has shown that a local hidden variables model can reproduce the correlations of a nonseparable mixed state for single *projective* measurements on each component system.

We show here that the conjecture made in Clifton & Halvorson (2000) is true. More precisely, we show that a bipartite system possesses a dense set of states violating the CHSH inequality for projective measurements if and only if $d_1 = d_2 = \infty$, and that the system possesses a dense set of states with nonlocal correlations if $d_1 < d_2 = \infty$. In the second case, we demonstrate that the states have nonlocal correlations for sequences of projective measurements: we do not exclude the possibility that they also violate a 'higher order' Bell inequality (Pitowsky 1989, Peres 1999, Gisin 1999) involving more than two measurement choices for each component system, nor do we exclude violations which involve positive operator valued measurements. Our results also yield an elementary proof of the main result of Clifton & Halvorson (2000).

1 Preliminaries

We first establish some basic facts about nonseparability and nonlocality necessary for the sequel.

Let $\mathcal{B}(\mathcal{H}_1 \otimes \mathcal{H}_2)$ denote the set of all (bounded) operators on $\mathcal{H}_1 \otimes \mathcal{H}_2$, and let $\mathcal{T} \equiv \mathcal{T}(\mathcal{H}_1 \otimes \mathcal{H}_2)$ be the subset of (positive, trace-

1) density operators. Throughout, we shall consider $\mathcal{T}$ as endowed with the metric (and corresponding topology) induced by the trace norm, $\|A\|_T \equiv \mathrm{Tr}((A^*A)^{1/2})$. We reserve the notation '$\|A\|$' for the standard operator norm. An operator A is called a *contraction* if $\|A\| \leq 1$. We denote the self-adjoint contractions acting on a Hilbert space $\mathcal{H}$ by $\mathcal{B}(\mathcal{H})_s$. The metric induced by the trace norm is appropriate physically for measuring the distance between quantum states, because (Schatten 1970, p. 46ff)

$$\|D - D'\|_T = \sup\Big\{|\mathrm{Tr}(DA) - \mathrm{Tr}(D'A)| : A \in \mathcal{B}(\mathcal{H}_1 \otimes \mathcal{H}_2)_s\Big\} \tag{1}$$

which implies that trace norm close states dictate close probabilities for the outcomes of measuring any observable.

For $D \in \mathcal{T}$, D is said to be a *product state* just in case there is a $D_1 \in \mathcal{T}(\mathcal{H}_1)$ and a $D_2 \in \mathcal{T}(\mathcal{H}_2)$ such that $D = D_1 \otimes D_2$. The *separable* density operators are then defined to be all members of $\mathcal{T}$ that may be approximated (in trace norm) by convex combinations of product states (Werner 1989). In other words, the separable density operators are those in the closed convex hull of the set of all product states in $\mathcal{T}$. By definition, then, the set of *non*separable density operators is open.

Let A_1, A_2 be self-adjoint contractions in $\mathcal{B}(\mathcal{H}_1)_s$, and, similarly, let $B_1, B_2 \in \mathcal{B}(\mathcal{H}_2)_s$. Then the corresponding operator

$$R \equiv \frac{1}{2}\Big(A_1 \otimes (B_1 + B_2) + A_2 \otimes (B_1 - B_2)\Big) \tag{2}$$

is called a *Bell operator* for the system $\mathcal{H}_1 \otimes \mathcal{H}_2$. Fix a density operator $D \in \mathcal{T}$. We can then define the *Bell coefficient* $\beta(D)$ of D by

$$\beta(D) \equiv \sup\Big\{|\mathrm{Tr}(DR)| : R \text{ is a Bell operator for } \mathcal{H}_1 \otimes \mathcal{H}_2\Big\}. \tag{3}$$

Bell's theorem, as elaborated by Bell (1987), Clauser *et al.* (1969), implies that for any state D and Bell operator R, a local hidden variable model of D's correlations is committed to predicting the *CHSH inequality* $|\mathrm{Tr}(DR)| \leq 1$. On the other hand, there are always states D for which $\beta(D) > 1$. We say such states are *CHSH violating*.

Convexity arguments entail that $\beta(D)$ is in fact equivalent to the supremum taken over all Bell operators where A_i, B_i are self-adjoint *unitary* operators satisfying $A_i^2 = B_i^2 = I$, i.e., generalized spin components (Summers & Werner 1995, Prop. 3.2). For completeness, we set out a detailed proof of this fact in Appendix A. Unless otherwise noted, we henceforth assume that all our Bell operators are constructed out of self-adjoint unitaries. Moreover, for such Bell operators we always have (Landau 1987)

$$R^2 = I \otimes I - \frac{1}{4}[A_1, A_2] \otimes [B_1, B_2], \tag{4}$$

from which it follows by an elementary calculation that $\|R\| \leq \sqrt{2}$. Thus, for any state D, $\beta(D) \leq \sqrt{2}$ since $|\mathrm{Tr}(DR)| \leq \|R\|$. Moreover, $\beta(D) \geq 1$, since we may always take $A_i = B_i = I$.

If any of the four operators A_i, B_i is $\pm I$, then (4) entails that $\|R\|^2 = \|R^2\| = 1$ and R cannot indicate any CHSH violation. Thus, we will find it convenient to define $\gamma(D)$ in analogy to the definition of $\beta(D)$, but with the added restriction that the supremum be taken over all Bell operators constructed from *nontrivial* (i.e., not $\pm I$) self-adjoint unitary operators. It immediately follows that for any $D \in \mathcal{T}$, $\gamma(D) \in [0, \sqrt{2}]$ and

$$\beta(D) = \max\{1, \gamma(D)\}. \tag{5}$$

Thus, any nonclassical CHSH violation indicated by $\beta(D) > 1$ is indicated just as well by $\gamma(D) > 1$.

Let $D, D' \in \mathcal{T}$ be such that $\|D - D'\|_T \leq \epsilon$. Then, for any Bell operator $R \in \mathcal{B}(\mathcal{H}_1 \otimes \mathcal{H}_2)$, it follows from (1) that

$$|\mathrm{Tr}(DR) - \mathrm{Tr}(D'R)| \leq \epsilon\|R\|. \tag{6}$$

In particular, since for any Bell operator R, $\|R\| \leq \sqrt{2}$,

$$|\mathrm{Tr}(DR)| \leq \epsilon\sqrt{2} + |\mathrm{Tr}(D'R)|. \tag{7}$$

Taking the supremum over nontrivial Bell operators R, first on the right-hand side of (7), and then on the left, we see that $\gamma(D) \leq \epsilon\sqrt{2} + \gamma(D')$. By symmetry, we have $\gamma(D') \leq \epsilon\sqrt{2} + \gamma(D)$, so that

$$|\gamma(D) - \gamma(D')| \leq \epsilon\sqrt{2} \tag{8}$$

and γ is a continuous function from $\mathcal{T}$ (in trace norm) into $[0, \sqrt{2}]$. It then follows from (5) that β is a continuous function from $\mathcal{T}$ into $[1, \sqrt{2}]$. Since the set of CHSH violating density operators is the pre-image of $(1, \sqrt{2}]$ under β, this set is open in the trace norm topology.

Suppose now that D is a convex combination $D = (1 - \lambda)W + \lambda W'$ where $W, W' \in \mathcal{T}$. Then, for any Bell operator R,

$$\begin{aligned} |\text{Tr}(DR)| &= |(1-\lambda)\text{Tr}(WR) + \lambda\text{Tr}(W'R)| \\ &\leq (1-\lambda)|\text{Tr}(WR)| + \lambda|\text{Tr}(W'R)|. \end{aligned} \tag{9}$$

Taking the supremum over nontrivial Bell operators first on the right-hand side of (9), and then on the left, we may conclude that

$$\gamma(D) \leq (1-\lambda)\gamma(W) + \lambda\gamma(W'). \tag{10}$$

Thus, γ is a convex function. It is easy to check that $\gamma(D) \leq 1$ for all product states D, and therefore the same holds for any separable state, by continuity and convexity of γ.

It follows from the work of Werner (1989) that when $d_1 = d_2 = n \geq 2$, there are *non*separable states that satisfy all CHSH inequalities. In the case where $d_1 = d_2 = 2$, the Werner state, which we shall denote by W_{22}, can be written as

$$W_{22} = \frac{1}{8}(I \otimes I) + \frac{1}{4}\Big[(I \otimes I) - U\Big], \tag{11}$$

where U is the (self-adjoint, unitary) permutation operator. Werner observed that for any separable density operator D, we must have $\text{Tr}(UD) \geq 0$. However, using the fact that $U^2 = I$ and $\text{Tr}(U) = 2$, we have

$$\text{Tr}(UW_{22}) = \frac{1}{8}\text{Tr}(U) + \frac{1}{4}\text{Tr}(U - I) = -\frac{1}{4} < 0. \tag{12}$$

Thus, W_{22} is nonseparable. Moreover, using the fact that $U = I \otimes I - 2P_s$, where P_s is the projection onto the singlet state, we may conveniently rewrite W_{22} in the form:

$$W_{22} = \frac{1}{8}(I \otimes I) + \frac{1}{2}P_s. \tag{13}$$

Since $\gamma[(1/4)(I \otimes I)] = 0$, and γ is convex,

$$\gamma(W_{22}) \leq \frac{1}{2}\gamma(P_s) = 2^{-1/2} < 1, \tag{14}$$

and W_{22} is not CHSH violating.

More generally, we define a state D to be *CHSH insensitive* whenever D is nonseparable yet not CHSH violating, i.e., $\gamma(D) \leq 1$. Such states may still violate Bell inequalities involving projective measurements of observables with spectral values lying outside $[-1, 1]$, or more than two pairs of projective measurements, or positive operator valued measurements. They may also contain "hidden" CHSH violations in the sense that they may violate *generalized* CHSH inequalities which involve performing consecutive projective measurements on each of the two subsystems. To make this precise, let $\mathcal{H}$ be an arbitrary Hilbert space, and let $\mathcal{T}(\mathcal{H})$ be the set of density operators on $\mathcal{H}$. For any $D \in \mathcal{T}(\mathcal{H})$ and $A \in \mathcal{B}(\mathcal{H})$ such that $ADA^* \neq 0$, we may define the new density operator D^A by

$$D^A \equiv \frac{ADA^*}{\mathrm{Tr}(ADA^*)}. \tag{15}$$

Then $D \in \mathcal{T}(\mathcal{H}_1 \otimes \mathcal{H}_2)(\equiv \mathcal{T})$ will violate a generalized CHSH inequality just in case there are projections Q_1 and Q_2 such that $D^{Q_1 \otimes Q_2}$ is CHSH violating. (In such a case, the violation is 'seen' after first performing a pair of selective measurements on the component systems.) For example, Popescu (1994, 1995) (see also Mermin 1996) has shown that when $n \geq 5$, the states constructed by Werner violate generalized CHSH inequalities. On the other hand, it is clear from (11) that W_{22} itself cannot violate a generalized CHSH inequality, since for nontrivial Q_1 or Q_2, $W_{22}^{Q_1 \otimes Q_2}$ is always a product state.

A state which violates *any* Bell inequality, including generalized inequalities, must be nonseparable. Moreover, since the correlations in such states — whether or not they are CHSH sensitive — cannot be reproduced by any local hidden variable theory, one is justified in terming them *nonlocal* states.

For example, while Werner has shown that the correlations dictated by W_{22} between the outcomes of projective measurements admit a local hidden variable model, this does not imply that W_{22} is nonlocal; for he left it as a conjecture that the same is true for positive operator valued measurements (Werner 1989, p. 4280).

2 CHSH violation and infinite-dimensional systems

In this section, we establish that a bipartite system has a dense set of nonlocal states when either component is infinite-dimensional.

We begin with an elementary observation about the action of A on D defined by (15). This action is a natural generalization of the action of an operator on unit vectors. Indeed, we may always add an ancillary Hilbert space $\mathcal{K}$ onto $\mathcal{H}$ (with $\dim \mathcal{K} \geq \dim \mathcal{H}$) such that D is the reduced density operator for a pure vector state $x \in \mathcal{H} \otimes \mathcal{K}$. In such a case, a straightforward verification shows that (when $(A \otimes I)x \neq 0$) the reduced density operator for $(A \otimes I)x / \|(A \otimes I)x\|$ is just D^A.

Let Φ be the map that assigns a unit vector $x \in \mathcal{H} \otimes \mathcal{K}$ its reduced density operator $\Phi(x)$ on $\mathcal{H}$. It is easy to see that Φ is trace-norm continuous (Clifton & Halvorson 2000). Let $\{P_n\}$ be any increasing sequence of projections in $\mathcal{B}(\mathcal{H})$ with least upper bound I. Then, $(P_n \otimes I)x \to x$ and

$$\begin{aligned} D^{P_n} &= \Phi[(P_n \otimes I)x / \|(P_n \otimes I)x\|] && (16) \\ &\to \Phi[x] = D, && (17) \end{aligned}$$

where the convergence is in trace norm. We make use of this convergence in our arguments below.

Proposition 1. *If $d_1 = d_2 = \infty$, then the set of CHSH violating states is trace norm dense in the set of all density operators on $\mathcal{H}_1 \otimes \mathcal{H}_2$.*

Proof: Fix an arbitrary density operator D on $\mathcal{H}_1 \otimes \mathcal{H}_2$, and fix orthonormal bases for the factor spaces $\{e_i\}$ and $\{f_j\}$. Let P_n be the projection onto the span of $\{e_i \otimes f_j\}_{i,j \leq n}$, and set

$$\psi_n = \frac{1}{\sqrt{2}}(|e_{n+1}\rangle|f_{n+1}\rangle + |e_{n+2}\rangle|f_{n+2}\rangle). \tag{18}$$

Consider the sequence of density operators $\{D_n\}$ defined by

$$D_n = (1 - \frac{1}{n})D^{P_n} + \frac{1}{n}P_{\psi_n} \tag{19}$$

where P_ψ projects onto the ray ψ generates. Since $\lim_{n\to\infty} D_n = D$ in trace norm, all that remains to show is that each D_n is CHSH violating. As ψ_n is the pure singlet state, there are "spin components" (i.e. self-adjoint unitaries) A_i^n, B_i^n ($i = 1, 2$) such that each A_i^n leaves the subspace generated by $|e_{n+1}\rangle, |e_{n+2}\rangle$ invariant and acts like the identity on the complement; similarly for each B_i^n and the subspace generated by $|f_{n+1}\rangle, |f_{n+2}\rangle$; and, moreover, the Bell operator

$$R_n \equiv \frac{1}{2}(A_1^n \otimes B_1^n + A_1^n \otimes B_2^n + A_2^n \otimes B_1^n - A_2^n \otimes B_2^n) \tag{20}$$

is such that $\mathrm{Tr}(P_{\psi_n} R_n) > 1$. Therefore, in view of (19), to show that $\mathrm{Tr}(D_n R_n) > 1$, and hence that D_n is CHSH violating, it suffices to observe that $\mathrm{Tr}(D^{P_n} R_n) = 1$. But this is immediate from the fact that R_n acts as the identity on P_n's range. *QED*

A similar argument shows that nonlocal states are dense in the case $d_1 < d_2 = \infty$.

Proposition 2. *If $d_1 < d_2 = \infty$, then the set of nonlocal states is trace norm dense in the set of all density operators on $\mathcal{H}_1 \otimes \mathcal{H}_2$.*

Proof: Fix an arbitrary density operator D on $\mathcal{H}_1 \otimes \mathcal{H}_2$, and fix orthonormal bases for the factor spaces $\{e_i\}_{i=1}^{d_1}$ and $\{f_j\}_{j=1}^{\infty}$. Let P'_n be the projection onto the span of $\{e_i \otimes f_j\}_{1 \leq i \leq d_1, 1 \leq j \leq n}$, and set

$$\psi'_n = \frac{1}{\sqrt{2}}(|e_1\rangle|f_{n+1}\rangle + |e_2\rangle|f_{n+2}\rangle). \tag{21}$$

Consider the sequence of density operators $\{D_n\}$ defined by

$$D_n = (1 - \frac{1}{n})D^{P'_n} + \frac{1}{n}P_{\psi'_n}. \tag{22}$$

As before, $\lim_{n\to\infty} D_n = D$ in trace norm, so it suffices to show that each D_n is nonlocal. Define the projections Q_1, Q_2 onto the spans of $\{e_i\}_{1\le i\le 2}$ and $\{f_j\}_{n+1\le j\le n+2}$, respectively. Then since $D_n^{Q_1\otimes Q_2} = P_{\psi_n'}$, D_n violates a generalized CHSH inequality. *QED*

Note that Prop. 2 entails that when $d_2 = \infty$, the set of nonseparable states is dense. This reproduces, by quite different methods, the main result of Clifton & Halvorson (2000).

3 Generic CHSH violation characterizes infinite-dimensional systems

As mentioned in the introduction, when both $d_1, d_2 < \infty$, there is always an open neighborhood of separable states (Życzkowski *et al.* 1998, Caves & Schack 2000, Życzkowski 1999). Since separable states cannot display any nonlocal correlations, it follows that in this case the CHSH violating states cannot be dense. Note, however, that this same method of argument could not establish an open CHSH nonviolating neighborhood in the case where $d_1 < d_2 = \infty$, for in that case we know that the separable states are nowhere dense. However, as we now show, such neighborhoods exist.

Let $D \in \mathcal{T}$ be a density operator with $\gamma(D) < 1$. It is not difficult to see that the distance from D to the set of CHSH violating states is bounded below by $2^{-1/2}(1-\gamma(D))$. Indeed, for any density operator D', if

$$\|D - D'\|_T \le 2^{-1/2}(1-\gamma(D)), \tag{23}$$

then from (8),

$$\gamma(D') \le 2^{1/2}\Big[2^{-1/2}(1-\gamma(D))\Big] + \gamma(D) = 1. \tag{24}$$

Thus any state D with $\gamma(D) < 1$ is surrounded by a neighborhood of states that are again not CHSH violators.

Proposition 3. *If $d_1 < \infty$ then, for any density operator $D_2 \in \mathcal{T}(\mathcal{H}_2)$, we have*

$$\gamma[d_1^{-1}(I\otimes D_2)] \le 1 - 2d_1^{-1} < 1. \tag{25}$$

Proof: Let A be a self-adjoint unitary operator (not $\pm I$) acting on $\mathcal{H}_1$. Then $A = P_1 - P_2$, where P_i is a projection ($i = 1, 2$). Since $A \neq \pm I$, $P_1 \neq 0$ and $P_2 \neq 0$. Thus,

$$\begin{aligned} |\mathrm{Tr}(d_1^{-1}A)| &= d_1^{-1}\big|\mathrm{Tr}(P_1) - \mathrm{Tr}(P_2)\big| & (26)\\ &\leq d_1^{-1}(d_1 - 2) = 1 - 2d_1^{-1}. & (27)\end{aligned}$$

Now let R be any Bell-operator for $\mathcal{H}_1 \otimes \mathcal{H}_2$, constructed from (non-trivial) self-adjoint unitary operators. Then,

$$\begin{aligned} &|\mathrm{Tr}(d_1^{-1}(I \otimes D_2)R)| \\ &= \frac{1}{2}\big|\mathrm{Tr}(d_1^{-1}(A_1 + A_2)) \cdot \mathrm{Tr}(D_2 B_1) \\ &\quad + \mathrm{Tr}(d_1^{-1}(A_1 - A_2)) \cdot \mathrm{Tr}(D_2 B_2)\big| & (28)\\ &\leq \frac{1}{2}\big|\mathrm{Tr}(d_1^{-1}A_1) + \mathrm{Tr}(d_1^{-1}A_2)\big| \\ &\quad + \frac{1}{2}\big|\mathrm{Tr}(d_1^{-1}A_1) - \mathrm{Tr}(d_1^{-1}A_2)\big| & (29)\\ &\leq 1 - 2d_1^{-1}. & (30)\end{aligned}$$

The last inequality follows since

$$|a_1 + a_2| + |a_1 - a_2| \leq 2\max\{|a_i|\}, \qquad (31)$$

for any two real numbers a_1, a_2. *QED*

Note that the considerations prior to this proposition entail that $d_1^{-1}(I \otimes D_2)$ lies in a neighborhood of CHSH nonviolating states of (trace norm) size at least $d_1^{-1}\sqrt{2}$. (Of course, this estimate could be improved if restrictions on D_2 were also taken into account.)

Proposition 4. *The set of CHSH violating density operators is trace norm dense in the set of all density operators on* $\mathcal{H}_1 \otimes \mathcal{H}_2$ *(and its complement is nowhere dense) if and only if* $\dim\mathcal{H}_1 = \dim\mathcal{H}_2 = \infty$.

Proof: Suppose that $d_1 = d_2 = \infty$. Then, from Prop. 1, the set of CHSH violating states is trace norm dense (and its *closed* complement must be nowhere dense). Conversely, if $d_1 < \infty$, then Prop. 3 (and the discussion preceding it) ensures the existence of many open neighborhoods of states that satisfy the CHSH inequality. *QED*

4 CHSH insensitive states

Props. 1–4 establish that CHSH insensitive states exist in the case when $d_1 < d_2 = \infty$. In particular, since there is at least one open set of states that do not violate the CHSH inequality, and since the nonseparable states are dense, there must be nonseparable states that are not CHSH violating. Indeed, Prop. 3 provides us with a class of states which we know have a surrounding neighborhood of states that are not CHSH violating, while Prop. 2 shows how, given any state, we may construct a sequence of nonseparable states which converges to that state. In Appendix B, we invoke the alternate method of constructing nonseparable states given in Clifton & Halvorson (2000) to construct a sequence of CHSH insensitive states that converges continuously to a product state. (We do so only for the simplest case of a bipartite system with exactly one two-dimensional component — such as a spin-1/2 particle, distinguishing its internal and external degrees of freedom.)

We have not so far shown that there are CHSH insensitive states in the cases $d_1 < d_2 < \infty$ and $d_1 = d_2 = \infty$. We now proceed to show that in all relevant cases, i.e., whenever $d_1, d_2 \geq 2$, CHSH insensitive states exist. Moreover, if $d_1 < \infty$, there is always an open neighborhood of CHSH insensitive states.

CHSH insensitive states can be constructed simply by embedding the 2×2 Werner state W_{22} into the higher-dimensional space. Let $\{e_i \otimes f_j\}$ denote an orthonormal product basis for $\mathcal{H}_1 \otimes \mathcal{H}_2$, and let $\mathcal{K}$ denote the 2×2 subspace spanned by $\{e_i \otimes f_j : i,j = 1,2\}$. Note that the projection onto $\mathcal{K}$ is just the product $P \otimes Q$ of the projections P onto $\{e_i : i = 1,2\}$ and Q onto $\{f_j : j = 1,2\}$. Corresponding to the permutation operator U of $\mathbb{C}^2 \otimes \mathbb{C}^2$, we let U' denote the (partial isometry) operator on $\mathcal{H}_1 \otimes \mathcal{H}_2$ which permutes the basis elements of $\mathcal{K}$ and maps $\mathcal{K}^\perp$ to 0. Then, by analogy with W_{22}, we may define

$$W'_{22} \equiv \frac{1}{8}(P \otimes Q) + \frac{1}{4}\Big[(P \otimes Q) - U'\Big]. \tag{32}$$

It is not difficult to see that $W'_{22} \in \mathcal{T}(\mathcal{H}_1 \otimes \mathcal{H}_2)$. We now verify that W'_{22}, *as a state of* $\mathcal{H}_1 \otimes \mathcal{H}_2$, is again CHSH insensitive.

For a density operator $D \in \mathcal{T}(\mathcal{H}_1 \otimes \mathcal{H}_2)$, let us say that D is $\mathcal{K}$-*separable* just in case D is in the closed convex hull of product states *all of whose ranges are contained in* $\mathcal{K}$.

Proposition 5. *Suppose that* $D \in \mathcal{T}(\mathcal{H}_1 \otimes \mathcal{H}_2)$ *and* D *has range contained in* $\mathcal{K}$. *If* D *is separable, then* D *is* $\mathcal{K}$*-separable.*

Before we give the proof of this proposition, we recall from Clifton & Halvorson (2000) some basic facts concerning the operation $D \to D^A$ on density operators defined in (15). Suppose that $D \in \mathcal{T}(\mathcal{H})$ is a convex combination of density operators

$$D = \sum_{i=1}^{n} \lambda_i D_i. \tag{33}$$

Then, for any $A \in \mathcal{B}(\mathcal{H})$, if $ADA^* \neq 0$, we may set

$$\lambda_i^A \equiv \lambda_i \frac{\mathrm{Tr}(AD_iA^*)}{\mathrm{Tr}(ADA^*)}, \tag{34}$$

and we have

$$D^A \equiv \frac{ADA^*}{\mathrm{Tr}(ADA^*)} = \sum_{i=1}^{n} \lambda_i^A D_i^A, \tag{35}$$

where $\sum_{i=1}^{n} \lambda_i^A = 1$. Thus, D^A is a convex combination of the D_i^A. Note, also, that when $ADA^* \neq 0$, the operation $D \to D^A$ is trace norm continuous at D (since multiplication by a fixed element in $\mathcal{B}(\mathcal{H})$ is trace norm continuous (Schatten 1970, p. 39)).

Further specializing to the case where $\mathcal{H} \equiv \mathcal{H}_1 \otimes \mathcal{H}_2$, note that if $D = D_1 \otimes D_2$ is a product state, and $A \in \mathcal{B}(\mathcal{H}_1)$, $B \in \mathcal{B}(\mathcal{H}_2)$ are arbitrary, then

$$D^{A\otimes B} = (D_1 \otimes D_2)^{A\otimes B} = D_1^A \otimes D_2^B. \tag{36}$$

Proof of the proposition: Let $P \otimes Q$ denote the projection onto $\mathcal{K}$. Since D has range contained in $\mathcal{K}$, we have $D^{P\otimes Q} = D$. Suppose now that D is separable. That is, $D = \lim_n D_n$ where each $D_n \in \mathcal{T}(\mathcal{H}_1 \otimes \mathcal{H}_2)$ is a convex combination of product states. Thus, by continuity we have

$$D = D^{P\otimes Q} = \lim_n D_n^{P\otimes Q}. \tag{37}$$

By the preceding considerations, each $D_n^{P\otimes Q}$ is a convex combination of product states, each of which has range contained in $\mathcal{K}$. Thus, D is $\mathcal{K}$-separable. *QED*

It is now straightforward to see that W'_{22} is nonseparable. Indeed, since W'_{22} has range contained in $\mathcal{K}$, if W'_{22} were separable, it would also be $\mathcal{K}$-separable. However, using the natural isomorphism between $\mathbb{C}^2 \otimes \mathbb{C}^2$ and $\mathcal{K}$, and the induced isomorphism between density operators on $\mathbb{C}^2 \otimes \mathbb{C}^2$ and density operators on $\mathcal{H}_1 \otimes \mathcal{H}_2$ with range in $\mathcal{K}$, it would follow that W_{22} is separable. Therefore, W'_{22} is nonseparable.

To see that W'_{22} is not CHSH violating, note that for any Bell operator R for $\mathcal{H}_1 \otimes \mathcal{H}_2$,

$$R' \equiv (P \otimes Q)R(P \otimes Q) \tag{38}$$

is again a Bell operator (constructed out of self-adjoint contractions PA_iP, QB_iQ that may not be unitary). Moreover,

$$\begin{aligned} |\mathrm{Tr}(W'_{22}R)| &= |\mathrm{Tr}((W'_{22})^{P\otimes Q}R)| && (39)\\ &= |\mathrm{Tr}(W'_{22}(P\otimes Q)R(P\otimes Q))| && (40)\\ &= |\mathrm{Tr}(W'_{22}R')|. && (41) \end{aligned}$$

Thus, if W'_{22} violates a CHSH inequality, it must violate a CHSH inequality with respect to some Bell operator R' whose range lies in $\mathcal{K}$. But any such R' has a counterpart in $\mathcal{B}(\mathbb{C}^2 \otimes \mathbb{C}^2)$ that would display a CHSH violation for W_{22}. Therefore, W'_{22} is not CHSH violating and is CHSH insensitive for $\mathcal{H}_1 \otimes \mathcal{H}_2$.

We end by combining the fact that there are always CHSH insensitive states with the results of the previous section to show that there are "many" CHSH insensitive states, unless both component spaces are infinite-dimensional.

Proposition 6. *There is an open set of CHSH insensitive density operators on* $\mathcal{H}_1 \otimes \mathcal{H}_2$ *if and only if* $\dim\mathcal{H}_1 < \infty$ *or* $\dim\mathcal{H}_2 < \infty$.

Proof: The "only if" follows immediately from Prop. 1. (If $d_1 = d_2 = \infty$, then the CHSH insensitive states are contained in the nowhere dense set of states which satisfy all CHSH inequalities.) To prove the converse, suppose that $d_1 < \infty$. It would suffice to show that there is a nonseparable state W with $\gamma(W) < 1$. For, in that case, we may use the continuity of γ to obtain an open neighborhood $\mathcal{O}$ of W which contains only states with no CHSH violations. Taking the intersection of $\mathcal{O}$ with the open set of nonseparable states would give the desired open set of CHSH insensitive states.

From considerations adduced above, there is always a CHSH insensitive state $W' \in \mathcal{T}(\mathcal{H}_1 \otimes \mathcal{H}_2)$. Since W' does not violate a CHSH inequality, we have $\gamma(W') \leq 1$. Moreover, from Prop. 3, there is a $D \in \mathcal{T}$ such that $\gamma(D) < 1$. For each n, let

$$W_n \equiv (1 - n^{-1})W' + n^{-1}D. \tag{42}$$

Clearly, $W_n \to W'$ in trace norm, and by the convexity of γ,

$$\begin{aligned} \gamma(W_n) &\leq (1 - n^{-1})\gamma(W') + n^{-1}\gamma(D) && (43) \\ &\leq (1 - n^{-1}) + n^{-1}\gamma(D) < 1, && (44) \end{aligned}$$

for all n. However, since W' is nonseparable, and the nonseparable states are open, there is an $m \in \mathbb{N}$ such that W_n is nonseparable for all $n \geq m$. Thus, setting $W \equiv W_m$ gives the desired nonseparable state with $\gamma(W) < 1$. *QED*

5 Conclusion

We have established the conjecture made in Clifton & Halvorson (2000) that bipartite systems whose components are both infinite-dimensional (e.g., a pair of particles, neglecting their spins) have states that generically violate the CHSH inequality. We also established that even if one of the components is finite-dimensional (e.g., a spin-1/2 particle), nonlocally correlated states remain dense. Finally, we have identified new classes of CHSH insensitive states for finite by infinite systems, and established that such states can only be neglected, for all practical purposes, in the infinite by infinite case.

Infinite-dimensional systems thus provide a resource of nonlocality which — practically speaking — cannot be completely destroyed by noise or by errors in preparation or measurement. In this they differ from finite-dimensional systems, where entangled mixed states can always be reduced to separable states by sufficient noise. One might naively conclude that, to the extent that it is practicable in quantum information and computation theory to exploit infinite-dimensional systems, it would be advantageous to do so. But in fact we can never exploit all the degrees of freedom in a infinite-dimensional system. So, though we hope the above results may be useful in developing the theory of entanglement in large finite-dimensional systems, we doubt that they themselves can lead to direct practical application.

Even in the case of large finite-dimensional systems, there is a potential pitfall. It may well be that nonlocality becomes harder and harder to destroy, by some sensible quantitative measure, as the size of the system becomes larger. However, the nonlocality results we have outlined give no indication of a general procedure for extracting or demonstrating nonlocality. Protecting some form of nonlocality is less useful if it is achieved at the cost of making it harder and harder to find. It would thus be very interesting to quantify the trade-offs which can usefully be made in this direction when large finite-dimensional systems are used to counter noise on a highly noisy channel.

Appendix A

We give here a self-contained version of Summers and Werner's (1995) argument that $\beta(D)$ is equal to the supremum of $|\mathrm{Tr}(DR)|$, where R only runs over the Bell operators for $\mathcal{H}_1 \otimes \mathcal{H}_2$ that are constructed from self-adjoint *unitary* operators.

Recall that the weak-operator topology on $\mathcal{B}(\mathcal{H})$ is the coarsest topology for which all functionals of the form

$$T \to |\langle Tx, y\rangle| \qquad x, y \in \mathcal{H}, \tag{45}$$

are continuous at 0. It then follows that the unit ball of $\mathcal{B}(\mathcal{H})$ is compact in the weak-operator topology (Kadison & Ringrose 1997, Thm. 5.1.3). (Of course, if dim $\mathcal{H} < \infty$, the unit ball of $\mathcal{B}(\mathcal{H})$ is also compact in the operator-norm topology.) Moreover, since the adjoint operation is weak-operator continuous, the set of self-adjoint operators is weak-operator closed in $\mathcal{B}(\mathcal{H})$, and $\mathcal{B}(\mathcal{H})_s$ is weak-operator compact (as well as convex).

Fix $A_2 \in \mathcal{B}(\mathcal{H}_1)_s$ and $B_1, B_2 \in \mathcal{B}(\mathcal{H}_2)_s$. We show that the map $\Psi_D : \mathcal{B}(\mathcal{H}_1)_s \to \mathbb{R}$ defined by

$$\Psi_D(A_1) \equiv \frac{1}{2}\mathrm{Tr}\Big(D(A_1 \otimes (B_1 + B_2) + A_2 \otimes (B_1 - B_2))\Big), \tag{46}$$

is affine and weak-operator continuous. From this it will follow that Ψ_D attains its extremal values on extreme points of $\mathcal{B}(\mathcal{H}_1)_s$ (Conway 1990, Prop. 7.9). These, however, consist precisely of the self-adjoint unitary operators (Kadison & Ringrose 1997, Prop. 7.4.6).

Now, to establish that Ψ_D is affine and weak-operator continuous, let $\Lambda_D : \mathcal{B}(\mathcal{H}_1)_s \to \mathbb{R}$ denote the linear functional defined by

$$\Lambda_D(A_1) \equiv \mathrm{Tr}\Big[D(A_1 \otimes (1/2)(B_1 + B_2))\Big]. \tag{47}$$

Then, Λ_D is the composition of the map

$$A_1 \to A_1 \otimes (1/2)(B_1 + B_2), \tag{48}$$

from $\mathcal{B}(\mathcal{H}_1)_s$ into $\mathcal{B}(\mathcal{H}_1 \otimes \mathcal{H}_2)_s$, with the functional $\mathrm{Tr}(D\ \cdot\ )$. However, the former is continuous (when both algebras are equipped with the weak-operator topology) since multiplication by a fixed operator is weakly continuous. Moreover, $\mathrm{Tr}(D\ \cdot\ )$ is weak-operator continuous on the unit ball of $\mathcal{B}(\mathcal{H}_1 \otimes \mathcal{H}_2)$. Thus, Λ_D is weak-operator continuous. Now, let

$$r_D \equiv \mathrm{Tr}\Big[D(A_2 \otimes (1/2)(B_1 - B_2))\Big].$$

Then, $\Psi_D = \Lambda_D + r_D$ is affine and weak-operator continuous.

From the above considerations it now follows that for every $A_1 \in \mathcal{B}(\mathcal{H}_1)_s$ and Bell operator R constructed using A_1, there is a Bell operator R' constructed from the same elements as R, except with A_1 replaced by a self-adjoint unitary operator, and such that $|\mathrm{Tr}(DR)| \leq |\mathrm{Tr}(DR')|$. By symmetry, the same conclusion applies to A_2, B_1 and B_2. Thus, for any given Bell operator R, there is a Bell operator R' constructed entirely from self-adjoint unitaries, and such that $|\mathrm{Tr}(DR)| \leq |\mathrm{Tr}(DR')|$.

Appendix B

In this appendix, we use the results of the current paper and of Clifton & Halvorson (2000) to construct a continuous 'path' of CHSH insensitive states with endpoint a product state. Reversing the convention $d_1 \leq d_2$ of the current paper (to align with that chosen in Clifton & Halvorson (2000)), we examine the case where $d_1 = \infty$ and $d_2 = 2$.

Let $\{e_i\} \subseteq \mathcal{H}_1$ and $\{f_1, f_2\} \subseteq \mathcal{H}_2$ be orthonormal bases. Attaching an ancillary Hilbert space $\mathcal{H}_3$, with infinite orthonormal basis $\{g_k\}$, we may define a unit vector $v_0 \in \mathcal{H}_1 \otimes \mathcal{H}_2 \otimes \mathcal{H}_3$ by

$$\begin{aligned} v_0 \quad \equiv \quad & \frac{1}{2}\Big(|e_1\rangle|f_1\rangle|g_1\rangle + |e_2\rangle|f_2\rangle|g_2\rangle \\ & + |e_2\rangle|f_1\rangle|g_3\rangle + |e_1\rangle|f_2\rangle|g_4\rangle \Big). \end{aligned} \tag{49}$$

Note that the reduced density operator $\Phi(v_0) \in \mathcal{T}(\mathcal{H}_1 \otimes \mathcal{H}_2)$ for v_0 is just $\frac{1}{2}P \otimes \frac{1}{2}I$, where P is the projection onto the subspace of $\mathcal{H}_1$ spanned by $\{e_1, e_2\}$. Thus, from Prop. 3 (interchanging 1 with 2), there is a CHSH nonviolating neighborhood surrounding $\Phi(v)$.

Now, for each $\lambda \in [0, 1]$, define the unit vector $v_\lambda \in \mathcal{H}_1 \otimes \mathcal{H}_2 \otimes \mathcal{H}_3$ by

$$v_\lambda \equiv (1 - \lambda)v_0 + [\lambda(2 - \lambda)]^{1/2} u \tag{50}$$

where u is the unit vector

$$u \equiv \sum_{n=1}^{\infty} 2^{-(n+1)/2} \Big(|e_{2n+1}\rangle|f_1\rangle|g_n\rangle + |e_{2n+2}\rangle|f_2\rangle|g_n\rangle \Big). \tag{51}$$

Clearly, $v_\lambda \to v_0$ as $\lambda \to 0$. Furthermore, by the continuity of Φ, $\Phi(v_\lambda) \to \Phi(v_0)$. It then follows that there is an $\epsilon > 0$ such that $\Phi(v_\lambda)$ is not CHSH violating for all $\lambda < \epsilon$. However, by construction v_λ is *separating* for the subalgebra $I \otimes \mathcal{B}(\mathcal{H}_2 \otimes \mathcal{H}_3)$, for all $\lambda \in (0, 1]$. That is, for any $A \in \mathcal{B}(\mathcal{H}_2 \otimes \mathcal{H}_3)$, if $(I \otimes A)v_\lambda = 0$, then $A = 0$. (To see this, observe that any such A would have to annihilate all the basis vectors $\{f_j \otimes g_k\}$ due to the orthogonality of the $\{e_i\}$.) Thus, invoking Clifton & Halvorson (2000, Lemmas 1,2), each $\Phi(v_\lambda)$ is nonseparable, and, for all $0 < \lambda < \epsilon$, CHSH insensitive.

Bibliography

Bell, J. S. (1987), *Speakable and unspeakable in quantum mechanics*, Cambridge University Press, Cambridge. Collected papers on quantum philosophy.

Braunstein, S. L., Caves, C. M., Jozsa, R., Linden, N., Popescu, S. & Schack, R. (1999), 'Separability of very noisy mixed states and implications for NMR quantum computing', *Physical Review Letters* **83**, 1054–7.

Caves, C. & Schack, R. (2000), 'Explicit product ensembles for separable quantum states', *Journal of Modern Optics* **47**, 387–99.

Clauser, J. F., Horne, M. A., Shimony, A. & Holt, R. A. (1969), 'Proposed experiment to test local hidden-variable theories', *Physical Review Letters* **23**, 880–4.

Clifton, R. & Halvorson, H. (2000), 'Bipartite mixed states of infinite-dimensional systems are generically nonseparable', *Physical Review A* **61**, 012108.

Conway, J. B. (1990), *A course in functional analysis*, 2nd ed., Springer-Verlag, New York.

Gisin, N. (1991), 'Bell's inequality holds for all nonproduct states', *Physics Letters A* **154**, 201–2.

Gisin, N. (1999), 'Bell inequality for arbitrary many settings of the analyzers', *Physics Letters A* **260**, 1–3.

Kadison, R. V. & Ringrose, J. R. (1997), *Fundamentals of the theory of operator algebras. Vol. I*, Vol. 15 of *Graduate Studies in Mathematics*, American Mathematical Society, Providence, RI.

Landau, L. J. (1987), 'On the violation of Bell's inequality in quantum theory', *Physics Letters A* **120**, 54–6.

Linden, N. & Popescu, S. (2001), 'Good dynamics versus bad kinematics. Is entanglement needed for quantum computation?', *Physical Review Letters* **87**, 047901.

Mermin, N. D. (1996), Hidden quantum non-locality, in R. Clifton, ed., 'Perspectives on Quantum Reality', Kluwer, Boston, pp. 57–71.

Peres, A. (1999), 'All the Bell inequalities', *Foundations of Physics* **29**, 589–614.

Pitowsky, I. (1989), *Quantum probability — quantum logic*, Springer-Verlag, Berlin.

Popescu, S. (1994), 'Bell's inequalities versus teleportation: What is nonlocality?', *Physical Review Letters* **72**, 797–9.

Popescu, S. (1995), 'Bell's inequalities and density matrices: Revealing "hidden" nonlocality', *Physical Review Letters* **74**, 2619–22.

Popescu, S. & Rohrlich, D. (1992), 'Generic quantum nonlocality', *Physics Letters A* **166**, 293–7.

Schatten, R. (1970), *Norm ideals of completely continuous operators*, Second printing. Ergebnisse der Mathematik und ihrer Grenzgebiete, Band 27, Springer-Verlag, Berlin.

Summers, S. J. & Werner, R. F. (1995), 'On Bell's inequalities and algebraic invariants', *Letters in Mathematical Physics* **33**, 321–34.

Werner, R. F. (1989), 'Quantum states with Einstein-Podolsky-Rosen correlations admitting a hidden-variable model', *Physical Review A* **40**, 4277–81.

Życzkowski, K. (1999), 'On the volume of the set of mixed entangled states ii', *Physical Review A* **60**, 3496.

Życzkowski, K., Horodecki, P., Sanpera, A. & Lewenstein, M. (1998), 'Volume of the set of separable states', *Physical Review A* **58**, 883–92.

Chapter 11

Complementarity between position and momentum as a consequence of Kochen-Specker arguments

1 Complementarity and Kochen-Specker

Complementarity is the idea that mutually exclusive pictures are needed for a complete description of quantum-mechanical reality. The paradigm example is the complementarity between particle and wave (or 'spacetime' and 'causal') pictures, which Bohr took to be reflected in the uncertainty relation $\Delta x \Delta p \geq \hbar$. Bohr saw this relation as defining the latitude of applicability of the concepts of position and momentum to a single system, not just as putting a limit on our ability to predict the values of both position and momentum within an ensemble of identically prepared systems. Furthermore, right at the start of his celebrated reply to the Einstein-Podolsky-Rosen (EPR) argument against the completeness of quantum the-

The author would like to thank All Souls College, Oxford for support, Mauro Dorato for his hospitality at the 3rd University of Rome, and Paul Busch, Jeremy Butterfield, Hans Halvorson, David Mermin, and Jason Zimba for helpful and encouraging discussions.

ory (1935), Bohr confidently asserted: "it is never possible, in the description of the state of a mechanical system, to attach definite values to both of two canonically conjugate variables" (Bohr 1935). Critics have often pointed out that complementarity does not logically follow from the uncertainty relation without making the positivistic assumption that position and momentum can only be simultaneously defined if their values can be simultaneously measured or predicted (Popper 1982). However, we shall show here how direct Kochen-Specker arguments for complementarity between position and momentum can be given that are entirely independent of the uncertainty relation and its interpretation.

The aim of a Kochen-Specker argument is to establish that a certain set of observables of a quantum system cannot have simultaneously definite values that respect the functional relations between compatible observables within the set (Kochen & Specker 1967, Bell 1966, Mermin 1993). Let $\mathcal{O}$ be a collection of bounded self-adjoint operators (acting on some Hilbert space) containing the identity I and both AB and $\lambda A + \mu B$ $(\lambda, \mu \in \mathbb{R})$, whenever $A, B \in \mathcal{O}$ and $[A, B] = 0$. Kochen and Specker called such a structure a *partial* algebra because there is no requirement that $\mathcal{O}$ contain *arbitrary* self-adjoint functions of its members (such as $i[A, B]$ or $A + B$, when $[A, B] \neq 0$). They then assumed that an assignment of values $[\cdot] : \mathcal{O} \to \mathbb{R}$ to the observables in $\mathcal{O}$ should at least be a *partial* homomorphism, respecting linear combinations and products of *compatible* observables in $\mathcal{O}$. That is, whenever $A, B \in \mathcal{O}$ and $[A, B] = 0$,

$$[AB] = [A]\,[B]\,, \; [\lambda A + \mu B] = \lambda\,[A] + \mu\,[B]\,, \; [I] = 1. \qquad (1)$$

Clearly these constraints are motivated by analogy with classical physics, in which all physical magnitudes (functions on phase space) trivially commute, and possess values (determined by points of phase space) that respect their functional relations. (The requirement that $[I] = 1$ is only needed to avoid triviality; for $[I^2] = [I] = [I]^2$ implies $[I] = 0$ or 1, and if we took $[I] = 0$, it would follow that $[A] = [AI] = [A][I] = 0$ for all $\mathcal{A} \in \mathcal{O}$.)

Constraints (1) are not entirely out of place in quantum theory.

For example, any common eigenstate Ψ of a collection of observables $\mathcal{O}$ automatically defines a partial homomorphism, given by assigning to each $A \in \mathcal{O}$ the eigenvalue A has in state Ψ. Difficulties — called Kochen-Specker contradictions or obstructions (Butterfield & Isham 1999, Hamilton *et al.* 2000) — arise when not all observables in $\mathcal{O}$ share a common eigenstate. In that case, there is no guarantee that value assignments on all the commutative subalgebras of $\mathcal{O}$ can be extended to a partial homomorphism on $\mathcal{O}$ as a whole. Should such an extension exist, one could be justified in thinking of the noncommuting observables in $\mathcal{O}$ as having simultaneously definite values, notwithstanding that a quantum state may not permit all their values to be predicted with certainty. But should some particular collection of observables $\mathcal{O}$ not possess *any* partial homomorphisms, the natural response would be to concede to Bohr that the observables in $\mathcal{O}$ "transcend the scope of classical physical explanation" and cannot be discussed using "unambiguous language with suitable application of the terminology of classical physics" (Bohr 1949). That is, one would have strong reasons for taking the noncommuting observables in $\mathcal{O}$ to be mutually complementary.

Bell (1966) has emphasized one other way to escape an obstruction with respect to some set of observables $\mathcal{O}$. One could still take all $\mathcal{O}$'s observables to have definite values by allowing the value of a particular $A \in \mathcal{O}$ to be a function of the context in which A is measured. Thus, suppose $\mathcal{O}_1, \mathcal{O}_2 \subseteq \mathcal{O}$ are two different commuting subalgebras both containing A, where $[\mathcal{O}_1, \mathcal{O}_2] \neq 0$. Then if $[\cdot]_1, [\cdot]_2$ are homomorphisms on these subalgebras such that $[A]_1 \neq [A]_2$, one could interpret this difference in values (the obstruction) as signifying that the measured result for A has to depend on whether it is measured along with the observables in $\mathcal{O}_1$ or those in $\mathcal{O}_2$. Such value assignments to the observables in $\mathcal{O}$ are called contextual, because the context in which an observable is measured is allowed to influence what outcome is obtained (Bell 1966). For example, Bohm's theory is contextual in exactly this sense (Bohm 1952, Pagonis & Clifton 1995). On the other hand (as Bell himself was quick to observe), complementarity also demands a kind of contextualism: in some contexts it is appropriate to assign a system a definite position, and in other contexts, a definite momentum. The difference from Bohm is that Bohr takes the definiteness of the values of observables *itself* to be a function of context. And this makes *all* the difference in cases where value contextualism can only be enforced by making the measured value of an observable nonlocally depend on whether an observable of another spacelike-separated system is measured. We shall see below that complementarity between position and momentum can only be avoided by embracing such nonlocality.

Numerous Kochen-Specker obstructions have been identified in the literature, and their practical and theoretical implications continue to be analyzed (see Bub 1997, Cabello & Garcia-Alcaíne 1998, Clifton & Kent 2000, Kent 1999, Meyer 1999, Pitowsky 1998). While obstructions cannot occur for observables sharing a common eigenstate, failure to possess a common eigenstate does not suffice for an obstruction. As Kochen and Specker themselves showed, the partial algebra generated by all components of a spin-1/2 particle possesses plenty of partial homomorphisms. But for particles with higher spin, or collections of more than one spin-1/2 particle, obstructions can occur, perhaps the simplest being those identified by Peres (Peres 1990, 1992, 1993) in the case of two spin-1/2 particles, and Mermin (Mermin 1990, Greenberger *et al.* 1990) in the case of three. Obstructions for sets that contain functions of position and momentum observables *have* been identified (Fleming 1995, Zimba 1998), but additional observables need to be invoked that weaken the case for complementarity between position and momentum alone. In the arguments below, we shall only need simple *continuous* functions of the individual position and momentum components of a system. Though all our observables have purely continuous spectra, obstructions arise in exactly the same way that they do in the arguments given by Peres and Mermin for the spin-1/2 case. And because our obstructions depend only on the structure of the Weyl algebra, they immediately extend to relativistic quantum field theories, which are constructed out of representations of the Weyl algebra (Wald 1994).

2 The Weyl algebra

Let $\vec{x} = (x_1, x_2, x_3)$ and $\vec{p} = (p_1, p_2, p_3)$ be the unbounded position and momentum operators for three degrees of freedom. We cannot extract a Kochen-Specker contradiction directly out of these operators, since domain questions prevent them from defining a simple algebraic structure. However, we may just as well consider the collection of all bounded, continuous, self-adjoint functions of x_1, and, similarly, the same set of functions in each of the variables x_2, x_3, p_1, p_2, p_3. Taking $\mathcal{O}$ to be the partial algebra of observables generated by all these functions (obtained by taking compatible products and linear combinations thereof), we shall show that $\mathcal{O}$ does not possess any partial homomorphisms.

Our arguments are greatly simplified by employing the following method, analogous to simplifying a problem in real analysis by passing to the complex plane. Assuming that $\mathcal{O}$ *does* possess a partial homomorphism $[\cdot] : \mathcal{O} \to \mathbb{R}$ (an assumption we shall eventually have to discharge), we can extend this mapping to the set $\mathcal{O}_{\mathbb{C}} \equiv \mathcal{O} + i\mathcal{O}$ in a well-defined manner, by taking $[X] \equiv [\Re(X)] + i[\Im(X)] \in \mathbb{C}$, where $\Re(X)$ and $\Im(X)$ are the unique real and imaginary parts of X. Now, if we consider any pair of commuting unitary operators $U, U' \in \mathcal{O}_{\mathbb{C}}$, then since U, U^*, U', U'^* pairwise commute, the four self-adjoint operators

$$\Re(U) = (U + U^*)/2, \qquad \Im(U) = i(U^* - U)/2, \tag{2}$$
$$\Re(U') = (U' + U'^*)/2, \qquad \Im(U') = i(U'^* - U')/2, \tag{3}$$

which must lie in $\mathcal{O}_{\mathbb{C}}$, also pairwise commute. Thus

$$\begin{aligned} UU' &= \Re(U)\Re(U') - \Im(U)\Im(U') \\ &\quad + i(\Re(U)\Im(U') + \Im(U)\Re(U')) \in \mathcal{O}_{\mathbb{C}}, \end{aligned} \tag{4}$$

using the fact that $\mathcal{O}$ is a partial algebra. In addition,

$$\begin{aligned}
[UU'] &= [\Re(U)\Re(U') - \Im(U)\Im(U')] \\
&\quad +i[\Re(U)\Im(U') + \Im(U)\Re(U')], && (5)\\
&= [\Re(U)][\Re(U')] - [\Im(U)][\Im(U')] \\
&\quad +i([\Re(U)][\Im(U')] + [\Im(U)][\Re(U')]), && (6)\\
&= ([\Re(U)] + i[\Im(U)])([\Re(U')] + i[\Im(U')]) && (7)\\
&= [U][U'], && (8)
\end{aligned}$$

using the fact that $[\cdot]$ is a partial homomorphism in step (6). So we have established that the following product rule must hold in $\mathcal{O}_{\mathbb{C}}$:

$$\begin{aligned}
& U, U' \in \mathcal{O}_{\mathbb{C}} \ \& \quad [U, U'] = 0 \\
\Rightarrow \quad & UU' \in \mathcal{O}_{\mathbb{C}} \ \& \quad [UU'] = [U][U']. && (9)
\end{aligned}$$

Henceforth, we shall only need this simple product rule, together with $[\pm I] = \pm 1$. Our obstructions will manifest themselves as contradictions obtained by applying the product rule to compatible unitary operators in $\mathcal{O}_{\mathbb{C}}$.

To see what operators those are, we first recall the definition of the Weyl algebra for three degrees of freedom. Consider the two families of unitary operators given by

$$U_{\vec{a}} = e^{-i\vec{a}\cdot\vec{x}/\hbar}, \ V_{\vec{b}} = e^{-i\vec{b}\cdot\vec{p}/\hbar}, \ \vec{a}, \vec{b} \in \mathbb{R}^3. \tag{10}$$

These operators act on any wavefunction $\Psi \in L_2(\mathbb{R}^3)$ as

$$(U_{\vec{a}}\Psi)(\vec{x}) = e^{-i\vec{a}\cdot\vec{x}/\hbar}\Psi(\vec{x}), \ (V_{\vec{b}}\Psi)(\vec{x}) = \Psi(\vec{x} - \vec{b}), \tag{11}$$

and satisfy the Weyl form of the canonical commutations relations $[x_j, p_k] = \delta_{jk} i\hbar I$,

$$U_{\vec{a}} V_{\vec{b}} = e^{-i\vec{a}\cdot\vec{b}/\hbar} V_{\vec{b}} U_{\vec{a}}. \tag{12}$$

The Weyl algebra (which is independent of the representation in (11)) is just the C^*-algebra generated by the two families of unitary operators in (10) subject to the commutation relation (12).

$\mathcal{O}_{\mathbb{C}}$ is properly contained in the Weyl algebra. Indeed, writing U_{a_j} ($\equiv e^{-ia_j x_j/\hbar}$) for the jth component of the operator $U_{\vec{a}}$, and similarly V_{b_k} ($\equiv e^{-ib_k p_k/\hbar}$), all nine of these component generators of the Weyl algebra lie in $\mathcal{O}_{\mathbb{C}}$, because their real and imaginary parts, cosine and sine functions of the x_j's and p_k's, lie in $\mathcal{O}$. By the product rule, $\mathcal{O}_{\mathbb{C}}$ also contains the products of compatible unitary operators for different degrees of freedom, as well as compatible products of those products. But, unlike the full Weyl algebra, $\mathcal{O}_{\mathbb{C}}$ does not contain incompatible products, like $U_{a_j} V_{b_j}$ when $a_j b_j \neq 2n\pi\hbar$ ($n \in \mathbb{Z}$). Nevertheless, $\mathcal{O}_{\mathbb{C}}$ is all we need to exhibit obstructions. The key is that we can choose values for the components of $\vec{a}, \vec{b}$ so that, for $j = 1$ to 3, $a_j b_j = (2n+1)\pi\hbar$. In that case, we immediately obtain from (12) the *anti*-commutation rule

$$[U_{\pm a_j}, V_{\pm b_j}]_+ = 0 = [U_{\mp a_j}, V_{\pm b_j}]_+, \tag{13}$$

which, together with the product rule, will generate the required obstructions.

3 Obstructions for two and three degrees of freedom

We first limit ourselves to continuous functions of the four observables x_1, x_2, p_1, p_2, extracting a contradiction in exactly the way Peres (1990) does for a pair of spin-1/2 particles. A first application of the product rule in $\mathcal{O}_{\mathbb{C}}$ yields

$$\begin{aligned} [U_{-a_1} U_{a_2}] &= [U_{-a_1}]\,[U_{a_2}], & (14)\\ [U_{a_1} V_{b_2}] &= [U_{a_1}]\,[V_{b_2}], & (15)\\ [V_{b_1} U_{-a_2}] &= [V_{b_1}]\,[U_{-a_2}], & (16)\\ [V_{-b_1} V_{-b_2}] &= [V_{-b_1}]\,[V_{-b_2}]. & (17) \end{aligned}$$

Multiplying equations (14)–(17) together, and using one further (trivial) application of the product rule

$$[U_{a_j}][U_{-a_j}] = [I] = 1 = [V_{b_k}][V_{-b_k}], \tag{18}$$

one obtains

$$[U_{-a_1}U_{a_2}]\,[V_{-b_1}V_{-b_2}]\,[U_{a_1}V_{b_2}]\,[V_{b_1}U_{-a_2}] = 1. \tag{19}$$

However, because of the anti-commutation rule (13), the first pair of product operators occurring in (19) actually *commute*, as do the second pair of product operators. Hence we may make a further application of the product rule to (19) to get

$$[U_{-a_1}U_{a_2}V_{-b_1}V_{-b_2}]\,[U_{a_1}V_{b_2}V_{b_1}U_{-a_2}] = 1. \tag{20}$$

Again, due to the anti-commutation rule, the two remaining (four-fold) product operators occurring in (20) commute, and their product is I. Thus, a final application of the product rule to (20) yields the contradiction $[-I] = -1 = 1$.

Notice that this obstruction remains for any given nonzero values for a_1 and a_2, provided only that we choose $b_{1,2} = (2n+1)\pi\hbar/a_{1,2}$. The obstruction would vanish if, instead, we chose any of the numbers a_1, a_2, b_1, b_2 to be zero. When $a_1 = a_2 = 0$ or $b_1 = b_2 = 0$, this is to be expected, since one would then no longer be attempting to assign values to nontrivial functions of *both* the positions and momenta. However, the breakdown of the argument when either a_2 or b_2 is zero does not necessarily mean that a more complicated argument could not be given for position-momentum complementarity by invoking only a *single* degree of freedom.

As Mermin (1990) has emphasized (for the spin-1/2 analogue of the above argument), one can get by without independently assuming the existence of values for the two commuting unitary operators occurring in (20), and thereby strengthen the argument. For we can suppose that the quantum state of the system is an eigenstate of these operators, with eigenvalues that *necessarily* multiply to -1. Using (11), a wavefunction Ψ will be an eigenstate of both products in (20) just in case

$$e^{i(a_1x_1-a_2x_2)/\hbar}\Psi(x_1+b_1, x_2+b_2) = c\Psi(x_1,x_2), \tag{21}$$

$$-e^{-i(a_1x_1-a_2x_2)/\hbar}\Psi(x_1-b_1, x_2-b_2) = c'\Psi(x_1,x_2), \tag{22}$$

for some $c, c' \in \mathbb{C}$. We should not expect there to be a *normalizable* wavefunction satisfying (21) and (22), because the commuting product operators in (20) have purely continuous spectra. But if we allow ourselves the idealization of using Dirac states (which can be approximated arbitrarily closely by elements of $L_2(\mathbb{R}^2)$), and just choose $a_1 = a_2$ for simplicity, then the two-dimensional delta function $\delta(x_1 - x_2 - x_0)$ — an improper eigenstate of the relative position operator $x_1 - x_2$ with 'eigenvalue' $x_0 \in \mathbb{R}$ — provides a simple solution to the above equations. However, this state cannot also be used to independently justify the assignment of values to the operators $U_{a_1}V_{b_2}$ and $V_{b_1}U_{-a_2}$ occurring in (19), which do not have $\delta(x_1 - x_2 - x_0)$ as an eigenstate.

It is ironic that $\delta(x_1 - x_2 - x_0) = \delta(p_1 + p_2)$ is exactly the state of two spacelike-separated particles that EPR invoked to argue *against* position-momentum complementarity. So in a sense the EPR argument carries the seeds of its own destruction. For suppose we follow their reasoning by invoking locality and the strict correlations entailed by the EPR state between x_1 and x_2, and between p_1 and p_2, to argue for the existence of noncontextual values for all four positions and momenta. Then all eight component unitary operators we employed above must have definite noncontextual values, since their real and imaginary parts are simple functions of those x's and p's. It is then a small step to conclude that the four product operators in (19) should also have definite noncontextual values satisfying the product rule, and from there contradiction follows. This final step cannot itself be justified by appeal to locality, for the four product observables in (19) do not pertain to either particle on its own and, hence, a measurement context for any one of these operators (i.e., their self-adjoint real and imaginary parts) necessarily requires a joint measurement undertaken on both particles (Mermin 1990). Still, the above argument sheds an entirely new light on the nonclassical features of the original EPR state, which have hitherto only been discussed from a statistical point of view (Bell 1986, Cohen 1997, Banaszek & Wódkiewicz 1998).

Our second argument employs all three degrees of freedom, extracting a contradiction in exactly the way Mermin (1990) does for three spin-1/2 particles. Again, a first application of the product rule in $\mathcal{O}_{\mathbb{C}}$ yields

$$\begin{aligned}
\left[U_{a_1}V_{-b_2}V_{-b_3}\right] &= \left[U_{a_1}\right]\left[V_{-b_2}\right]\left[V_{-b_3}\right], & (23)\\
\left[V_{-b_1}U_{a_2}V_{b_3}\right] &= \left[V_{-b_1}\right]\left[U_{a_2}\right]\left[V_{b_3}\right], & (24)\\
\left[V_{b_1}V_{b_2}U_{a_3}\right] &= \left[V_{b_1}\right]\left[V_{b_2}\right]\left[U_{a_3}\right], & (25)\\
\left[U_{-a_1}U_{-a_2}U_{-a_3}\right] &= \left[U_{-a_1}\right]\left[U_{-a_2}\right]\left[U_{-a_3}\right]. & (26)
\end{aligned}$$

Multiplying (23)–(26) together, again using (18), yields

$$\begin{aligned}
&\left[U_{a_1}V_{-b_2}V_{-b_3}\right]\left[V_{-b_1}U_{a_2}V_{b_3}\right]\\
&\left[V_{b_1}V_{b_2}U_{a_3}\right]\left[U_{-a_1}U_{-a_2}U_{-a_3}\right] = 1. \qquad (27)
\end{aligned}$$

But now, exploiting the anti-commutation rule once again, the four product operators occurring in square brackets in (27) pairwise commute, and their product is easily seen to be $-I$. So one final application of the product rule to (27) once more yields the contradiction $[-I] = -1 = 1$.

As before, we may interpret the x's and p's as the positions and momenta of three spacelike-separated particles. And we can avoid independently assuming values for the four products in (27) by taking the state of the particles to be a simultaneous (improper) eigenstate of these operators — exploiting that state's strict correlations and EPR-type reasoning from locality to motivate values for all the component operators. (The reader is invited to use (11) to determine the set of all such common eigenstates, which are new position-momentum analogues of the Greenberger-Horne-Zeilinger state (Mermin 1990).) This time, the *only* way to prevent contradiction is to introduce contextualism to distinguish, for example, the value of U_{a_1} as it occurs in (23) from the value this operator (or rather its inverse) receives in (26) in the context of different operators for particles 2 and 3 — forcing the values of $\sin a_1x_1$ and $\cos a_1x_1$ to *nonlocally* depend on whether position or momentum observables for 1 and 2 are measured. Bohr of course denied that there could be any such nonlocal "mechanical" influence, but only "an influence on the very conditions which define" which of the two mutually complementary pictures available for each system can be unambiguously employed (Bohr 1935).

Bibliography

Banaszek, K. & Wódkiewicz, K. (1998), 'Nonlocality of the Einstein-Podolsky-Rosen state in the Wigner representation', *Physical Review A* **58**, 4345–7.

Bell, J. S. (1966), 'On the problem of hidden variables in quantum mechanics', *Review of Modern Physics* **38**, 447–52.

Bell, J. S. (1986), 'EPR correlations and EPW distributions', *Annals of the (N. Y.) Academy of Science* **480**, 263–6.

Bohm, D. (1952), 'A suggested interpretation of the quantum theory in terms of "hidden" variables. I', *Physical Review* **85**, 166–79.

Bohr, N. (1935), 'Can quantum-mechanical description of physical reality be considered complete?', *Physical Review* **48**, 696–702.

Bohr, N. (1949), Discussion with Einstein on epistemological problems in atomic physics, in P. A. Schilpp, ed., 'Albert Einstein, Philosopher-Scientist', Library of Living Philosophers, Evanston, pp. 201–41.

Bub, J. (1997), *Interpreting the Quantum World*, Cambridge University Press, Cambridge.

Butterfield, J. & Isham, C. J. (1999), 'A topos perspective on the Kochen-Specker theorem. II. Conceptual aspects and classical analogues', *International Journal of Theoretical Physics* **38**, 827–59.

Cabello, A. & Garcia-Alcaíne, G. (1998), 'Proposed experimental tests of the Bell-Kochen-Specker theorem', *Physical Review Letters* **80**, 1797–99.

Clifton, R. & Kent, A. (2000), 'Simulating quantum mechanics by non-contextual hidden variables', *The Royal Society of London. Proceedings. Series A* **456**, 2101–14.

Cohen, O. (1997), 'Nonlocality of the original Einstein-Podolsky-Rosen state', *Physical Review A* **56**, 3484–92.

Einstein, A., Podolsky, B. & Rosen, N. (1935), 'Can quantum-mechanical description of physical reality be considered complete?', *Physical Review* **47**, 777–80.

Fleming, G. N. (1995), 'A GHZ argument for a single spinless particle', *Annals of the (N. Y.) Academy of Sciences* **755**, 646–53.

Greenberger, D. M., Horne, M. A., Shimony, A. & Zeilinger, A. (1990), 'Bell's theorem without inequalities', *American Journal of Physics* **58**, 1131–43.

Hamilton, J., Isham, C. J. & Butterfield, J. (2000), 'Topos perspective on the Kochen-Specker theorem. III. Von Neumann algebras as the base category', *International Journal of Theoretical Physics* **39**, 1413–36.

Kent, A. (1999), 'Noncontextual hidden variables and physical measurements', *Physical Review Letters* **83**, 3755–7.

Kochen, S. & Specker, E. P. (1967), 'The problem of hidden variables in quantum mechanics', *Journal of Mathematics and Mechanics* **17**, 59–87.

Mermin, N. D. (1990), 'Simple unified form for the major no-hidden-variables theorems', *Physical Review Letters* **65**, 3373–6.

Mermin, N. D. (1993), 'Hidden variables and the two theorems of John Bell', *Reviews of Modern Physics* **65**, 803–15.

Meyer, D. (1999), 'Finite precision measurement nullifies the Kochen-Specker theorem', *Physical Review Letters* **83**, 3751–4.

Pagonis, C. & Clifton, R. (1995), 'Unremarkable contextualism: dispositions in the Bohm theory', *Foundations of Physics* **25**, 281–96.

Peres, A. (1990), 'Incompatible results of quantum measurements', *Physics Letters A* **151**, 107–8.

Peres, A. (1992), 'Recursive definition for elements of reality', **22**, 357–61.

Peres, A. (1993), *Quantum Theory: Concepts and Methods*, Kluwer, Dordrecht.

Pitowsky, I. (1998), 'Infinite and finite Gleason's theorems and the logic of indeterminacy', *Journal of Mathematical Physics* **39**, 218–28.

Popper, K. R. (1982), *Quantum Theory and the Schism in Physics*, Hutchinson, London.

Wald, R. (1994), *Quantum Field Theory in Curved Spacetime and Black Hole Thermodynamics*, University of Chicago Press, Chicago.

Zimba, J. (1998), 'Simple realism and canonically conjugate observables in non-relativistic quantum mechanics', *Foundations of Physics Letters* **11**, 503–33.

Chapter 12

Reconsidering Bohr's reply to EPR

with Hans Halvorson

1 Introduction

The past few decades have seen tremendous growth in our understanding of interpretations of quantum mechanics. For example, a number of 'no-go' results have been obtained which show that some or other interpretation violates constraints that we would expect any plausible interpretation of quantum mechanics to satisfy. Thus, although there is no immediate hope of convergence of opinion on interpretive issues, we certainly have an increased understanding of the technical and conceptual issues at stake. Perhaps, then, we can make use of this increased technical awareness to shed some new light on the great old episodes in the conceptual development of quantum mechanics.

One historical episode of enduring philosophical interest is the debate between Bohr and Einstein (along with Podolsky and Rosen)

We would like to thank Jeremy Butterfield for a close reading of an earlier draft.

over the completeness of quantum mechanics. Although folklore has it that Bohr was the victor in this debate, Fine & Beller (1994) have recently claimed that Bohr's reply to the EPR argument of 1935 is basically a failure. In particular, Fine and Beller claim that '... as a result of EPR, Bohr eventually turned from his original concept of disturbance, to make a final — and somewhat forced — landing in positivism' (Fine & Beller 1994, p. 29). They also make the stronger philosophical claim that '... a positivistic shift is the only salvageable version of Bohr's reply' (Fine & Beller 1994, p. 9). Unfortunately, Fine and Beller do not devote much attention to establishing this philosophical claim. (Nor does it seem to us that Beller's more extended treatment (Beller 1999, Chap. 7) goes any further towards establishing the philosophical claim.) Even if we concede — for purposes of argument — that the later Bohr embraced positivism, we are *not* willing to concede that he was rationally compelled to do so. In fact, we will argue that Bohr's defense of the completeness of quantum mechanics does not depend in any way on questionable philosophical doctrines. To this end, we will supply a formal reconstruction of Bohr's reply to EPR, showing that his reply is dictated by the dual requirements that any description of experimental data must be *classical* and *objective*.

The structure of this paper is as follows. In Section 2, we provide an informal preliminary account of the EPR argument and of Bohr's reply. In Section 3, we consider some salient features of Bohr's general outlook on quantum theory. We then return to Bohr's reply to EPR in Sections 4 and 5. In Section 4, we reconstruct Bohr's reply to EPR in the case of Bohm's simplified spin version of the EPR experiment. Finally, in Section 5, we reconstruct Bohr's reply to EPR in the case of the original (position-momentum) version of the EPR experiment.

2 Informal preview

In classical mechanics, a state description for a point particle includes a precise specification of both its position and its momentum. In contrast, a quantum-mechanical state description supplies only a statistical distribution over various position and momentum values. It would be quite natural, then, to regard the quantum-mechanical description as *incomplete* — i.e. as providing less than the full amount of information about the particle. Bohr, however, insists that the imprecision in the quantum-mechanical state description reflects a fundamental indeterminacy in nature rather than the incompleteness of the theory. The EPR argument attempts to directly rebut this completeness claim by showing that quantum mechanics (in conjunction with plausible extra-theoretical constraints) entails that particles always have both a precise position and a precise momentum.

EPR ask us to consider a system consisting of a pair of space-like separated particles. They then note that, according to quantum mechanics, there is a state ψ_{epr} in which the positions of the two particles are strictly correlated, *and* the momenta of the two particles are strictly correlated. It follows then that if we were to measure the position of the first particle, we could predict with certainty the outcome of a position measurement on the second particle; *and* if we were to measure the momentum of the first particle, we could predict with certainty the outcome of a momentum measurement on the second particle.

EPR then claim that our ability to predict with certainty the outcomes of these measurements on the second particle shows that each such measurement reveals a pre-existing 'element of reality'. In what has come to be known as the 'EPR reality criterion', they say:

> If, without in any way disturbing a system, we can predict with certainty (i.e. with probability equal to unity) the value of a physical quantity, then there exists an element of physical reality corresponding to this physical quantity. (Einstein *et al.* 1935, p. 77)

In particular, if we determine the position of the first particle in this strictly correlated state, then we can conclude that the second particle also has a definite position. And if we determine the momentum of the first particle in this strictly correlated state, then the second particle must also have a definite momentum.

Of course, it does not immediately follow that there is any single situation in which both the position and the momentum of the second particle are elements of reality. However, EPR also make the (*prima facie* plausible) assumption that what counts as an element of reality for the second particle should be independent of which measurement is performed on the first particle. In other words, a measurement on the first particle can play a *probative*, but not a *constitutive*, role with respect to the elements of reality for the second particle. Consequently, EPR conclude that both the position and the momentum of the second particle are elements of reality, regardless of which measurement is performed on the first particle.

2.1 Bohr's reply

According to Bohr, the EPR argument somehow misses the point about the nature of quantum-mechanical description. Unfortunately, though, not much scholarly work has been done attempting to reconstruct Bohr's reply in a cogent fashion.

We should begin by noting that Bohr most certainly does not maintain the 'hyperpositivist' position according to which no possessed properties or reality should be attributed to an unmeasured system. (For example, Ruark claims that, for Bohr, 'a given system has reality only when it is actually measured' (Ruark 1935, p. 466).) Quite to the contrary, Bohr explicitly claims that when the position of the first particle is measured, '... we obtain a basis for conclusions about the initial position of the other particle relative to the rest of the apparatus' (Bohr 1935*a*, p. 148). Thus, Bohr agrees with EPR that once the position (respectively, momentum) of the first particle is actually measured, the position of the second particle is an element of reality — *whether or not* its position is actually empirically determined. In other words, Bohr accepts the outcome of an application of the EPR reality criterion, so long as its application is restricted to individual measurement contexts (i.e. the results of its application in different contexts are not combined).

In order, then, to rationally reject EPR's conclusion, Bohr must reject the claim that elements of reality for the second particle cannot be constituted by measurements carried out on the first particle.

In other words, Bohr believes that a measurement on the first particle *can* serve to constitute elements of reality for the second, spacelike separated, particle.

To this point, we have not said anything particularly novel about Bohr's reply to EPR. It is relatively well known that his reply amounts to claiming — what EPR thought was absurd (Einstein *et al.* 1935, p. 480) — that what is real with respect to the second particle can depend in a nontrivial way on which measurement is performed on the first particle. However, where previous defenders of Bohr have uniformly stumbled is in giving a coherent account of *how* a measurement on one system can influence what is real for some spacelike separated system.

Unfortunately, Bohr's statements on this issue are brief and obscure. For example, he says,

> It is true that in the measurements under consideration any direct mechanical interaction of the [second] system and the measuring agencies is excluded, but a closer examination reveals that the procedure of measurement has an essential influence on the conditions on which the very definition of the physical quantities in question rests. (Bohr 1935*b*, p. 65)

That is, a measurement on the first system influences the conditions which must obtain in order for us to 'define' elements of reality for the second system. Moreover, this influence is of such a sort that a position (momentum) measurement on the first particle supplies the conditions needed to define the position (momentum) of the second particle.

Before we proceed to our positive account, we need first to dismiss one *prima facie* plausible, but nonetheless mistaken, explication of Bohr's notion of defining a quantity. In particular, some have claimed that, according to Bohr, an observable of a system comes to have a definite value when the wavefunction of the system collapses onto one of that observable's eigenstates. This amounts to attributing to Bohr the claim that:

Eigenstate-Eigenvalue Link: A quantity Q is defined in state ψ iff ψ is an eigenvector for Q;

along with the claim that by measuring an observable, we can cause the quantum state to collapse onto an eigenstate of that observable. In that case, Bohr would claim that by measuring the position of the first particle, we collapse the EPR state onto an eigenstate of position for the second particle — and thereby 'cause' the second particle to have a definite position. Similarly, if we were to measure the momentum of the first particle, we would 'cause' the second particle to have a definite momentum. In either case, the measurement on the first particle would be the cause of the reality associated with the second particle.

However, there are at least two good reasons for rejecting this reading of Bohr. First, Bohr explicitly claims that a measurement of the first particle cannot bring about a 'mechanical' change in the second particle. In philosophical terms, we might say that Bohr does not believe that the position measurement on the first particle *causes* the second particle to have a position, at least not in the same sense that a brick can *cause* a window to shatter. Thus, if Bohr does believe in a collapse of the wavefunction, it is as some sort of *non-physical* (perhaps epistemic) process. However, it is our firm opinion that, unless the quantum state can be taken to represent our ignorance of the 'true' hidden state of the system, there is no coherent non-physical interpretation of collapse. (We doubt the coherence of recent attempts to maintain both a subjectivist interpretation of quantum probabilities, and the claim that 'there are no unknown quantum states' (Caves *et al.* 2002).) Thus, if Bohr endorses collapse, then he is already committed to the incompleteness of quantum mechanics, and the EPR argument is superfluous.

The second, and more important, reason for resisting this reading of Bohr is the complete lack of textual evidence supporting the claim that Bohr believed in wavefunction collapse (see Howard 2000). Thus, there is no good reason to think that Bohr's reply to the EPR argument depends in any way on the notion of wavefunction collapse.

3 Classical description and appropriate mixtures

In order to do justice to Bohr's reply to EPR, it is essential that we avoid caricatured views of Bohr's general philosophical outlook, and of his interpretation of quantum mechanics. This is particularly difficult, because there has been a long history of misinterpretation of Bohr. For example, in terms of general philosophical themes, one might find Bohr associated with anti-realism, idealism, and subjectivism. Moreover, in terms of the specific features of an interpretation of quantum mechanics, Bohr is often associated with wavefunction collapse, creation of properties/attributes upon measurement, and 'cuts' between the microscopic and macroscopic realms. However, these characterizations of Bohr are pure distortion, and can find no justification in his published work. Indeed, Bohr's philosophical commitments, and the picture of quantum mechanics that arises from these commitments, are radically different from the mythical version that we have received from his critics and from his well-intended (but mistaken) followers. (Our own understanding of Bohr has its most immediate precedent in recent work on 'no collapse' interpretations of quantum mechanics (see Bub 1995, 1997, Halvorson & Clifton 1999, and Chapter 2 of this volume). However, this sort of analysis of Bohr's interpretation was suggested independently, and much earlier, by Howard (1979). See also (Howard 1994, 2000).)

According to Bohr, the phenomena investigated by quantum theory cannot be accounted for within the confines of classical physics. Nonetheless, he claims that '... however far the phenomena transcend the scope of classical physical explanation, the account of all evidence must be expressed in classical terms' (Bohr 1949, p. 209). That is, classical physics embodies a standard of intelligibility that should be exemplified by any description of the empirical evidence. In particular, although the various sources of evidence cannot be reconciled into a single classical picture, the description of any single source of evidence must be classical.

Bohr's statements about the notion of 'classical description' have been horribly misunderstood. For a catalog of these misunderstandings and for evidence that they are indeed mistaken, we refer the reader to Howard (1979, 1994, 2000). On the positive side, we will follow Howard (1994) in the claim that the notion of classical description is best explicated via the notion of an 'appropriate mixture'.

> ... we make the clearest sense out of Bohr's stress on the importance of a classical account of experimental arrangements and of the results of observation, if we understand a classical description to be one in terms of appropriate mixtures. (Howard 1994, p. 222)

As Howard (1979) shows, the notion of an appropriate mixture can be developed in such a way that Bohr's (sometimes obscure) statements about the possibilities of classical description become mathematically clear statements about the possibility of treating the quantum state as a classical probability measure. In order to see this, we first collect some terminology.

Let $\mathcal{H}$ be a finite-dimensional vector space with inner-product $\langle \cdot, \cdot \rangle$, and let $\mathfrak{B}(\mathcal{H})$ denote the family of linear operators on $\mathcal{H}$. We say that a self-adjoint operator W on $\mathcal{H}$ is a *density operator* just in case W has non-negative eigenvalues that sum to 1. If ψ is a vector in $\mathcal{H}$, we let $|\psi\rangle\langle\psi|$ denote the projection onto the ray in $\mathcal{H}$ generated by ψ. Thus, if Tr denotes the trace on $\mathfrak{B}(\mathcal{H})$, then $\mathrm{Tr}(|\psi\rangle\langle\psi|A) = \langle \psi, A\psi \rangle$ for any operator A on $\mathcal{H}$. A *measurement context* can be represented by a pair (ψ, R), where ψ is a unit vector (representing the quantum state), and R is a self-adjoint operator (representing the measured observable).

Following Howard (1979), we say that a 'mixture', represented by a density operator W, is *appropriate* for (ψ, R) just in case $W = \sum_{i=1}^{n} \lambda_i |\phi_i\rangle\langle\phi_i|$, $(n \leq \dim\mathcal{H})$, where each ϕ_i is an eigenvector for R, and $\lambda_i = |\langle \psi, \phi_i \rangle|^2$ for $i = 1, \ldots, n$. In other words, W is a mixture of eigenstates for R, and it reproduces the probability distribution that ψ assigns to the values of R. Thus, an appropriate mixture for (ψ, R) can be taken to represent our ignorance of the value of R in the state ψ.

Once again, we emphasize that Bohr never explicitly invokes wavefunction collapse, nor does he need to. Indeed, the idea of a 'measurement problem' was foreign to Bohr, who seems to take it as a brute empirical fact — needing no further explanation from within quantum theory — that an observable possesses a value when it is measured. Of course, we now know that if Bohr rejects collapse, then he would also have to reject the claim that an observable possesses a value only if the system is in an eigenstate for that observable (i.e., the eigenstate←eigenvalue link) (Clifton 1996). But there is little reason to believe that Bohr would have been tempted to endorse this suspect claim in the first place.

3.1 Appropriate mixtures and elements of reality

An appropriate mixture is supposed to give a description in which the measured observable is an 'element of reality'. However, the connection between an appropriate mixture (i.e. some density operator) and elements of reality is not completely clear. Clearly, the intent of writing the appropriate mixture as $W = \sum_{i=1}^{n} \lambda_i |\phi_i\rangle\langle\phi_i|$ is that each 'proposition' $|\phi_i\rangle\langle\phi_i|$ has a truth value. However, if W is degenerate then W has infinitely many distinct expansions as a linear combination of orthogonal one-dimensional projections. Thus, W itself does not determine the elements of reality; rather, it is some expansion of W into a linear combination of one-dimensional projections that determines the elements of reality.

In this case, however, we might as well focus on the one-dimensional projections themselves. Thus, we will say that the set $S = \{|\phi_i\rangle\langle\phi_i| : i = 1, \dots, n\}$ is an *appropriate event space* for the measurement context (ψ, R) just in case S is maximal relative to the following three conditions: (1) Each ϕ_i is an eigenvector of R; (2) If $i \neq j$ then ϕ_i and ϕ_j are orthogonal; (3) Each ϕ_i is *non*orthogonal to ψ. Each of these conditions has a natural interpretation. The first condition states that each proposition in S attributes some value to R (viz. the eigenvalue r_i satisfying $R\phi_i = r_i\phi_i$); the second condition states that the propositions in S are mutually exclusive; and the third condition states that each proposition in S is possible relative to ψ. Note, moreover, that every appropriate event space S can be obtained by taking the projection operators in some orthogonal expansion $W = \sum_{i=1}^{n} \lambda_i |\phi_i\rangle\langle\phi_i|$ of an appropriate mixture for (ψ, R), and then eliminating those projections with coefficient 0.

If we suppose that R possesses a definite value in the context (ψ, R) (and that it is *False* that it possesses any other value) then an appropriate event space S gives a *minimal* list of truth-valued propositions in the context (ψ, R). However, Bohr himself is *not* an ontological minimalist; rather, he claims that 'we must strive continually to extend the scope of our description, but in such a way that our messages do not thereby lose their objective and unambiguous character' (Petersen 1963, p. 10). Thus, we should look for the *maximal* set of propositions that can be consistently supposed to have a truth-value in the context (ψ, R).

It has been pointed out (in relation to the modal interpretation of quantum mechanics (see Chapter 1 of this volume)) that we can consistently assume that all projections in $S^{\perp}$ are *False*. Moreover, if we do so, then our lattice of truth-valued propositions will be maximal; i.e. we cannot add further elements of reality without violating the requirement of classical description. Thus, given an appropriate event space S, we will take the full family of truth-valued propositions to be those in the set (cf. Chapter 1):

$$\mathbf{Def}(S) := \left\{ P^2 = P = P^* : \forall Q \in S \left[Q \leq P \text{ or } QP = 0 \right] \right\}.$$

It is straightforward to verify that $\mathbf{Def}(S)$ is a sublattice of the lattice of all projection operators on $\mathcal{H}$. Moreover, it can be shown that $\mathbf{Def}(S)$ is maximal in the following sense: If $\mathcal{L}$ is a lattice of projections such that $\mathbf{Def}(S) \subset \mathcal{L}$, then ψ *cannot* be represented as a classical probability distribution over all elements in $\mathcal{L}$. (In this section and the next, we state results without proof. Each of these results is a corollary of the results proved in Halvorson & Clifton (1999).)

4 Bohr's reply: spin case

We can now make use of the appropriate mixtures account to reconstruct Bohr's reply to EPR. For the sake of mathematical simplicity, we first consider Bohm's spin version of the EPR experiment. We return to the original EPR experiment in the final section.

Suppose that we have prepared an ensemble of spin-1/2 particles in the singlet state:

$$\psi = \frac{1}{\sqrt{2}}(\,|x+\rangle|x-\rangle - |x-\rangle|x+\rangle\,), \tag{1}$$

where $\sigma_x|x\pm\rangle = \pm|x\pm\rangle$. Then, $\sigma_x \otimes I$ is strictly anticorrelated with $I \otimes \sigma_x$, and $\sigma_y \otimes I$ is strictly anticorrelated with $I \otimes \sigma_y$. Thus, the outcome of a measurement of $\sigma_x \otimes I$ would permit us to predict with certainty the outcome of a measurement of $I \otimes \sigma_x$; and the outcome of a measurement of $\sigma_y \otimes I$ would permit us to predict with certainty the outcome of a measurement of $I \otimes \sigma_y$.

For any orthonormal basis $\{\phi_i\}$ of eigenvectors for $\sigma_x \otimes I$, the event space

$$\{\,|\phi_i\rangle\langle\phi_i|\;:\;|\langle\psi,\phi_i\rangle|^2 \neq 0\}, \tag{2}$$

is appropriate for $(\psi, \sigma_x \otimes I)$. However, since $\sigma_x \otimes I$ is degenerate, there are infinitely many distinct orthonormal bases of eigenvectors for $\sigma_x \otimes I$. Moreover, each basis gives rise to a distinct event space, and each distinct event space permits us to attribute *different* elements of reality to the second (unmeasured) particle.

More concretely, let $P^x_\pm$ denote the projection onto the ray generated by $|x\pm\rangle$, and similarly for $P^y_\pm$ and $P^z_\pm$. Then, each of the following event spaces is appropriate for $(\psi, \sigma_x \otimes I)$:

$$S_{xx} = \{P^x_+ \otimes P^x_-\,,\; P^x_- \otimes P^x_+\}.$$

$$S_{xy} = \{P^x_+ \otimes P^y_+\,,\; P^x_+ \otimes P^y_-\,,\; P^x_- \otimes P^y_+\,,\; P^x_- \otimes P^y_-\}.$$

$$S_{xz} = \{P^x_+ \otimes P^z_+\,,\; P^x_+ \otimes P^z_-\,,\; P^x_- \otimes P^z_+\,,\; P^x_- \otimes P^z_-\}.$$

Clearly, though, these event spaces give *theoretically inequivalent* descriptions of the measurement context. While S_{xx} gives a description in which the second particle has spin-x values that are perfectly anticorrelated with the spin-x values of the first particle, S_{xy} gives a description in which the second particle has spin-y values that are uncorrelated with the spin-x values of the first particle. How do we determine which description is the *correct* one?

One might be inclined to argue that it is an advantage to have more than one 'interpretation' (i.e. empirically adequate description) of the same measurement context. That is, one might argue that there is no single correct description of the second particle in this context; rather, there are several incompatible, but individually acceptable, descriptions of the second particle. However — despite his otherwise unorthodox philosophical stance — Bohr is not a pluralist about descriptions of measurement contexts. Indeed, he claims that a measurement context uniquely dictates an interpretation.

> ... we are not dealing with an incomplete description characterized by the arbitrary picking out of different elements of reality at the cost of sacrificing other such elements, but with a rational discrimination between essentially different experimental arrangements and procedures ... . (Bohr 1935*a*, p. 148)

Thus, the theorist is not free to make a willy-nilly choice of which elements of reality to ascribe to the second particle; rather, her choice is to be fixed (in some, yet to be explicated, way) by the measurement context.

For Bohr, the correct description of the present measurement context (in which spin-x is measured on the first particle and no measurement is performed on the second particle) is S_{xx}, where the two particles have perfectly anticorrelated spin-x values. However, we do not yet have any explanation for *why* Bohr thinks that this description is privileged. In the next two sections, we shall provide an explanation for Bohr's preference.

4.1 The EPR reality criterion

Isn't it *obvious* that S_{xx} is the correct description of the context in which $\sigma_x \otimes I$ is measured in the EPR state? In particular, if we know that $\sigma_x \otimes I$ has some value (either $+1$ or -1) can we not infer immediately that $I \otimes \sigma_x$ has the opposite value? But what reason do we have to think that $I \otimes \sigma_x$ has any value at all? Since we are refusing to invoke wavefunction collapse, it does not help to note that Lüders' rule entails that a measurement of $\sigma_x \otimes I$ collapses ψ onto either $|x+\rangle|x-\rangle$ or $|x-\rangle|x+\rangle$. Perhaps then our intuition that $I \otimes \sigma_x$ has a value is based on some variant of the EPR reality criterion:

If we can predict with certainty the outcome of a measurement of $I \otimes \sigma_x$, then it must possess a value.

According to Howard, it is a contextualized version of the EPR reality criterion that dictates which properties Bohr attributes to the second (unmeasured) particle. Howard says, '... there is no obvious reason why, with the added necessary condition of a restriction to specific experimental contexts, [Bohr] could not accept the EPR reality criterion as it stands' (Howard 1979, p. 256). He then spells out concretely what such a contextualized version of the reality criterion would require.

> Once the experimental context is stipulated, which amounts to the specification of the candidates for real status, our decision as to which particular properties to consider as real will turn on the question of predictability with certainty. (Howard 1979, p. 256)

We will now give a formal description of this notion of a contextualized reality criterion.

First, when Howard says that the experimental context specifies the 'candidates' for real status, he presumably means that an observable must be compatible with the measured observable in order to be such a candidate. For example, if we measure $\sigma_x \otimes I$, then $\sigma_y \otimes I$ is not even a candidate for real status. However, in order for the quantum state ψ to be representable as a classical probability distribution over two projections P and P', it is *not* necessary for P and P' to be compatible. Rather, ψ can be represented as a classical probability distribution over P and P' if and only if $[P, P']\psi = 0$. Thus, since we wish to maintain that each spectral projection of R has a truth-value in the context (ψ, R), we will say that a property P is a *candidate for real status* just in case $[P, P']\psi = 0$ for every spectral projection P' of R.

However, since compatibility (or compatibility relative to a state) is not transitive, not every observable that is compatible with the measured observable can be an element of reality. For example, both $I \otimes \sigma_x$ and $I \otimes \sigma_y$ are compatible with $\sigma_x \otimes I$, but it is not possible for both $I \otimes \sigma_x$ and $I \otimes \sigma_y$ to be elements of reality. Thus, we need a criterion that will permit us to choose among the candidates for real status in such a way that we do not end up with a set of properties that cannot be described classically.

According to Howard, 'our decision as to which particular properties to consider as real will turn on the question of predictability with certainty'. In other words, P is real only if it is strictly correlated with one of the possible outcomes of a measurement of R; i.e. there is some spectral projection P' of R such that P and P' are strictly correlated in the state ψ. That is, $\langle \psi, (P - P')^2 \psi \rangle = 0$, which is equivalent to $P\psi = P'\psi$.

Let $\mathcal{R}$ denote the family of spectral projections for R. Then Howard's proposal amounts to attributing reality to the following set of properties in the context (ψ, R):

$$\mathcal{L}(\psi, R) := \Big\{ P^2 = P = P^* : [P, R]\psi = 0 \;\; \& \;\; \exists P' \in \mathcal{R} \text{ s.t. } P'\psi = P\psi \Big\}.$$

We leave the following straightforward verifications to the reader: (1) $\mathcal{L}(\psi, R)$ is a sublattice of the lattice of all projections on $\mathcal{H}$. (2) The quantum state $|\psi\rangle\langle\psi|$ is a mixture of dispersion-free states on $\mathcal{L}(\psi, R)$. (For this, recall that it is sufficient to show that $[P, Q]\psi = 0$ for all $P, Q \in \mathcal{L}(\psi, R)$.) (3) $I \otimes P^x_\pm \in \mathcal{L}(\psi, \sigma_x \otimes I)$ and $I \otimes P^y_\pm \notin \mathcal{L}(\psi, \sigma_x \otimes I)$; and similarly with the roles of x and y interchanged.

Thus, the contextualized reality criterion accurately reproduces Bohr's pronouncements on the EPR experiment. However, there is a serious difficulty with this analysis of Bohr's reply. In particular, the EPR reality criterion is best construed as a version of 'inference to the best explanation' (cf. Redhead 1989, p. 72): The best explanation of our ability to predict the outcome of a measurement with certainty is that the system has some pre-existing feature that we are detecting. However, since Bohr is not a classical scientific realist (see, e.g., Howard 1979), we cannot expect him to be persuaded by such inferences to the best explanation. Thus, although Howard's contextualized reality criterion gives the right answers, it fails to give a plausible explanation of why Bohr gave the answers he did.

4.2 Objectivity and invariance

Despite Bohr's rejection of classical scientific realism, he maintains that our descriptions of experimental phenomena must be 'objective'. Presumably, Bohr's notion of objectivity is to some extent derivative from the idealist philosophical tradition, and therefore has philosophical subtleties that go far beyond the scope of this paper. For our present purposes, however, it will suffice to use a straightforward and clear notion of objectivity that Bohr might have endorsed: For a feature of a system to be objective, that feature must be invariant under the 'relevant' group of symmetries. We now explicate this notion, and we show that it dictates a unique classical description of the EPR experiment.

Recall that the event space $S_{xy} = \{P^x_+ \otimes P^y_+, P^x_+ \otimes P^y_-, P^x_- \otimes P^y_+, P^x_- \otimes P^y_-\}$ allows us to describe an ensemble in which the first particle has spin-x values, and the second particle has (uncorrelated) spin-y values. Now, consider the symmetry U of the system defined by the following mapping of orthonormal bases:

$$\begin{aligned}
|y+\rangle|y+\rangle &\longmapsto +|z-\rangle|z-\rangle \\
|y+\rangle|y-\rangle &\longmapsto -|z-\rangle|z+\rangle \\
|y+\rangle|y-\rangle &\longmapsto -|z-\rangle|z+\rangle \\
|y-\rangle|y-\rangle &\longmapsto +|z+\rangle|z+\rangle .
\end{aligned}$$

Then, $U^*(\sigma_x \otimes I)U = \sigma_x \otimes I$, and $U\psi = \psi$. That is, U leaves both the state and the measured observable of the context invariant. However,

$$U^*(P^x_\pm \otimes P^y_\pm)U \;=\; P^x_\pm \otimes P^z_\pm .$$

That is, U does not leave the individual elements of S_{xy}, nor even the set as a whole, invariant. In fact, there is no quantum state that is dispersion-free on both $P^x_+ \otimes P^y_+$ and on its transform $U^*(P^x_+ \otimes P^y_+)U = P^x_+ \otimes P^z_+$. Thus, the candidate elements of reality in S_{xy} are not left invariant by the relevant class of symmetries.

In general, let us say that a set S of projections on $\mathcal{H}$ is *definable* in terms of ψ and R just in case: For any unitary operator U on $\mathcal{H}$, if $U\psi = \psi$ and $U^*RU = R$ then $U^*PU = P$ for all $P \in S$. It is straightforward to verify that the set $S_{xx} = \{P^x_+ \otimes P^x_-, P^x_- \otimes P^x_+\}$ is definable in terms of ψ and R. In fact, it is the only such appropriate event space for this context.

Theorem 1. *$\{P^x_+ \otimes P^x_-, P^x_- \otimes P^x_+\}$ is the unique appropriate event space for $(\psi, \sigma_x \otimes I)$ that is definable in terms of ψ and $\sigma_x \otimes I$.*

Proof. Suppose that S is an appropriate event space for (ψ, R) that is definable in terms of ψ and R, and let $|\phi\rangle\langle\phi| \in S$. Let $P_1 = (P^x_+ \otimes P^x_-) + (P^x_- \otimes P^x_+)$ and let $P_2 = (P^x_+ \otimes P^x_+) + (P^x_- \otimes P^x_-)$. Then $U := P_1 - P_2$ is a unitary operator. It is obvious that $U^*(\sigma_x \otimes I)U = \sigma_x \otimes I$ and $U\psi = \psi$. Thus, definability entails that $|\phi\rangle\langle\phi|$ commutes with U; and therefore $|\phi\rangle\langle\phi|$ is either a subprojection of P_1 or is a subprojection of P_2. However, the latter is not possible since $P_2\psi = 0$. Thus, $|\phi\rangle\langle\phi|$ is a subprojection of P_1. However, there are only two one-dimensional subprojections of P_1 that are compatible with R, namely $P^x_+ \otimes P^x_-$ and $P^x_- \otimes P^x_+$. Since $|\phi\rangle\langle\phi|$ must be compatible with R it follows that either $|\phi\rangle\langle\phi| = P^x_+ \otimes P^x_-$ or $|\phi\rangle\langle\phi| = P^x_- \otimes P^x_+$. □

Thus, we have a situation analogous to simultaneity relative to an inertial frame in relativity theory. In that case, there is only one simultaneity relation that is invariant under all symmetries that preserve an inertial observer's worldline (Malament 1977). Thus, we might wish to regard this simultaneity relation as the correct one relative to that observer, and the others as spurious. In the quantum-mechanical case, there is only one set of properties that is invariant under all the symmetries that preserve the quantum state and the measured observable. So, we should regard these properties as those that possess values relative to that measurement context.

It is easy to see that $\mathcal{L}(\psi, \sigma_x \otimes I) = \mathbf{Def}(S_{xx})$. Thus Howard's suggestion of applying a contextualized reality criterion turns out to be (extensionally) equivalent to requiring that the elements of reality be definable in terms of ψ and R. It follows that those attracted by Howard's analysis of Bohr's response to EPR now have independent grounds to think that $\mathcal{L}(\psi, \sigma_x \otimes I)$ gives the correct list of elements of reality in the context $(\psi, \sigma_x \otimes I)$.

5 Bohr's reply: position-momentum case

There are a couple of formal obstacles that we encounter in attempting to reconstruct Bohr's reply to the *original* EPR argument. First, there is an obstacle in describing the EPR experiment itself: The EPR state supposedly assigns dispersion-free values to the relative position $Q_1 - Q_2$ and to the total momentum $P_1 + P_2$ of the two particles. However, $Q_1 - Q_2$ and $P_1 + P_2$ are continuous spectrum observables, and no standard quantum state (i.e., density operator) assigns a dispersion-free value to a continuous spectrum observable. Thus, in terms of the standard mathematical formalism for quantum mechanics, *the EPR state does not exist*. Second, there is an obstacle in applying the account of appropriate mixtures to the EPR experiment: Since the position (or momentum) observable of the first particle has a continuous spectrum, no density operator W is a convex combination of dispersion-free states of the measured observable. Thus, there are no appropriate mixtures (in our earlier sense) for this measurement context.

We can overcome both of these obstacles by expanding the state space of our system so that it includes eigenstates for continuous spectrum observables. To do this rigorously, we will employ the C^*-algebraic formalism of quantum theory. We first recall the basic elements of this formalism.

A C^*-algebra $\mathcal{A}$ is a complex Banach space with norm $A \mapsto \|A\|$, involution $A \mapsto A^*$, and a product $A, B \mapsto AB$ satisfying:

$$(AB)^* = B^*A^*, \qquad \|A^*A\| = \|A\|^2, \qquad \|AB\| \leq \|A\|\|B\|. \tag{3}$$

We assume that $\mathcal{A}$ has a two-sided identity I. Let ω be a linear functional on $\mathcal{A}$. We say that ω is a *state* just in case ω is positive [i.e. $\omega(A^*A) \geq 0$ for all $A \in \mathcal{A}$], and ω is normalized [i.e. $\omega(I) = 1$]. A state ω is said to be *pure* just in case: If $\omega = \lambda\rho + (1-\lambda)\tau$ where ρ, τ are states of $\mathcal{A}$ and $\lambda \in (0,1)$, then $\omega = \rho = \tau$. A state ω is said to be *dispersion-free* on $A \in \mathcal{A}$ just in case $\omega(A^*A) = |\omega(A)|^2$. If ω is dispersion-free on A for all $A \in \mathcal{A}$, we say that ω is *dispersion-free* on the algebra $\mathcal{A}$.

We represent a measurement context by a pair $(\omega, \mathcal{R})$, where ω

is a state of $\mathcal{A}$, and $\mathcal{R}$ is a mutually commuting family of operators in $\mathcal{A}$ (representing the measured observables). We are interested now in determining which families of observables can be described classically as possessing values in the state ω. Thus, if $\mathcal{B}$ is a C^*-subalgebra of $\mathcal{A}$, we say that $\omega|_{\mathcal{B}}$ (i.e., the restriction of ω to $\mathcal{B}$) is a *classical probability measure* (or more briefly, *classical*) just in case

$$\omega(A) = \int \omega_\lambda(A) d\mu(\lambda), \qquad A \in \mathcal{B}, \tag{4}$$

where each ω_λ is a dispersion-free state of $\mathcal{B}$.

We now construct a specific C^*-algebra that provides the model for a single particle with one degree of freedom. In the standard Hilbert space description of a single particle, we can take our state space to be the Hilbert space $L_2(\mathbb{R})$ of (equivalence classes of) square-integrable functions from $\mathbb{R}$ into $\mathbb{C}$. The position observable can be represented by the self-adjoint operator Q defined by $Q\psi(x) = x \cdot \psi(x)$ (on a dense domain in $L_2(\mathbb{R})$), and the momentum observable can be represented by the self-adjoint operator $P = -i(d/dx)$ (also defined on a dense domain in $L_2(\mathbb{R})$). (We set $\hbar = 1$ throughout.) We may then define one-parameter unitary groups by setting $U_a := \exp\{iaQ\}$ for $a \in \mathbb{R}$, and $V_b := \exp\{ibP\}$ for $b \in \mathbb{R}$. Let $\mathcal{A}[\mathbb{R}^2]$ denote the C^*-subalgebra of operators on $L_2(\mathbb{R})$ generated by $\{U_a : a \in \mathbb{R}\} \cup \{V_b : b \in \mathbb{R}\}$. We call $\mathcal{A}[\mathbb{R}^2]$ the *Weyl algebra* for one degree of freedom.

Of course, $\mathcal{A}[\mathbb{R}^2]$ itself does not contain either Q or P. However, the group $\{U_a : a \in \mathbb{R}\}$ can be thought of as a surrogate for Q, in the sense that a state ω of $\mathcal{A}[\mathbb{R}^2]$ should be thought of as an 'eigenstate' for Q just in case ω is dispersion-free on the set $\{U_a : a \in \mathbb{R}\}$. Similarly, the group $\{V_b : b \in \mathbb{R}\}$ can be thought of as a surrogate for P. (Moreover, the indeterminacy relation between Q and P can be formulated rigorously as follows: There is no state of $\mathcal{A}[\mathbb{R}^2]$ that is simultaneously dispersion-free on both $\{U_a : a \in \mathbb{R}\}$ and $\{V_b : b \in \mathbb{R}\}$ (see Chapter 9, p. 314.)

5.1 Formal model of the EPR experiment

In the standard formalism, the state space of a pair of particles (each with one degree of freedom) can be taken as the tensor product Hilbert space $L_2(\mathbb{R}) \otimes L_2(\mathbb{R})$. Similarly, in the C^*-algebraic formalism, the algebra of observables for a pair of particles (each with one degree of freedom) can be represented as the tensor product $\mathcal{A}[\mathbb{R}^2] \otimes \mathcal{A}[\mathbb{R}^2]$.

The EPR state is supposed to be that state in which $Q_1 - Q_2$ has the value λ, and $P_1 + P_2$ has the value μ. (At present, we have no guarantee of either existence or uniqueness.) Since $\exp\{ia(Q_1 - Q_2)\} = U_a \otimes U_{-a}$ and $\exp\{ib(P_1 + P_2)\} = V_b \otimes V_b$, and since dispersion-free states preserve functional relations, the EPR state should assign the (dispersion-free) value $e^{ia\lambda}$ to $U_a \otimes U_{-a}$ and the value $e^{ib\mu}$ to $V_b \otimes V_b$. Fortunately, for a fixed pair (λ, μ) of real numbers, there is a unique pure state ω of $\mathcal{A}[\mathbb{R}^2] \otimes \mathcal{A}[\mathbb{R}^2]$ that satisfies these two conditions (Halvorson 2000, Theorem 1). We will simply call ω the *EPR state.*

Suppose then that we are in a context in which all elements in $\mathcal{Q}_1 := \{U_a \otimes I : a \in \mathbb{R}\}$ can be assigned definite numerical (complex) values (e.g. a context in which the position of the first particle has been determined). We can then ask: Which observables can be consistently described, along with the elements of $\mathcal{Q}_1$, as possessing values in the state ω? Since the elements of $\mathcal{Q}_2 := \{I \otimes U_a : a \in \mathbb{R}\}$ commute pairwise with the elements of $\mathcal{Q}_1$, we could provide a consistent description in which the second particle has a definite position (that is strictly correlated with the first particle's position). However, since the elements of $\mathcal{P}_2 := \{I \otimes V_a : a \in \mathbb{R}\}$ also commute pairwise with the elements of $\mathcal{Q}_1$, we could provide a consistent description in which the second particle has a definite momentum (which is uncorrelated with the position of the first particle). The requirement of consistency does not itself tell us which of these descriptions is the correct one. In order to find a basis for choosing between the descriptions, we turn again to symmetry considerations.

Let $\mathcal{A}, \mathcal{B}$ be C^*-algebras, and let π be a mapping of $\mathcal{A}$ into $\mathcal{B}$. We say that π is a $*$-*homomorphism* just in case π is linear, multiplicative, and preserves adjoints. If π is also a bijection, we say that π is a $*$-*isomorphism*; and we say that π is a $*$-*automorphism* when we wish to indicate that $\mathcal{B}$ was already assumed to be isomorphic to $\mathcal{A}$. Finally, let ω be a state of $\mathcal{A}$, let $\mathcal{R}$ be a mutually commuting family of operators in $\mathcal{A}$, and let $\mathcal{B}$ be a C^*-subalgebra of $\mathcal{A}$. We say that $\mathcal{B}$ is *definable* in terms of ω and $\mathcal{R}$ just in case: For any $*$-automorphism α of $\mathcal{A}$, if $\alpha(\mathcal{R}) = \mathcal{R}$ and $\omega \circ \alpha = \omega$, then $\alpha(\mathcal{B}) = \mathcal{B}$. Thus, in our present circumstance, we wish to determine which (if any) of the candidate algebras of 'elements of reality' identified above is definable in terms of the EPR state and the measured observables $\mathcal{Q}_1$.

5.2 The reconstruction theorem

We turn now to the main technical result of our paper. Our main result shows that if the position observable of the first particle is assumed to be an element of reality, and if the elements of reality are invariant under symmetries that leave the EPR state and the position of the first particle invariant, then (a) we can consistently assume that the position of the second particle is an element of reality, but (b) we cannot consistently assume that the momentum of the second particle is an element of reality.

For the statement and the proof of our result, it will be convenient to pass to the 'GNS representation' of $\mathcal{A}[\mathbb{R}^2] \otimes \mathcal{A}[\mathbb{R}^2]$ induced by the EPR state (see Kadison & Ringrose 1997, Thm. 4.5.2). That is, there is a Hilbert space $\mathcal{H}$, a unit vector $\Omega \in \mathcal{H}$, and a $*$-homomorphism π from $\mathcal{A}[\mathbb{R}^2] \otimes \mathcal{A}[\mathbb{R}^2]$ into the algebra of bounded operators on $\mathcal{H}$ such that

$$\omega(A) = \langle \Omega, \pi(A)\Omega \rangle, \qquad A \in \mathcal{A}[\mathbb{R}^2] \otimes \mathcal{A}[\mathbb{R}^2]. \tag{5}$$

Since $\mathcal{A}[\mathbb{R}^2] \otimes \mathcal{A}[\mathbb{R}^2]$ is simple, π is a $*$-isomorphism. Thus, we can suppress reference to π, and suppose that $\mathcal{A}[\mathbb{R}^2] \otimes \mathcal{A}[\mathbb{R}^2]$ is given concretely as a C^*-algebra of operators acting on $\mathcal{H}$, and that the EPR state ω is given by the unit vector $\Omega \in \mathcal{H}$.

Theorem 2. *Let ω be the EPR state. Suppose that $\mathcal{B}$ is a subalgebra of $\mathcal{A}[\mathbb{R}^2] \otimes \mathcal{A}[\mathbb{R}^2]$ such that:*

1. *$\omega|_{\mathcal{B}}$ is a classical probability distribution;*

2. *$\{U_a \otimes I : a \in \mathbb{R}\} \subseteq \mathcal{B}$;*

3. *$\mathcal{B}$ is definable in terms of ω and $\{U_a \otimes I : a \in \mathbb{R}\}$.*

Then $[I \otimes U_a, Z]\Omega = 0$ for all $a \in \mathbb{R}$ and $Z \in \mathcal{B}$.

This theorem guarantees that we can suppose that $\{I \otimes U_a : a \in \mathbb{R}\}$ are elements of reality without violating the requirement of classical describability. (We could then invoke Bohr's demand for 'maximizing the scope of our description' to justify the claim that the position of the second particle *is* an element of reality.) Moreover, since there is no $b \in \mathbb{R}\backslash\{0\}$ such that $[(I \otimes U_a), (I \otimes V_b)]\Omega = 0$ for all $a \in \mathbb{R}$, the theorem excludes the possibility that the momentum of the second particle is an element of reality.

For the proof of the theorem, we will need to invoke a technical lemma. First, let $\mathfrak{B}(\mathcal{H})$ denote the algebra of bounded linear operators on the Hilbert space $\mathcal{H}$. If $\mathcal{B}$ is a subset of $\mathfrak{B}(\mathcal{H})$, we let $\mathcal{B}'$ denote the set of all operators in $\mathfrak{B}(\mathcal{H})$ that commute with each operator in $\mathcal{B}$, and we let $\mathcal{B}'' = (\mathcal{B}')'$.

Lemma 1. *Let $\mathcal{B}$ be a C^*-algebra of operators acting on $\mathcal{H}$. Let $V_t = \exp\{-itH\}$, where H is a bounded self-adjoint operator acting on $\mathcal{H}$. If $V_t \mathcal{B} V_{-t} = \mathcal{B}$ for all $t \in \mathbb{R}$, then there is a one-parameter unitary group $\{W_t : t \in \mathbb{R}\} \subseteq \mathcal{B}''$ such that $V_t A V_{-t} = W_t A W_{-t}$ for all $A \in \mathcal{B}$ and $t \in \mathbb{R}$.*

Proof. See Theorem 4.1.15 of Sakai (1971). □

Proof of the theorem. Suppose that $\mathcal{B}$ is a C^*-subalgebra of $\mathcal{A}[\mathbb{R}^2] \otimes \mathcal{A}[\mathbb{R}^2]$ that satisfies conditions 1.–3. of the theorem. We wish to show that

$$[I \otimes U_a, Z]\Omega = 0. \tag{6}$$

for all $Z \in \mathcal{B}$ and $a \in \mathbb{R}$. Fix $Z \in \mathcal{B}$ and $a \in \mathbb{R}$. Define the real and imaginary parts of $I \otimes U_a$ by setting

$$A := (1/2)\big[(I \otimes U_a) + (I \otimes U_{-a})\big], \tag{7}$$

$$B := (i/2)\big[(I \otimes U_{-a}) - (I \otimes U_a)\big]. \tag{8}$$

Thus, $A + iB = I \otimes U_a$ and it will suffice to show that $[A, Z]\Omega = 0$ and $[B, Z]\Omega = 0$. We will treat the case of A; the case of B can be dealt with by a similar argument. Define the real part of $e^{-ia\lambda}(U_a \otimes I)$ by setting

$$A' \quad := \quad (1/2)[e^{-ia\lambda}(U_a \otimes I) + e^{ia\lambda}(U_{-a} \otimes I)]. \tag{9}$$

Let $V_t = \exp\{-it(A' - A)\}$ for all $t \in \mathbb{R}$. Thus, in order to show that $[A, Z]\Omega = 0$ it will suffice to show that $[V_t, Z]\Omega = 0$ for all $t \in \mathbb{R}$.

For each $t \in \mathbb{R}$, define a $*$-automorphism α_t of $\mathcal{A}[\mathbb{R}^2] \otimes \mathcal{A}[\mathbb{R}^2]$ by setting $\alpha_t(X) = V_t X V_{-t}$ for all $X \in \mathcal{A}[\mathbb{R}^2] \otimes \mathcal{A}[\mathbb{R}^2]$. Since the EPR state ω is dispersion-free on $A' - A$ [use the fact that $\omega(U_a \otimes U_b) = 0$ when $b \neq -a$], it follows that

$$\omega(\alpha_t(X)) = \omega(V_t X V_{-t}) = \omega(X), \tag{10}$$

for all $X \in \mathcal{A}[\mathbb{R}^2] \otimes \mathcal{A}[\mathbb{R}^2]$ and $t \in \mathbb{R}$. Moreover, $\alpha_t(U_a \otimes I) = V_t(U_a \otimes I)V_{-t} = U_a \otimes I$ for all $a \in \mathbb{R}$. By hypothesis, $\mathcal{B}$ is definable in terms of $\{U_a \otimes I : a \in \mathbb{R}\}$ and ω. Thus, $V_t \mathcal{B} V_{-t} = \alpha_t(\mathcal{B}) = \mathcal{B}$ for all $t \in \mathbb{R}$, and Lemma 1 entails that there is a unitary group $\{W_t : t \in \mathbb{R}\} \subseteq \mathcal{B}''$ such that $V_t X V_{-t} = W_t X W_{-t}$ for all $X \in \mathcal{B}$ and $t \in \mathbb{R}$. Since $\omega|_{\mathcal{B}''}$ is classical (see Halvorson & Clifton 1999, Cor. 2.9), and since $Z, W_{-t} \in \mathcal{B}''$, it follows that

$$V_t Z V_{-t}\Omega = W_t Z W_{-t}\Omega = W_t W_{-t} Z\Omega = Z\Omega. \tag{11}$$

Therefore, $[V_t, Z]\Omega = 0$ for all $t \in \mathbb{R}$. □

We conjecture that there is a unique subalgebra $\mathcal{B}$ of $\mathcal{A}[\mathbb{R}^2] \otimes \mathcal{A}[\mathbb{R}^2]$ that maximally satisfies conditions 1.–3. of the previous theorem. In fact, this stronger result would follow if it could be established that the von Neumann algebra $\pi(\{U_a \otimes U_b : a, b \in \mathbb{R}\})''$ is maximal abelian in $\mathfrak{B}(\mathcal{H})$. (Moreover, this latter fact would follow from Theorem I.6 of (Fannes *et al.* 1974). Unfortunately, the proof of this theorem is invalid (Araki 1976).)

6 Conclusion

We have shown that Bohr's reply to EPR is a logical consequence of four requirements: (1) *Empirical Adequacy:* When an observable is measured, it possesses some value in accordance with the probabilities determined by the quantum state; (2) *Classical Description:* Properties P and P' can be simultaneously real in a quantum state only if that state can be represented as a joint classical probability distribution over P and P'; (3) *Objectivity:* Elements of reality must be invariants of those symmetries that preserve the defining features of the measurement context; (4.) *Maximality:* Our description should be maximal, subject to the prior three constraints. Obviously, these requirements have nothing to do with the verifiability criterion of meaning or with similar positivistic doctrines. Thus, Bohr's reply to EPR does not require a shift towards positivism.

Nonetheless, our reconstruction of Bohr's reply does not in itself constitute an argument for the superiority of Bohr's point of view over EPR's more 'realist' point of view, which rejects the claim that the reality of a system can be constituted 'from a distance'. However, we wish to emphasize that Bohr is not so much concerned with what is *truly* real for the distant system as he is with the question of what we would be *warranted in asserting* about the distant system from the standpoint of classical description. In particular, Bohr argues that in certain measurement contexts we are warranted in attributing certain elements of reality to distant (unmeasured) systems. He also claims, however, that if we attempt to make *context-independent* attributions of reality to these distant systems, then we will come into conflict with the experimental record.

Moreover, as Bohr himself might have claimed, a similar sort of context-dependence already arises in special relativity. In particular, an inertial observer is warranted in saying that any two events that are orthogonal to his worldline at some worldpoint are simultaneous. However, if we attempt to make *context-independent* attributions of simultaneity to distant events — where the 'context' is now set by the observer's frame of reference — then we will run into conflicts with the experimental record.

Of course, a proper defense of Bohr's point of view would require much more space than we have here. However, we have supplied ample justification for the claim that Bohr's reply to EPR — and his philosophy of quantum theory in general — deserves a more fair treatment than it has recently received.

Bibliography

Araki, H. (1976), 'Review of Fannes *et al.* (1974)', *Mathematical Reviews* **52**, 12601.

Beller, M. (1999), *Quantum Dialogue*, University of Chicago Press, Chicago.

Bohr, N. (1935*a*), 'Can quantum-mechanical description of physical reality be considered complete?', *Physical Review* **48**, 696–702.

Bohr, N. (1935*b*), 'Quantum mechanics and physical reality', *Nature* **136**, 65.

Bohr, N. (1949), Discussion with Einstein on epistemological problems in atomic physics, in P. Schilpp, ed., 'Albert Einstein: Philosopher-Scientist', Tudor, NY, pp. 201–41.

Bub, J. (1995), Complementarity and the orthodox (Dirac-von Neumann) interpretation of quantum mechanics, in R. Clifton, ed., 'Perspectives on Quantum Reality', Kluwer, NY, pp. 211–26.

Bub, J. (1997), *Interpreting the Quantum World*, Cambridge University Press, NY.

Caves, C., Fuchs, C. & Schack, R. (2002), 'Quantum probabilities as Bayesian probabilities', *Physical Review A* **65**, 022305.

Clifton, R. (1996), 'The properties of modal interpretations of quantum mechanics', *British Journal for the Philosophy of Science* **47**, 371–98.

Einstein, A., Podolsky, B. & Rosen, N. (1935), 'Can quantum-mechanical description of physical reality be considered complete?', *Physical Review* **47**, 777–80.

Fannes, M., Verbeure, A. & Weder, R. (1974), 'On momentum states in quantum mechanics', *Ann. Inst. Henri Poincaré* **20**, 291–6.

Fine, A. & Beller, M. (1994), Bohr's response to EPR, in J. Faye & H. Folse, eds, 'Niels Bohr and Contemporary Philosophy', Kluwer, NY, pp. 1–31.

Halvorson, H. (2000), 'The Einstein-Podolsky-Rosen state maximally violates Bell's inequalities', *Letters in Mathematical Physics* **53**, 321–9.

Halvorson, H. & Clifton, R. (1999), 'Maximal beable subalgebras of quantum mechanical observables', *International Journal of Theoretical Physics* **38**, 2441–84.

Howard, D. (1979), Complementarity and Ontology: Niels Bohr and the problem of scientific realism in quantum physics, PhD thesis, Boston University.

Howard, D. (1994), What makes a classical concept classical?, in J. Faye & H. Folse, eds, 'Niels Bohr and Contemporary Philosophy', Kluwer, NY, pp. 201–29.

Howard, D. (2000), A brief on behalf of Bohr, University of Notre Dame.

Kadison, R. & Ringrose, J. (1997), *Fundamentals of the Theory of Operator Algebras*, American Mathematical Society, Providence, RI.

Malament, D. (1977), 'Causal theories of time and the conventionality of simultaneity', *Noûs* **11**, 293–300.

Petersen, A. (1963), 'The philosophy of Niels Bohr', *Bulletin of the Atomic Scientists* **19**, 8–14.

Redhead, M. L. G. (1989), *Incompleteness, Nonlocality, and Realism*, Oxford University Press, NY.

Ruark, A. (1935), 'Is the quantum-mechanical description of physical reality complete?', *Physical Review* **48**, 466–7.

Sakai, S. (1971), *C^*-algebras and W^*-algebras*, Springer, NY.

Chapter 13

Simulating quantum mechanics by non-contextual hidden variables

with Adrian Kent

1 Introduction

Bell's theorem (Bell 1964) establishes that local hidden variable theories are committed to Bell inequalities violated by quantum correlations. Since violations of Bell inequalities can be verified without requiring that the observables whose correlations figure in the inequalities be measured with arbitrarily high precision, Bell's theorem yields a method of falsifying local hidden variable theories. Moreover, the evidence for violations of Bell inequalities

We thank J. Finklestein, N. D. Mermin, and an anonymous referee for helpful comments. A. K. was supported by a Royal Society University Research Fellowship and thanks the Oxford Centre for Quantum Computation for much appreciated hospitality.

very strongly suggests that such theories have indeed been falsified (Clauser *et al.* 1969, Aspect *et al.* 1981, Tittel *et al.* 1998, Weihs *et al.* 1998).

We are concerned here, however, not with locality but with non-contextuality. Non-contextual hidden variable theories ascribe definite truth values to projections or, in the case of generalized measurements, any positive operators, so that the truth values predict the outcome of any measurement involving the relevant operator and are independent of the other projections or positive operators involved in the measurement. Of course, non-contextual hidden variable theories that reproduce the quantum correlations between spatially separated systems must, by Bell's theorem, be non-local. However, our interest here is in determining whether *non-relativistic* quantum theory can be simulated classically via non-contextual hidden variables. Since non-relativistic classical mechanics does not presuppose a light cone structure, non-locality is not a meaningful constraint on hidden variables in this context, and we shall henceforth ignore questions of non-locality altogether.

Unlike the arguments against local hidden variables, the known arguments against non-contextual hidden variables require observables to be measurable with perfect precision. These arguments derive from the work of Gleason (1957), Bell (1966) and Kochen & Specker (1967). (For more recent discussions see Redhead 1989, Mermin 1993, Zimba & Penrose 1993, Peres 1995, Bub 1997.)

Pitowsky (1983, 1985) argued some time ago that these no-go arguments could be evaded by restricting attention to appropriately chosen subsets of the space of observables. While ingenious, Pitowsky's models, which are constructed via the axiom of choice and the continuum hypothesis, have the defect that they rely for their interpretation on a radically non-standard version of probability theory, according to which (for example) the conjunction of two probability one events can have probability zero.

More recently, Meyer (1999) has emphasized that the fact that all physical measurements are of finite precision leaves a loophole in the arguments against non-contextual hidden variables. One could hypothesize that the class of possible physical measurements is only a dense subset of the full set of von Neumann or positive operator valued measurements. That is, in any given finite precision measurement there is a fact of the matter, unknown to us, as to which precise measurement is being carried out, and these realized measurements always belong to some particular dense subset, which again need not necessarily be known to us. Under this hypothesis, the arguments against non-contextual hidden variables, which rely on ascribing definite values to all projections in a real three-dimensional space (or to certain well chosen finite subsets of projections), have been shown not to go through (Meyer 1999), nor can any similar arguments be constructed for either projections or positive operator valued measurements in Hilbert spaces of three or higher dimensions (Kent 1999). These recent counterexamples rely only on constructive set theory.

The aim of the present paper is more ambitious. We shall show that all the predictions of non-relativistic quantum mechanics that are verifiable to within any finite precision *can* be simulated classically by non-contextual hidden variable theories. That is, there are non-contextual hidden variable models whose predictions are practically indistinguishable from those of non-relativistic quantum mechanics for either projection valued or positive operator valued measurements. We give explicit examples, whose construction requires only constructive set theory and whose interpretation needs only elementary (standard) probability theory.

Before giving details, we should explain why we find the question interesting. We have no particular interest in advocating non-contextual hidden variable theories. However, we believe that it is important to distinguish strongly held theoretical beliefs from rigorously established facts in analyzing the ways in which quantum theory is demonstrably non-classical. We share, too, with Meyer (1999) another motivation: questions about the viability of hidden variable models for a particular quantum process translate into questions about the classical simulability of some particular aspect of quantum behavior, and are interesting independently of the physical plausibility of the relevant models. In particular, from the point of view of quantum computation, the precision attainable in a measurement is a computational resource. Specifying infinite precision requires infinite resources and prevents any useful comparison with discrete classical computation. It is interesting to see that, once the assumption of infinite precision is relaxed, the outcomes and probabilities of quantum measurements can indeed be simulated classically and non-contextually.

2 Outline of results

We begin by reviewing the standard theoretical argument against non-contextual hidden variable models, which is based on infinite precision von Neumann, i.e., projection valued, measurements performed on a system represented by an n-dimensional Hilbert space H_n, where n is finite.

Determining a unique value for some measured observable O, with spectral projections $\{P_i\}$, is equivalent to distinguishing exactly one member of the set $\{P_i\}$ and assigning it value '1' (to signify that the corresponding eigenvalue of O is determined to occur on measurement), while assigning all the other projections in $\{P_i\}$ value '0'. Let $\mathcal{P}$ denote a set of projections, and $\overline{\mathcal{P}}$ denote the set of all observables whose spectral projections lie in $\mathcal{P}$. Then, whether there can be hidden variables that uniquely determine the measured values of all the observables in $\overline{\mathcal{P}}$ is equivalent to asking whether there exists a *truth function* $t : \mathcal{P} \to \{0,1\}$ satisfying:

$$\sum_i t(P_i) = 1 \text{ whenever } \sum_i P_i = I \text{ and } \{P_i\} \subseteq \mathcal{P}. \qquad (1)$$

If $n > 2$, and we take $\overline{\mathcal{P}}$ to be all observables of the system, it is an immediate consequence of Gleason's theorem (Gleason 1957) that $\mathcal{P}$ admits no truth function. A simpler proof of the relevant part of Gleason's result was given by Bell (1966), who discussed its implications for hidden variable models.

Kochen & Specker (1967) exhibited a finite set of spin-1 observables $\overline{\mathcal{P}}$ such that $\mathcal{P}$ admits no truth function, and arguments for the nonexistence of truth functions on finite sets of projections are often (summarily) called the 'Kochen-Specker Theorem'. Simpler examples of finite sets of observables admitting no truth function have since been given by Peres (1995) and Zimba & Penrose (1993), among others.

Clearly, then, even before statistical considerations enter, for a hidden variable theory to remain viable, its hidden variables cannot determine unique values for all observables, or even certain finite subsets of observables. There is a simple intuitive reason why contradictions can arise when one attempts to construct a truth function on some set of projections $\mathcal{P}$. Recall that two projections P, P' are said to be compatible if $[P, P'] = 0$. Call two resolutions of the identity operator $\sum_i P_i = I$ and $\sum_j P'_j = I$ *compatible* if $[P_i, P'_j] = 0$ for all i, j. If the various resolutions of the identity generated by $\mathcal{P}$ are all mutually compatible, then closing $\mathcal{P}$ under products of projections and complements yields a Boolean algebra under the operations $P \wedge Q = PQ$ and $P^{\perp} = I - P$, and Boolean algebras always possess truth functions. However, in dimension $n > 2$, one can choose $\mathcal{P}$ so that it generates resolutions of the identity with many projections in common between different *incompatible* resolutions (in particular, such $\mathcal{P}$'s fail to generate a Boolean algebra). This can drastically reduce the number of 'unknowns' relative to equations in (1) to the point where they have no solution.

Conversely, one *can* have a viable hidden variable theory (still assuming infinite measurement precision) only if one is prepared to assign a value to a projection that is not simply a function of the hidden variables, but also of the resolution of the identity that the projection is considered to be a member of. Physically, this would mean that the measured value of the projection would be allowed to depend upon the context in which it is measured, i.e., on the complete commuting set of observables that the projection is jointly measured with — hence the phrase 'contextual hidden variable theory'. Alternatively, one can adopt the approach Kochen and Specker themselves advocated, and view the nonexistence of truth functions as an argument for a quantum-logical conception of a system's properties.

However, when it comes to considering practical experiments, we are not actually forced either towards contextual hidden variables or quantum logic by the Kochen-Specker theorem. The Pitowsky models (Pitowsky 1983) mentioned above provide one possible alternative approach. Unfortunately, both the axiom of choice and the continuum hypothesis (or some weaker axiom in this direction) are needed to define Pitowsky's models. Moreover, the non-standard version of probability theory required for their interpretation has such bizarre properties that we doubt whether these models can reasonably be said to constitute a classical explanation or simulation of quantum theory. At the very least, these features decrease the value of the comparison with classical physics and classical computation.

However, the more recent constructive arguments by Meyer (1999) and Kent (1999) show that one can always find a subset $\mathcal{P}_d$ of projections on H_n, for any finite n, such that $\mathcal{P}_d$ admits a truth function t and generates a countable dense set of resolutions of the identity. By the latter, we mean that for any k-length resolution $\sum_{i=1}^{k} P_i = I$ (where $k \leq n$) and any $\epsilon > 0$, one can always find another k-length resolution $\sum_{i=1}^{k} P_i' = I$ such that $|P_i - P_i'| < \epsilon$ for all i and $\{P_i'\} \subseteq \mathcal{P}_d$. In particular, for any self-adjoint operator O and any $\epsilon > 0$, there is an $O' \in \overline{\mathcal{P}}_d$, with the same eigenspectrum as O, such that the probabilities for measurement outcomes of O' lie within ϵ of the corresponding probabilities for O.

Given this, the non-contextual hidden variable theorist is free to adopt the hypothesis that any *finite* precision measurement by which we attempt an approximate measurement of an observable O with spectral decomposition $\sum_{i=1}^{k} a_i P_i$ actually corresponds to a measurement of some other observable $O' = \sum_{i=1}^{k} a_i P_i'$ lying in the precision range and belonging to $\overline{\mathcal{P}}_d$. Moreover, one can allow this observable O' to be specified by the state and characteristics of the measuring device alone: O' need not depend either on the quantum state of the system or on the hidden variables of the particle being measured. The hypothesis, in other words, is that when we set up an experiment to carry out a finite precision measurement there is some fact of the matter as to which precise observable is actually being measured, and the measured observable in fact belongs to $\overline{\mathcal{P}}_d$. The precise observable being measured is presumably specified by hidden variables associated with the measuring apparatus. It is not known to us in any given experiment; we know only the precision range.

This, it should be stressed, does *not* amount to reintroducing contextualism into the hidden variable theory. While the distribution of hidden variables associated with the measured system needs to be related to the quantum state of that system in order to recover its quantum statistics, these variables are supposed to be independent of both the hidden variables associated with the measuring apparatus and *its* quantum state; in particular, they are not correlated with the unknown choice of O'. Note too that the fact that measurements are finite precision is not sufficient in itself to overthrow the standard analysis. If we assumed, as above, that some precise O' was always specified, and we allowed the possibility that O' could be chosen to be any observable sufficiently close to O, the Kochen-Specker theorem would again threaten. It is the restriction to observables in $\overline{\mathcal{P}}_d$, that has a set of projections *admitting* a truth function, that creates a loophole.

Thus, a finite precision attempt to measure O can have its outcome specified, in a non-contextual way, by the value picked out for O' by the truth function on $\mathcal{P}_d$. Invoking this toy non-contextual hidden variable model, the practical import of the theoretical argument against such models supplied by the Kochen-Specker theorem can be regarded as 'nullified' (to use Meyer's term).

Meyer and Kent leave open the question as to whether we can actually construct a non-contextual hidden variable model that is consistent with all the *statistical* predictions of quantum theory. Let us call a collection of truth functions on a set $\mathcal{P}$ *full* if for any two distinct $P, P' \in \mathcal{P}$, there exists a truth function t on $\mathcal{P}$ such that $t(P) \neq t(P')$. For example, if $\mathcal{P}$ generates a Boolean algebra, then it will possess a full set of truth valuations (one for each minimal nonzero projection of $\mathcal{P}$), but the converse fails. Unless $\mathcal{P}_d$ possesses a full set of truth valuations, the previous paragraph's toy model cannot satisfy the statistical predictions of quantum theory. For suppose that two distinct $P, P' \in \mathcal{P}_d$ are always mapped to the same truth value under all truth valuations. Then every set of hidden variables must dictate the same values for P and P', and therefore the theory will have to predict the same expectation values for P and P' in every quantum state of the system. But this is absurd: since P and P' are distinct, there is certainly some quantum state of the system (which we need only assume is prepared to within finite precision) in which the expectations of P and P' differ.

There certainly are sets of projections $\mathcal{P}$ that admit truth valuations, but not a full set (see the 'Kochen-Specker diagram' on p. 70 of Kochen and Specker 1967, involving only 17 projections). Moreover, it is not obvious that any $\mathcal{P}_d$ with the above properties — for example those given by Meyer (1999) and Kent (1999) — must necessarily possess a full set of truth valuations. Hence the unfalsifiability of non-contextual hidden variables in the face of quantum statistics has yet to be established. The goal of the present paper is to establish this unfalsifiability.

In Section **3**, we shall prove the following result:

> **Theorem 1** *There exists a set of projections $\hat{\mathcal{P}}_d$ on H_n, closed under products of compatible projections and complements, that generates a countable dense set of resolutions of the identity with the property that no two compatible projections in $\hat{\mathcal{P}}_d$ are members of incompatible resolutions.*

Consider, first, the structure of $\hat{\mathcal{P}}_d$. Regard two resolutions of the identity in $\hat{\mathcal{P}}_d$ as equivalent if they are compatible. Then since compatible projections in $\hat{\mathcal{P}}_d$ can only figure in compatible resolutions, we obtain an equivalence relation. To see its transitivity, suppose $\{P_i\}$, $\{P'_j\}$ and $\{P'_j\}$, $\{P''_k\}$ are compatible pairs of resolutions in $\hat{\mathcal{P}}_d$, fix some arbitrary indices i, j, k, and consider the resolution $I = P_iP'_j + (I - P_i)P'_j + (I - P'_j)$. Since $[I - P'_j, P''_k] = 0$, we must have $[P_iP'_j, P''_k] = 0$, which in turn entails

$$P''_k P_i P'_j = P_i P'_j P''_k = P_i P''_k P'_j. \tag{2}$$

Summing this equation over j yields $[P''_k, P_i] = 0$, and hence the resolutions $\{P_i\}$, $\{P''_k\}$ must be compatible as well.

It follows that the relation of compatibility between resolutions of the identity generated by $\hat{\mathcal{P}}_d$ partitions that set into a collection of Boolean algebras that share only the projections 0 and I in common. In particular, the truth valuations on $\hat{\mathcal{P}}_d$ are full, since each of its Boolean subalgebras possesses a full set of truth valuations, and any collection of assignments of truth values to all the Boolean subalgebras of $\hat{\mathcal{P}}_d$ extends trivially to a truth valuation on the whole of $\hat{\mathcal{P}}_d$.

Moreover, it should already be clear that the set of truth valuations on $\hat{\mathcal{P}}_d$ will be sufficiently rich to recover the statistics of any quantum state by averaging over the values of the hidden variables that determine the various truth valuations.

First, given any state D, and nonzero projection $P \in \hat{\mathcal{P}}_d$, there will always be many truth valuations that map $t(P) = 1$, and we may assign the subset of hidden variables for which $t(P) = 1$ measure $\mathrm{Tr}(DP)$. This prescription also works for calculating joint probabilities of compatible projections $P, P' \in \hat{\mathcal{P}}_d$, since

$$\mathrm{Prob}_D(P = 1, P' = 1) = \mathrm{Prob}_D(PP' = 1) = \mathrm{Tr}(DPP'), \text{ and } PP' \in \hat{\mathcal{P}}_d. \tag{3}$$

So, given any Boolean subalgebra B of $\hat{\mathcal{P}}_d$, we can define a measure on the hidden variables such that

$$\mathrm{Prob}_D(P = 1) = \mathrm{Tr}(DP), \text{ for all } P \in B. \tag{4}$$

Since the Boolean subalgebras of $\hat{\mathcal{P}}_d$ are disjoint apart from the elements 0 and I, which have measure 0 and 1 respectively under this definition, the product measure defined by (4) and

$$\begin{aligned}\mathrm{Prob}_D(P = 1, P' = 1) &= \mathrm{Prob}_D(P = 1)\mathrm{Prob}_D(P' = 1),\\ &\qquad \text{for all incompatible } P, P' \in \hat{\mathcal{P}}_d \quad (5)\end{aligned}$$

gives a consistent definition of a measure on the hidden variables and reproduces all the predictions of quantum mechanics concerning projection valued measurements.

To sum up, then, the existence of the set $\hat{\mathcal{P}}_d$ defeats the practical possibility of falsifying non-contextual hidden variables on either nonstatistical or statistical grounds.

The same goes for falsifying classical logic. Following Kochen and Specker, let us call a set of projections $\mathcal{P}$ a *partial Boolean algebra* if it is closed under products of compatible projections and complements, and a truth valuation $t : \mathcal{P} \to \{0, 1\}$ will be called a *two-valued homomorphism* if t also preserves compatible products and complements (i.e., $t(PP') = t(P)t(P')$ and $t(P^{\perp}) = 1 - t(P)$). In their Theorem 0, Kochen and Specker establish that if $\mathcal{P}$ possess a full set of two-valued homomorphisms, then $\mathcal{P}$ is imbeddable into a Boolean algebra. Since $\hat{\mathcal{P}}_d$ is closed under the relevant operations, it is a partial Boolean algebra, and it is clear that all its truth valuations will in fact be two-valued homomorphisms. Thus $\hat{\mathcal{P}}_d$ can always be imbedded into a Boolean algebra, and this, from a practical point of view, nullifies any possible argument for a quantum logical conception of properties.

In Section **4**, we rule out falsifications of non-contextual models based on generalized observables represented by POV measures. Let $\mathcal{A}$ be a set of positive operators on H_n, and consider all the positive operator (PO) decompositions of the identity that $\mathcal{A}$ generates, i.e., decompositions $\sum_i A_i = I$ with $\{A_i\} \subseteq \mathcal{A}$. Since $\sum_i A_i = I$ does not entail $A_i A_j = 0$ for $i \neq j$, there can be more members of $\{A_i\}$ than the dimension, n, of the space, and the POs in a resolution of the identity need not bc mutually compatible. Still, we can ask the analogous question: does there exist a truth function $t : \mathcal{A} \to \{0, 1\}$ satisfying

$$\sum_i t(A_i) = 1 \text{ whenever } \sum_i A_i = I \text{ and } \{A_i\} \subseteq \mathcal{A}? \qquad (6)$$

Moreover: does $\mathcal{A}$ possess enough truth valuations to recover the statistical predictions, prescribed by any state D, that pertain to the members of $\mathcal{A}$? Again, we show the answer is 'Yes' for some sets $\mathcal{A}$ containing countable dense sets of *finite* PO resolutions of the identity. By this we mean (just as in the projective resolution case) that for any k-length PO decomposition $\sum_i^k A_i = I$ and $\epsilon > 0$, there is another k-length PO decomposition $\sum_i^k A_i' = I$ such that $|A_i - A_i'| < \epsilon$ for all i and $\{A_i'\} \subseteq \mathcal{A}$.

Note that from the point of view of practically performable measurements, there is no need to consider infinite PO resolutions of the identity $\sum_i^{\infty} A_i = I$ (which, of course, exist even in finite dimensions). The reason is that, for any state D of the system, a

POV measurement amounts to ascertaining the values of the numbers $\mathrm{Tr}(DA_i)$. Since these numbers normalize to unity, only a finite subset of them are measurable to within any finite precision. Thus any infinite POV measurement is always practically equivalent to a finite one. Similarly, any von Neumann measurement of an observable is always practically equivalent to the measurement of an observable with finite spectrum. For this reason, we have not sought to generalize any of our results to infinite-dimensional Hilbert spaces, though it might well be of theoretical interest to do so.

Specifically, in Section **4** we shall establish:

> **Theorem 2** *There exists a set of POs $\hat{A}_d$ on H_n that generates a countable dense set of finite PO resolutions of the identity with the property that no two resolutions share a common PO.*

Though this result is weaker than its analogue for projections in Theorem 1, it still insulates non-contextual models of POV measurements from falsification. Because the resolutions generated by $\hat{A}_d$ fail to overlap, the truth values for POs within a resolution may be set quite independently of the values assigned to POs in other resolutions. Thus, it is clear that there will be sufficiently many such valuations to recover the statistics of any density operator D. And, as before, we can suppose that any purported POV measurement, of any length k, actually corresponds to a POV measurement corresponding to a k-length PO resolution (within the precision range of the measurement) that lies in $\hat{A}_d$.

3 Non-contextual hidden variables for PV measures

Our goal in this section is to establish Theorem 1. Let H_n be an n-dimensional Hilbert space, and denote an (ordered) orthonormal basis of H_n by

$$\langle e_i \rangle = \{e_1, e_2, \ldots, e_n\}. \tag{7}$$

Let M be the metric space whose points are orthonormal bases, with the distance between two bases $\langle e_i \rangle$ and $\langle e_i' \rangle$ given by $|I - U|$, where

U is the unitary operator mapping $e_i \mapsto e'_i$ for all i. Next, consider $U(n)$, the unitary group on H_n (endowed with the operator norm topology), and recall that $U(n)$ is compact. Fix a reference point $\langle e_i \rangle \in M$, and consider the mapping $\varphi : M \mapsto U(n)$ defined by $\varphi(\langle e'_i \rangle) = U$, where U is the unitary operator that maps $e_i \mapsto e'_i$ for all i. Evidently φ is a homeomorphism, thus M is a compact, complete, separable metric space.

Call two points $\langle e_i \rangle, \langle e'_i \rangle \in M$ *totally incompatible* whenever *every* projection onto a subspace generated by a nonempty proper subset of the vectors $\langle e_i \rangle$ is incompatible with *every* projection onto a subspace generated by a nonempty proper subset of the vectors $\langle e'_i \rangle$. We shall need to make use of the following:

Lemma 1: For any finite sequence

$$\langle e_i^{(1)} \rangle, \ldots, \langle e_i^{(m-1)} \rangle \in M, \tag{8}$$

the subset of points, $T^{(m)}$, that are totally incompatible with all members of the sequence, is dense in M.

Proof: Let the indices k and l range over the values of some fixed enumeration of the (proper) subsets of $\{1, \ldots, n\}$, let $P_k^{(j)}$ denote the projection onto the subspace generated by the kth subset of $\langle e_i^{(j)} \rangle$ (for $j = 1, ..., m-1$), and for an arbitrary unlabeled point $\langle f_i \rangle \in M$, let P_l be the projection onto the subspace generated by the lth subset of $\langle f_i \rangle$. Define:

$$I_{kl}^{(j)} \stackrel{\text{def}}{=} \left\{ \langle f_i \rangle \in M : [P_l, P_k^{(j)}] \neq 0 \right\}. \tag{9}$$

Clearly $T^{(m)}$ is just the finite intersection of all the sets of form $I_{kl}^{(j)}$ over all j, k, l. Now the intersection of any two open dense sets in M is again an open dense set. So if we can argue that each $I_{kl}^{(j)}$ is both open and dense, then it will follow that $T^{(m)}$ is dense in M.

So fix j, k, l once and for all. To see that $I_{kl}^{(j)}$ is open, pass to its complement $\overline{I}_{kl}^{(j)}$, and consider any Cauchy sequence $\{\langle f_i^{(p)} \rangle\}_{p=1}^{\infty} \subseteq \overline{I}_{kl}^{(j)}$ with limit $\langle f_i \rangle \in M$. We must show $\langle f_i \rangle \in \overline{I}_{kl}^{(j)}$, i.e., that $[P_l, P_k^{(j)}] = 0$. By hypothesis, $[P_l^{(p)}, P_k^{(j)}] = 0$ for all p, and $\langle f_i^{(p)} \rangle \rightarrow$

$\langle f_i \rangle$. Let U_p be the unitary operator mapping $f_i \mapsto f_i^{(p)}$ for all i. Then $P_l^{(p)} = U_p P_l U_p^{-1}$ for all p, $U_p \to I$ (in operator norm), and we have:

$$\begin{aligned} 0 &= \lim_{p\to\infty} [P_l^{(p)}, P_k^{(j)}] \\ &= \lim_{p\to\infty} \left[U_p P_l U_p^{-1}, P_k^{(j)}\right] = [P_l, P_k^{(j)}]. \end{aligned} \tag{10}$$

To see that $I_{kl}^{(j)}$ is dense, fix an arbitrary point $\langle f_i \rangle \in M$, and arbitrary $\epsilon > 0$. We must show that one can always find a unitary U such that:

$$|I - U| < \epsilon \text{ and } [UP_l U^{-1}, P_k^{(j)}] \neq 0; \tag{11}$$

for then the point $\langle Uf_i \rangle$ must lie inside $I_{kl}^{(j)}$ and within ϵ of $\langle f_i \rangle$. Since j, k, l are all fixed, we are free to set $P = P_l$ and $Q = P_k^{(j)}$ for simplicity, bearing in mind that $P, Q \neq 0$ or I. To establish (11), then, all we need to show is that assuming

$$|I - U| < \epsilon \;\; \Rightarrow \;\; [UPU^{-1}, Q] = 0 \tag{12}$$

leads to a contradiction.

First, we dispense with the case $P = Q$. Consider the one parameter group of unitaries $U_t \equiv e^{itH}$ where H is self-adjoint. Since $Q \neq 0, I$, we may suppose that $[H, Q] \neq 0$. By (12), $U_t Q U_{-t}$ and Q commute for all sufficiently small t, in which case we may write

$$Q = A_t + B_t, \;\; U_t Q U_{-t} = A_t + C_t, \tag{13}$$

where A_t, B_t, and C_t are pairwise orthogonal projections. Then,

$$0 = \lim_{t\to 0} |U_t Q U_{-t} - Q| = \lim_{t\to 0} |B_t - C_t|. \tag{14}$$

By (14), we may choose a $\delta > 0$ so that $|B_t - C_t| < \frac{1}{2}$ for all $t < \delta$. If B_t were nonzero for some $t < \delta$, then we could choose a unit vector e in the range of B_t, and in that case we would have $\|(B_t - C_t)e\| = \|e\| = 1$. However, this contradicts the fact that $|B_t - C_t| < \frac{1}{2}$. Thus, in fact $B_t = 0$ for all $t < \delta$, and by symmetry $C_t = 0$ for all $t < \delta$. Hence, for all $t < \delta$, $U_t Q U_{-t} = A_t = Q$, i.e., $[U_t, Q] = 0$. However, since $\lim_{t \to 0} t^{-1}(U_t - I) = iH$, any operator that commutes with all U_t in a neighborhood of the identity must commute with H. Thus $[H, Q] = 0$, contrary to hypothesis.

Next, consider the case of general P and Q ($\neq 0, I$). Since $[P, Q] = 0$, we may write $Q = A + B$, $P = A + C$, where A, B and C are pairwise orthogonal projections. Without loss of generality, we may assume $A \neq 0$ (i.e., $PQ \neq 0$); for if not, then we may replace P by $I - P$ (in order to guarantee $PQ = A \neq 0$), and under that replacement (12) continues to hold. Similarly, we may assume that $A + B + C \neq I$; for if not, we could replace Q by $I - Q$. Since neither $A + B + C$ nor A equals 0 or I, there is a self-adjoint H' such that $[H', B] = [H', C] = 0$ but $[H', A] \neq 0$. Defining $U_t \equiv e^{itH'}$, we have that $[U_t P U_{-t}, Q] = [U_t A U_{-t}, A]$ for all t. Thus (12) implies:

$$|I - U_t| < \epsilon \;\Rightarrow\; [U_t A U_{-t}, A] = 0, \tag{15}$$

which, in turn, entails the contradiction $[H', A] = 0$ by the argument of the previous paragraph (with H' in place of H, and A in place of Q). *QED*.

Proposition 1: There is a countable dense subset of M whose members are pairwise totally incompatible.

Proof: Since M is separable, it possesses a countable dense set $T = \{\langle e_i^{(1)}\rangle, \ldots, \langle e_i^{(m-1)}\rangle, \ldots\}$. If, in moving along this sequence, some point $\langle e_i^{(m)}\rangle$ were found to be not totally incompatible with all previous members of the sequence, then, by Lemma 1, we could always discard $\langle e_i^{(m)}\rangle$ and replace it with a new point $\langle \hat{e}_i^{(m)}\rangle \in T^{(m)}$. The replacement point $\langle \hat{e}_i^{(m)}\rangle$ *will* be totally incompatible with all previous members, and (since $T^{(m)}$ is dense) it can always be chosen to lie within a distance 2^{-m} of $\langle e_i^{(m)}\rangle$. Moving down the sequence T, and replacing points in this way as many times as needed, we obtain a new countably infinite sequence $\hat{T}$ that is pairwise totally incompatible. And $\hat{T}$ is itself dense. For in every ball $B(p, \epsilon) = \{q : d(p, q) < \epsilon\}$ around any point $p \in M$ there must be infinitely many members of the dense set T. If one of those members did not need replacing, then clearly $B(p,\epsilon)$ contains an element of $\hat{T}$. But this must also be true if all of them needed replacing, because the replacement points $\langle \hat{e}_i^m \rangle$ lie closer and closer to $\langle e_i^m \rangle$ as $m \to \infty$. *QED*.

Now we can complete the proof of Theorem 1. Let

$$\{\langle \hat{e}_i^{(1)} \rangle, \langle \hat{e}_i^{(2)} \rangle, \ldots, \}$$

be the dense subset of Proposition 1. Let the k in $\hat{P}_k^{(m)}$ (the projection onto the span of the kth subset of $\langle \hat{e}_i^{(m)} \rangle$) now range over an enumeration of *all* the subsets of $\{1, \ldots, n\}$ (including the empty set, corresponding to $k = 0$ and $\hat{P}_0^{(m)} = 0$, and the entire set, corresponding to $k = 2^n$ and $\hat{P}_{2^n}^{(m)} = I$). Define:

$$B_m \stackrel{\text{def}}{=} \left\{ \hat{P}_k^{(m)} : k = 1, \ldots, 2^n \right\}, \quad \hat{\mathcal{P}}_d \stackrel{\text{def}}{=} \bigcup_{m=1}^{\infty} B_m. \tag{16}$$

Clearly each B_m is a maximal Boolean algebra. Moreover, Proposition 1 assures us that any two (nontrivial) compatible projections $P, P' \in \hat{\mathcal{P}}_d$ must lie in the same B_m, as well as all the resolutions of the identity in which P, P' figure. Thus compatible projections in $\hat{\mathcal{P}}_d$ only appear in compatible resolutions. Trivially, $\hat{\mathcal{P}}_d$ is closed under compatible products and complements, since each B_m is so closed, and the projections contained in different B_m's (excepting 0 and I) are all incompatible. Finally, since $\{\langle \hat{e}_i^{(1)} \rangle, \langle \hat{e}_i^{(2)} \rangle, \ldots, \}$ is dense in M, it is immediate that $\hat{\mathcal{P}}_d$ generates a dense set of resolutions of the identity, and Theorem 1 is proved. Note, finally, that by specifying from the outset a particular countable dense subset of $U(n)$, all our arguments in the proof can be made constructively. (In particular, the argument for (9) could have been given directly, rather than via a *reductio ad absurdum* from (10).)

4 Non-contextual hidden variables for POV measures

We turn, next, to establish Theorem 2. Let $\mathcal{O}$ be the set of all operators on H_n, endowed with the operator norm topology, and let $\mathcal{O}^+$ denote the PO's on H_n, a closed subset of $\mathcal{O}$. Fix, once and for all, some basis $B \subseteq H_n$, and consider the matrix representations of all operators in $\mathcal{O}$ relative to B. Define $\mathcal{Q}_C$ to be the set of all operators with complex rational matrix entries (relative to B), and $\mathcal{Q}_C^+$ to be the subset of positive operators therein. Clearly $\mathcal{Q}_C$ is dense in $\mathcal{O}$. It follows that $\mathcal{Q}_C^+$ is dense in $\mathcal{O}^+$. For consider any $A \geq 0$. Then there is an $X \in \mathcal{O}$ such that $A = X^*X$. Since there is a sequence $\{X_m\}_{m=1}^{\infty} \subseteq \mathcal{Q}_C$ converging to X, $\{X_m^* X_m\}_{m=1}^{\infty}$ is a sequence in $\mathcal{Q}_C^+$ converging to A. Next, define $\mathcal{A}$ to be the subset of positive operators in $\mathcal{Q}_C^+$ whose (complex rational) matrix entries are all *nonzero*. To see that $\mathcal{A}$ is also dense in $\mathcal{O}^+$, it suffices to observe that it is dense in $\mathcal{Q}_C^+$. So consider any $A \in \mathcal{Q}_C^+$, choose any $A' \in \mathcal{A}$, and let $\{t_m\}_{m=1}^{\infty}$ be a sequence of positive rationals tending to 0. Then clearly $A + t_m A' \to A$, each $A + t_m A' \in \mathcal{Q}_C^+$ (since the positive operators form a convex cone, and the rationals a field), and at most n^2 of the operators $\{A + t_m A'\}_{m=1}^{\infty}$ can have a zero matrix entry (since the operators A and A' are fixed).

We now need to establish:

Lemma 2: For any k, $\mathcal{A}$ generates a dense set of k-length PO resolutions of the identity.

Proof: Let $\sum_{i=1}^{k} A_i = I$ be any k-length PO resolution of the identity, and fix $\epsilon > 0$. Choose a rational $r \in (0, \epsilon)$ and set $\delta = r/(5+k) > 0$. Then, since $\mathcal{A}$ is dense in $\mathcal{O}^+$, we may choose k POs $\{A_i'\}_{i=1}^{k} \subseteq \mathcal{A}$ such that $|A_i - A_i'| < \delta$ for all i. With $H = \sum_{i=1}^{k} A_i'$, observe that

$$-\delta I < -|A_i' - A_i| I \leq A_i' - A_i \leq |A_i' - A_i| I < \delta I, \tag{17}$$

which, summed over $i = 1, \ldots, k$, yields

$$-\delta k I < H - I < \delta k I. \tag{18}$$

Next, introduce the new positive operators:

$$A_i'' = (1+k\delta)^{-1}\left[A_i' + t_i\left((1+k\delta)I - H\right)\right], \tag{19}$$

for all i. Here the t_i are positive rationals obeying $\sum_{i=1}^{k} t_i = 1$ and $t_i < 2/k$, and are chosen so that all the matrix entries of all the A_i'' are nonzero. Such a choice can always be made, since for each fixed i there are only finitely many choices one can make for t_i (in fact, at most n^2) such that a matrix entry of A_i'' can vanish. Each A_i'' is positive, since $A_i' \geq 0$, $(1+k\delta)I - H > 0$ (by (18)), and the positive operators form a convex cone. Each A_i'' is also complex rational (hence lies in $\mathcal{Q}_C^+$), since each A_i' and, therefore, H is, as is I (obviously), and all of k, δ, t_i are rational. Thus, $\{A_i''\}_{i=1}^{k} \subseteq \mathcal{A}$. Summing (19) over $i = 1, \ldots, k$ reveals that $\sum_{i=1}^{k} A_i'' = I$. Finally, note that since

$$|H - I| = \inf\{\lambda > 0 : -\lambda I \leq H - I \leq \lambda I\}, \tag{20}$$

(18) entails $|H - I| \leq \delta k$. And, since $\sum_{i=1}^{k} A_i = I$, $|A_i| \leq 1$ for all i. With latter inequalities, the triangle inequality, $t_i < 2/k$, and the inequalities of the previous paragraph, we obtain:

$$\begin{aligned}
|A_i - A_i''| &\leq \left|(1+k\delta)^{-1}\left[A_i' + t_i\left((1+k\delta)I - H\right)\right] - A_i\right| \\
&\leq (1+k\delta)^{-1}|A_i' - (1+k\delta)A_i| \\
&\quad + t_i(1+k\delta)^{-1}|(1+k\delta)I - H| \\
&\leq (1+k\delta)^{-1}|A_i' - A_i - k\delta A_i| \\
&\quad + t_i(1+k\delta)^{-1}|I - H + k\delta I| \\
&\leq (1+k\delta)^{-1}(\delta + k\delta) + t_i(1+k\delta)^{-1}(\delta k + k\delta) \\
&\leq \delta(5+k)/(1+k\delta) < \delta(5+k) = r < \epsilon,
\end{aligned} \tag{21}$$

for all i. *QED.*

Now since $\mathcal{A}$ consists only of complex rational POs, it is countable. The set of all finite subsets of a countable set is itself countable. Thus $\mathcal{A}$ can include at most countably many finite PO decompositions of the identity. Let the variable m range over an enumeration of these resolutions, denoting the mth resolution, which will have some length k_m, by $\{A_i^{(m)}\}_{i=1}^{k_m}$. For each m, define the unitary operator U_m to be given by the diagonal matrix $\mathrm{diag}(e^{i\theta_m}, 1, \ldots, 1)$ relative to the basis B, where

$$\sin\theta_m = (\pi/4)^m, \ \ \cos\theta_m = (1 - (\pi/4)^{2m})^{1/2}, \tag{22}$$

and for definiteness we take the positive square root. For each m, define a new k_m-length PO resolution $\{\hat{A}_i^{(m)}\}_{i=1}^{k_m}$ by

$$\hat{A}_i^{(m)} = U_m A_i^{(m)} U_m^{-1}, \text{ for all } i = 1, \ldots, k_m. \tag{23}$$

Then we have:

Proposition 2: No two of the PO resolutions $\{\hat{A}_i^{(m)}\}_{i=1}^{k_m}$ (for different m) share a common PO.

Proof: Assuming that for some i, j and m, p we have $\hat{A}_i^{(m)} = \hat{A}_j^{(p)}$, we must show that $m = p$. By (23),

$$U_m A_i^{(m)} U_m^{-1} = U_p A_j^{(p)} U_p^{-1}. \tag{24}$$

Let (x_{ab}) and (y_{ab}) be the nonzero complex rational matrix coefficients (relative to B) of $A_i^{(m)}$ and $A_j^{(p)}$, respectively. Using the definition of U_m, (24) entails, in particular, that $x_{12}e^{i\theta_m} = y_{12}e^{i\theta_p}$. Since $x_{12} \neq 0$, we may write:

$$e^{i(\theta_m - \theta_p)} = c, \tag{25}$$

where c is a complex rational. Equating the real parts of (25), one obtains

$$\cos\theta_m \cos\theta_p = r - \sin\theta_m \sin\theta_p, \tag{26}$$

where r $(= \Re(c))$ is rational. Squaring both sides of (26), inserting the expressions (22), and rearranging, yields:

$$2r\left(\frac{\pi}{4}\right)^{m+p} - \left(\frac{\pi}{4}\right)^{2p} - \left(\frac{\pi}{4}\right)^{2m} = r^2 - 1. \tag{27}$$

The transcendentality of π requires that the coefficient of any given power of π in this equation must be zero. This cannot happen unless $m = p$, since if $m \neq p$, then π^{m+p}, π^{2p}, and π^{2m} are three distinct powers of π, and the latter two powers occur with nonzero coefficient. *QED.*

We finish with the argument for Theorem 2. All that remains to show is that the countable collection of finite PO resolutions contained in

$$\hat{\mathcal{A}}_d \stackrel{\text{def}}{=} \bigcup_{m=1}^{\infty} \{\hat{A}_i^{(m)}\}_{i=1}^{k_m} \tag{28}$$

is dense. Fix k and let $\{A_i\}_{i=1}^{k}$ be an arbitrary k-length resolution. By Lemma 2, we can find k-length resolutions in $\mathcal{A}$ arbitrarily close to $\{A_i\}_{i=1}^{k}$. Since the unitary operators defined by (22) satisfy $U_m \to I$, by Proposition 2 we can find k-length resolutions in $\hat{\mathcal{A}}_d$ arbitrarily close to any k-length resolution in $\mathcal{A}$. This completes the proof.

5 Discussion

To many physicists, the fact that quantum theory can most elegantly be expressed in a radically non-classical language makes explicit demonstrations of its non-classicality essentially redundant. This paper is addressed not to them, but to those who, like us, are interested in what we can establish for certain on the question. As we have already stressed, we hold no particular brief for non-contextual hidden variable theories. However, our conclusion, in the light of the results of Meyer (1999) and Kent (1999) and the constructions above, is that there is no truly compelling argument establishing that non-relativistic quantum mechanics describes classically inexplicable physics. Only when quantum theory and relativity are combined can really compelling arguments be mounted against the possibility of classical simulation, and then — so far as is presently known — only against the particular class of simulations defined by local hidden variable theories.

One feature of our toy models is that the non-contextual hidden variables at any given time define outcomes only for one measurement, not for a sequence of measurements at separate times. This is adequate for our purposes, since we are interested only in establishing the point that, despite the Kochen-Specker theorem, non-contextual hidden variables can simulate the measurement process in non-relativistic quantum mechanics.

As our model has no dynamics, it cannot supply a proper account of processes extended in time. However, our discussion could be extended to a treatment of sequential projective or positive operator valued measurements assuming there is no intervening evolution, simply by assuming that the hidden variables, like the state vector, undergo a discontinuous change after a measurement, so that the probability distribution of the post-measurement hidden variables corresponds to that defined by the new state vector. If such postulates were adopted, they would, in the spirit of the hidden variables program, need to be seen as approximations to be justified by a more fundamental theory with complete dynamics. A complete theory would also, of course, need to describe successive measurements in which the intervening evolution of the quantum state is non-trivial.

Constructing a dynamical non-contextual hidden variable theory with these properties goes beyond our ambitions here, and it should be stressed that we have not shown that our toy models, or any non-contextual hidden variable theories, can in fact be extended to a viable dynamical non-contextual hidden variable theory. It is sufficient for our argument here to note that Kochen-Specker constraints *per se* give no reason to think that such a theory cannot be constructed. In fact, though, we see no argument against the possibility of such a theory, given that it is possible to give a precise collapse dynamics for quantum states (Ghirardi *et al.* 1986).

From another viewpoint, it might be argued that our proofs and constructions establish more than is strictly necessary. While we have defined a hidden variable simulation of quantum theory by constructing a dense subset of the set of projective decompositions with the property that no two compatible projections belong to incompatible resolutions, this last property is not necessary. To give a trivial example, one could also define a simulation on the basis of a dense set of projections exactly one of which belongs to two (or more) incompatible resolutions. Similar comments apply to the analysis for positive operator decompositions.

Our response would simply be that the constructions we give do the job. They are certainly not unique, and there may well be other constructions which differ in interesting ways, or even perhaps define more natural non-contextual hidden variable models.

A related argument that might also be made is that the constructions of Meyer (1999) and Kent (1999) already imply the simulability of quantum theory by non-contextual hidden variables, since they each describe a truth valuation t on a dense set in which both truth values occur densely (i.e., for any projection or positive operator mapped to 1 by t, there is another one arbitrarily close to it mapped to 0). One could imagine a model in which the particle, confronted by a measuring apparatus set with a particular precision range, first calculates the approximate probability that it should produce a '1' by evaluating the expectation in its quantum state of some projection randomly chosen from within the precision range, and then goes through some deterministic algorithm to *choose* a (generally different) projection from the range for which it will reveal its predetermined measurement value. In this picture, it is left to the deterministic algorithm to simulate quantum statistics by choosing projections of value '0' or '1' in the right proportions.

Such models cannot be logically excluded, but they seem to us overly baroque: hidden variable theories in which the effects of the system and apparatus hidden variables can be separated seem to us cleaner and simpler than theories which rely on some conspiratorial interaction between those variables at the point of measurement.

Bibliography

Aspect, A. *et al.* (1981), 'Experimental tests of realistic local theories via Bell's theorem', *Physical Review Letters* **47**, 460–3.

Bell, J. (1964), 'On the Einstein-Podolsky-Rosen paradox', *Physics* **1**, 195–200.

Bell, J. S. (1966), 'On the problem of hidden variables in quantum mechanics', *Reviews of Modern Physics* **38**, 447–52.

Bub, J. (1997), *Interpreting the Quantum World*, Cambridge Univ. Press, Cambridge. Chap. 3.

Clauser, J. F. *et al.* (1969), 'Proposed experiment to test local hidden-variable theories', *Physical Review Letters* **23**, 880–4.

Ghirardi, G. *et al.* (1986), 'Unified dynamics for microscopic and macroscopic systems', *Physical Review D* **34**, 470–91.

Gleason, A. M. (1957), 'Measures on the closed subspaces of a Hilbert space', *Journal of Mathematics and Mechanics* **6**, 885–93.

Kent, A. (1999), 'Non-contextual hidden variables and physical measurements', *Physical Review Letters* **83**, 3755–7.

Kochen, S. & Specker, E. P. (1967), 'The problem of hidden variables in quantum mechanics', *Journal of Mathematics and Mechanics* **17**, 59–87.

Mermin, N. D. (1993), 'Hidden variables and the two theorems of John Bell', *Review of Modern Physics* **65**, 803–15.

Meyer, D. (1999), 'Finite precision measurement nullifies the Kochen-Specker theorem', *Physical Review Letters* **83**, 3751–4.

Peres, A. (1995), *Quantum Theory: Concepts and Methods*, Kluwer, Boston. Chap. 7, pp. 187–211.

Pitowsky, I. (1983), 'Deterministic model of spin and statistics', *Physical Review D* **27**, 2316–26.

Pitowsky, I. (1985), 'Quantum mechanics and value definiteness', *Philosophy of Science* **52**, 154–6.

Redhead, M. L. G. (1989), *Incompleteness, Nonlocality, and Realism*, Clarendon Press, Oxford. Chap. 5.

Tittel, W. *et al.* (1998), 'Violation of Bell inequalities by photons more than 10 km apart', *Physical Review Letters* **81**, 3563–6.

Weihs, G. *et al.* (1998), 'Violation of Bell's inequality under strict Einstein locality conditions', *Physical Review Letters* **81**, 5039–43.

Zimba, J. & Penrose, R. (1993), 'On Bell non-locality without probabilities: more curious geometry', *Studies in History and Philosophy of Science* **24**, 697–720.

Chapter 14

The subtleties of entanglement and its role in quantum information theory

1 Introduction

For years, since the EPR 'paradox' (1935) and Schrödinger's (1935, 1936) work on entanglement in the 1930s, philosophers of science have struggled to understand quantum correlations. By the end of the 1960s, Bell's theorem (1964) became widely recognized as establishing that these correlations cannot be explained through the operation of local common causes. Yet it was also clear that, by themselves, quantum correlations cannot be exploited to transmit information between the locations of entangled systems. Many metaphors were developed by philosophers in the 1970s and 1980s to characterize these apparently nonlocal yet curiously benign correlations. They were variously taken to involve 'passion-at-a-distance', 'nonrobustness', 'relational holism', and 'randomness in harmony' (see, respectively, the contributions of Shimony, Redhead, Teller and Fine to Cushing and McMullin (1989)). However, the impact of these philosophical discussions on the consciousness of the practicing physicist was virtually negligible. That quantum nonlocality was a fact of nature worth reckoning with was treated at about the same level of seriousness that scientists regard evidence for telepathy. In his *Will to Believe*, William James described the scientist's attitude towards telepathy thus:

> Why do so few 'scientists' even look at the evidence for telepathy, so called? Because they think, as a leading biologist, now dead, once said to me, that even if such a thing were true, scientists ought to band together to keep it suppressed and concealed. It would undo the uniformity of Nature and all sorts of other things without which scientists cannot carry on their pursuits. But if this very man had been shown something which as a scientist he might *do* with telepathy, he might not only have examined the evidence, but even have found it good enough.

Since the beginning of the 1990s, when the interest in quantum information theory began to explode, physicists' attitudes have changed dramatically, and James' observations about telepathy now seem downright prophetic! Witness the following passage from the introduction to Popescu and Rohrlich's (1998) 'The Joy of Entanglement':

> ... today, the EPR paradox is more paradoxical than ever and generations of physicists have broken their heads over it. Here we explain what makes entanglement so baffling and surprising. But we do not break our heads over it; we take a more positive approach to entanglement. After decades in which everyone talked about entanglement but no one did anything about it, physicists have begun to *do* things with entanglement.

What entanglement is now known to *do* (among other things) is increase the capacity of classical communication channels — so-called 'entanglement-assisted communication'. No longer is entanglement the *deus ex machina* philosophers' metaphors would lead us to believe. Indeed, physicists have now developed a rich theory of entanglement storage and retrieval with deep analogies to the behavior of heat as a physical resource in classical thermodynamics.

My aim in this paper is a modest one. I do not have any particular thesis to advance about the nature of entanglement, nor can I claim novelty for any of the material I shall discuss (much of which is now readily accessible through excellent texts: Preskill (1998), Lo *et al.* (1998), Bouwmeester *et al.* (2000), and Nielsen & Chuang (2000). My aim is simply to raise some questions about entanglement that spring naturally from certain developments in quantum information theory and are, I believe, worthy of serious consideration by philosophers of science. In Section 2, I shall discuss different senses in which a bipartite quantum state can be said to be 'nonlocal'. All the different senses collapse into a single concept when the state is pure, but conceptually novel questions arise when the state at issue is mixed. In Section 3, I will limit my discussion to the two paradigm cases of entanglement-assisted communication: dense coding and teleportation. Finally, in Section 4, I shall discuss different kinds/degrees of entanglement and give a whirlwind tour of the basics of 'entanglement thermodynamics'. Space limitations force me to assume some prior acquaintance with elementary quantum mechanics, though I have endeavored to keep my discussion as self-contained and nontechnical as possible.

2 Different manifestations of nonlocality

Suppose we have two spatially separated observers, Alice and Bob, and a source that creates identically prepared pairs of particles. One member of each pair goes to Alice and the other to Bob. If Alice and Bob measure various observables on their particles, there will, in general, be correlations between the measurement results they obtain. The quantum state of each particle pair will be represented by a density operator ρ on a tensor product $H^A \otimes H^B$ of two Hilbert spaces, with the predicted correlation between any given Alice observable A and Bob observable B given by $\langle A \otimes B \rangle_\rho = \mathrm{Tr}[\rho(A \otimes B)]$. Call a state ρ *Bell correlated* if there are spin-type observables $A_i =$

$A_i^* = A_i^{-1}, B_i = B_i^* = B_i^{-1}, i = 1, 2$, such that

$$|\langle A_1 \otimes B_1 \rangle_\rho + \langle A_1 \otimes B_2 \rangle_\rho + \langle A_2 \otimes B_1 \rangle_\rho - \langle A_2 \otimes B_2 \rangle_\rho| > 2.$$

Then Bell's theorem tells us that if ρ is Bell correlated, its correlations cannot be simulated in any local hidden variables (LHV) model. Being Bell-correlated is, therefore, one sense in which a state — understood simply as a catalog of correlations — can be said to be 'nonlocal'.

I am well aware that some quantum physicists remain committed to the view that the absence of an LHV model for correlations does not entail that they are nonlocal, but simply 'nonclassical' (e.g., Fuchs and Peres 2000). Their escape typically involves arguing that assuming the existence of hidden variables presupposes a form of realism inappropriate in quantum theory. On the other hand, other physicists have recently taken to calculating the average number of classical bits that would need to be 'communicated between the particles' to simulate their Bell correlations (Steiner 2000, Brassard *et al.* 1999, Massar *et al.* 2001). This has apparently been done in ignorance of similar earlier work by Maudlin (1994), and it would be interesting to determine how all these results are related. But the philosophical payoff seems clear: If we can show in *purely information-theoretic terms* that no local account of Bell correlations is possible, wouldn't that show that realism is a red-herring? Or is the use of *classical* information theory as a benchmark inappropriate in the quantum context? Since in 'quantum information' theory, it is typically the medium and not the method of calculating the information content of the message that is 'quantum', I shall proceed, tentatively, on the assumption that the answer to the first question, but not the second, is 'Yes'.

A state ρ is called a *product state* just in case there are density operators ρ^A and ρ^B on H^A and H^B, respectively, such that $\rho = \rho^A \otimes \rho^B$. More generally, a state ρ is called *separable* just in case it can be written as a convex combination of product states

$$\rho = \sum_i p_i \rho_i^A \otimes \rho_i^B, \quad p_i \in [0, 1], \quad \sum_i p_i = 1.$$

Clearly the correlations of separable states can be simulated locally, because we may think of the separate quantum states ρ_i^A and ρ_i^B as *themselves* supplying the required local hidden variables, with the probabilities p_i representing our ignorance about which hidden variables obtain in a given measurement trial. Thus, a state is separable if and only if its correlations admit a *quantum* LHV model.

So Bell correlated states are nonlocal while separable states are local. What about nonseparable, or *entangled*, states: Are they always nonlocal? It certainly does not automatically follow from the fact that a state admits no quantum LHV model that it does not admit any LHV model whatsoever. However, if an entangled state is pure, it can be shown that it must be Bell correlated (Popescu and Rohrlich 1992), and so the answer is 'Yes' for pure entangled states. Not so in the mixed case. Consider the following 'Werner state' in 2×2 dimensions (Werner 1989):

$$\rho_W = \left(1 - \frac{1}{\sqrt{2}}\right)\frac{1}{4} I \otimes I + \frac{1}{\sqrt{2}} |\Psi_-\rangle\langle\Psi_-|,$$

where $|\Psi_-\rangle$ is the infamous EPR-Bohm singlet state of two spin-1/2 particles. Since spin operators are traceless, the maximum possible Bell correlation in state ρ_W is determined by $\frac{1}{\sqrt{2}}$ times the maximum Bell correlation achievable in the singlet state, which is $2\sqrt{2}$. Thus ρ_W is *not* Bell correlated. Nevertheless, ρ_W *is* entangled — not because the particular convex combination used above to define ρ_W involves the singlet state (which is, of course, not a product state), but because there is no other way to *rewrite* ρ_W as a convex combination of product states. Nevertheless, while ρ_W does not admit a quantum LHV model, Werner (1989) *was* able to construct an LHV model for its correlations.

It would be natural to conclude that since an entangled state need not be Bell correlated — and can even admit an LHV model — some entangled states are surely local. One might object to this by observing that every convex decomposition of a mixed entangled state into pure states must involve at least one entangled pure state (else the mixed state would be separable); and, therefore, since pure entangled states are nonlocal, mixed entangled states should be thought of as nonlocal too. But this reasoning runs afoul of well-known difficulties with the ignorance interpretation of mixtures. A mixed density operator like ρ_W need not result from physically mixing, in known proportions, pure states of two spin-1/2 particles. It could result, instead, from reducing the density matrix of a larger system in a pure entangled state with the spin-1/2 particles — a state in which there is no fact of the matter about what pure state the particles occupy for anyone to be ignorant about! Moreover, even if we knew that a mixed state has been produced by physically mixing pure states, there are many inequivalent ways to mix and produce the same density matrix. For example, the 'maximally mixed' product state $(I \otimes I)/4$ of two spin-1/2 particles is surely local, and can be produced by an equal mix of the four product pure states

$$|\downarrow_z\rangle \otimes |\downarrow_z\rangle,\ |\downarrow_z\rangle \otimes |\uparrow_z\rangle,\ |\uparrow_z\rangle \otimes |\downarrow_z\rangle,\ |\uparrow_z\rangle \otimes |\uparrow_z\rangle.$$

Yet $(I \otimes I)/4$ can also be produced by physically mixing, in equal proportions, the following four entangled eigenstates of the 'Bell operator':

$$\begin{aligned}|\Psi^{\pm}\rangle &= \frac{1}{\sqrt{2}}(|\downarrow_z\rangle \otimes |\uparrow_z\rangle \pm |\uparrow_z\rangle \otimes |\downarrow_z\rangle),\\ |\Phi^{\pm}\rangle &= \frac{1}{\sqrt{2}}(|\downarrow_z\rangle \otimes |\downarrow_z\rangle \pm |\uparrow_z\rangle \otimes |\uparrow_z\rangle).\end{aligned}$$

In the absence of further knowledge of the mixing process, the possibility of producing $(I \otimes I)/4$ by mixing eigenstates of the Bell operator shows only that there is a nonlocal quantum hidden variables model of the state $(I \otimes I)/4$, not that there can be no local one! So we see that it is in general fallacious to take the locality properties of the pure components of a mixture to be indicative of the locality or nonlocality of the mixture itself.

There is, however, a more interesting objection to the claim that certain entangled states, like the Werner state, are local. What if we require more of an LHV model than that it simply reproduce the correlations between Alice's and Bob's measurement outcomes? Suppose prior to making the measurements on their particles from which they calculate their correlations, we allow Alice and Bob to 'preprocess' the particles by performing arbitrary local operations on them and communicating about the results they obtain. By this is meant the following.

A nonselective local operation on Alice's particle is any 'completely positive, linear, and trace-preserving' transformation of its density operator that leaves unchanged the state of Bob's particle. Such a transformation need not be unitary. For example, it could be brought about by Alice first combining her particle with an ancilla system which has no prior entanglement with either Alice's or Bob's particle, then executing a unitary transformation on the combined particle+ancilla system, and, finally, tracing out the ancilla to get a reduced state for Alice's particle again. Alice could also perform local measurements on her particle (whose outcome statistics will, in general, only be representable by positive operators rather than projections), and communicate to Bob her results so that he can coordinate the local operations he performs on his particle with the outcomes Alice gets from hers. In particular, we can allow Alice and Bob to perform non-trace-preserving *selective* local operations, in which they drop from further consideration certain members of their initial ensemble — on the basis of certain measurements results they obtain — and communicate classically between them to ensure agreement about which particle pairs are dropped.

The remarkable thing is that it is possible, after Alice and Bob avail themselves of local operations and classical communication on an initially 'local' but entangled particle pair, for the final state of the surviving ensemble of particle pairs to be Bell correlated! (See Popescu (1995) and Mermin (1996).) As it happens, this only works for higher-dimensional Werner states, not the simple 2×2 state ρ_W we have considered. In order to 'display' ρ_W's nonlocality, we need to allow Alice, and similarly Bob, the additional freedom to perform *collective* operations on all of her particles at once (Peres 1996). These operations will still be local, insofar as they do not affect the states of Bob's particles (even if they might introduce entanglement amongst Alice's). Conventional wisdom then dictates that, after all is said (between Alice and Bob) and done (by their collective local operations), an initial entangled state that, as a result of 'preprocessing' changes to a Bell correlated one, should be understood as nonlocal. For example, Gisin (2000, p. 271) writes 'A state that is explicitly nonlocal after some local operations does not deserve the qualification of local... because reproducing all correlations [of the initial state] is not enough'. Similarly, Popescu & Rohrlich (1998, p. 41) state: 'Alice and Bob could never obtain nonlocal states from local states by using only local interactions'.

What are we to make of this conventional wisdom? It is certainly true that if we demand there exist an *extended* LHV model, not just for the correlations of the initial state, but for all of Alice and Bob's preprocessing and the Bell correlations of their final state, then the model cannot be local. Moreover, it appears we are not putting an LHV model at a disadvantage by requiring it to model all of this; for in confining Alice and Bob to the use of *local* operations on their particles and normal *causal* communication, we are not adding any 'extra nonlocality' between the particles for the model to explain. Still, what *precisely* is the argument for taking the original unprocessed entangled state to intrinsically possess nonlocal (as opposed to merely entangled) correlations?

The only way I can see to justify such a claim is if the following conjecture could be proved: If an HV model M of the correlations of an initial entangled state has an extension to — or is the restriction of — a more encompassing HV model of all the collective local operations and classical communication needed to produce a Bell correlated state, then the model M of the initial state *must already have contained nonlocal elements*. If true, this would entail that Werner's LHV model for ρ_W's correlations does not extend to any adequate model — *local or not* — of the production of the final state after (collective) preprocessing. Similarly, the truth of my conjecture would mean that Bohm's theory, which is certainly capable of reproducing all the predictions of quantum theory associated with Alice and Bob's preprocessing, yields a nonlocal dynamics between two particles in the Werner state (understood as the reduced state of a larger three particle pure state, if necessary), notwithstanding its lack of Bell correlation. However, even if my conjecture is true, I am unaware of any general argument that every entangled state must yield to local operations and classical communication and eventually produce a Bell correlated state. Moreover, we shall see in Section 4 that certain entangled states definitely resist preprocessing into *singlet* states.

Figure 14.1 summarizes the different senses of nonlocality I have discussed, and their logical relations. Note that for pure states, all distinctions in the diagram collapse. It can also be shown that the Bell correlated states of infinite-dimensional systems are generic (see Chapter 10), and therefore, for all practical purposes, the diagram collapses in that case as well.

3 Entanglement-assisted communication

I turn now to discuss two novel and conceptually puzzling practical uses of entanglement: dense coding (Bennett & Wiesner 1992) and teleportation (Bennett *et al.* 1993). Both involve forms of communication between Alice and Bob in which it *looks* like Bob is able to receive more information from Alice than she actually sent him.

In dense coding (see Figure 14.2), the aim is for Alice to convey to Bob two classical bits of information, or 'cbits' for short. They initially share a pair of spin-1/2 particles, or 'qubits', in the singlet state $|\Psi^-\rangle$. While Bob holds on to his particle, Alice has the option of doing nothing to hers, or performing one of three unitary operations given by the Pauli spin operators σ_x, σ_y, or σ_z. After exercising her option, the final joint state of the particles will either be $|\Psi^-\rangle$, $(\sigma_x \otimes I)|\Psi^-\rangle$, $(\sigma_y \otimes I)|\Psi^-\rangle$, or $(\sigma_z \otimes I)|\Psi^-\rangle$. These four states are easily seen to be mutually orthogonal (use $\sigma_x\sigma_y = i\sigma_z$ + cyclic permutations, and $\langle\Psi^-|(\sigma_{x,y,z} \otimes I)|\Psi^-\rangle = 0$) and are, in fact, the four Bell operator eigenstates (up to phase). Thus, Alice can now send her qubit to Bob, who can combine it with his own qubit and measure the Bell operator on them jointly to perfectly distinguish which of the four options Alice took.

What is 'dense' about this coding is that it *apparently* exceeds a well-known limit — called the Holevo bound — on the capacity of qubits to carry classical information. For a consequence of this

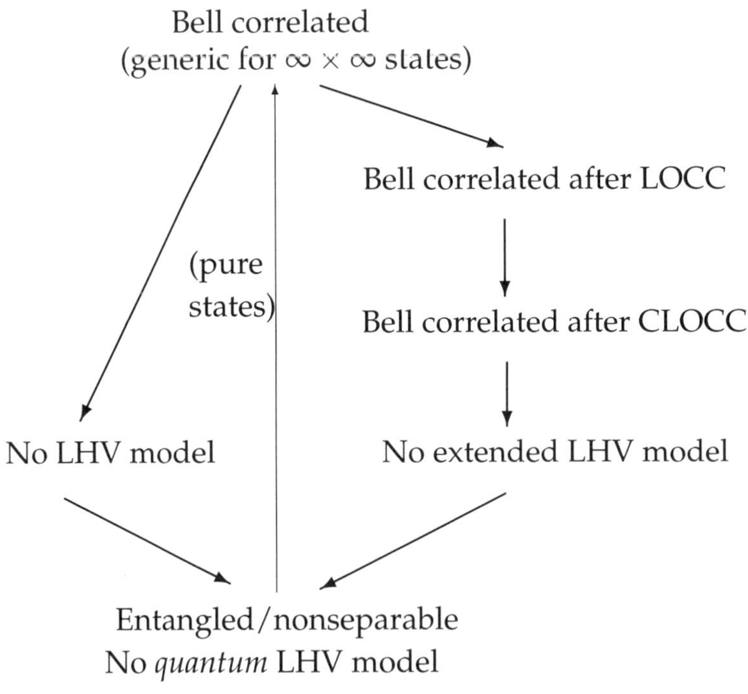

Figure 14.1: Different manifestations of nonlocality (LHV = local hidden variables, LO = local operations, CC = classical communication, LOCC = LO + CC, CLOCC = *collective* LOCC)

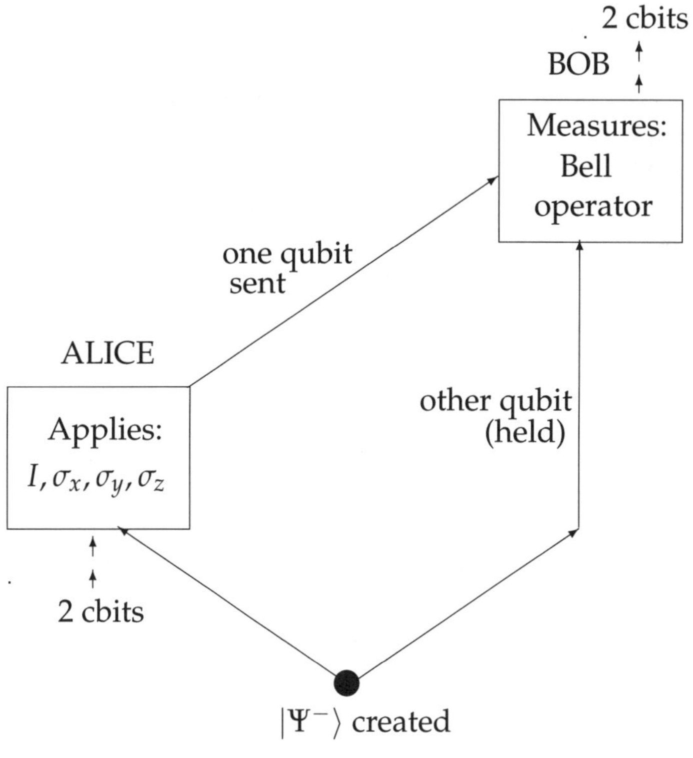

Figure 14.2: Dense coding

bound is that a state of n qubits can be used between a sender and receiver to transmit at most n cbits (Nielsen and Chuang 2000, Exercise 12.3). Yet Bob receives two cbits with only *one* qubit changing hands! On closer inspection, however, there is no contradiction. The setup presupposed for Holevo's bound says nothing about which qubits must 'change hands' but, rather, assumes only that the sender encodes her information by performing operations on a collection of n qubits, and the receiver retrieves as much of that information as possible by making measurements on that same collection. Nothing precludes the sender from operating locally only on some subset of the n 'information carrying qubits' which, if that subset is entangled with the rest, can still produce a nontrivial change in their total state. Thus, from the point of view of Holevo's bound, what Alice actually does in the dense coding is to code her two cbits into the state of *two* qubits, and Bob chooses a Bell operator measurement that allows him to retrieve as much information as he possibly can. Note, also, that Alice's trick wouldn't work if her qubit weren't entangled with Bob's. For example, replacing $|\Psi^-\rangle$ by $|\uparrow_z\rangle \otimes |\downarrow_z\rangle$ would only leave Bob, at the end of the protocol, with one of the *two* orthogonal states $|\uparrow_z\rangle \otimes |\downarrow_z\rangle, |\downarrow_z\rangle \otimes |\downarrow_z\rangle$ (up to phase), from which he can distinguish only 1 cbit about Alice's choice — consistent with the Holevo bound on the information transmittable using only a single qubit.

Though there is no outright contradiction with the Holevo bound in the entangled case, it is hard to escape the sense that the qubit in Bob's possession could not possibly carry *any* of the information he finally receives. After all, Bob has his qubit in his hands well in advance of Alice deciding what information she wants to send by choosing between the four unitary operations. The Holevo bound is cleverly silent about this, leaving us the option of filling in how the n cbits the sender encodes get 'distributed amongst' the n encoding qubits. Moreover, while the bound allows us to say that if Bob looks at either qubit in isolation he can get *up to* 1 cbit of information, it is easy to see that, in fact, absolutely no information *about Alice's unitary operation* can be obtained in that way. Should we, then, resist the temptation to fill in the story and marvel in the fact that quantum holism gives merely the *appearance* of nonlocal information transfer between the qubits without actually instantiating it?

Consider the following classical analogue of dense-coding (suggested to me by Arthur Cunningham). A source produces a pair of cards. One card, which is blank, goes to Alice; the other, on which is (randomly) written one of four ordered pairs $(x_2, y_2) = (0,0), (0,1), (10),$ or $(1,1)$, goes to Bob. Alice chooses one of these same four ordered pairs, call it (x_1, y_1), to write on her card, and then sends it to Bob. Finally, Bob combines the information from the two cards and computes the binary sum $(x_1 \oplus x_2, y_1 \oplus y_2) = (0,0), (0,1), (10),$ or $(1,1)$. In this way, Alice can communicate to Bob one of the latter four pairs, i.e., 2 cbits, by writing on her card the appropriate pair (x_1, y_1). Note that by looking at either card on its own, Bob can infer *nothing* about the 2 cbits Alice is trying to send him — no (relevant) information is carried by either card in isolation. Thus, even in the absence of quantum holism, there is no reason to think that the information carried by a communication channel must be parceled out amongst the component physical systems making up the channel. It seems that dense coding does not *mandate* a holistic metaphysics after all.

Unfortunately, this classical communication protocol is disanalogous to dense coding in one important respect. In order for Alice to choose the appropriate pair (x_1, y_1) with which to convey her 2 cbits, *she must know in advance what is written on Bob's card.* What would be the mechanism by which she could get this sort of knowledge in the quantum case? For example, Alice can learn nothing from the prior EPR correlation between the qubits, because she does not perform any measurements on her qubit! One is, therefore, tempted to conjecture that any classical simulation of dense coding, if it does not involve holistic elements, must still involve nonlocal effects of one sort or another.

In teleportation (see Figure 14.3), Alice and Bob again share a pair of spin-1/2 particles in the singlet state $|\Psi^-\rangle$, but now they swap their equipment: Alice first measures the Bell operator, and then Bob applies one of the four unitary transformations I, σ_x, σ_y, or σ_z. The aim is for Alice to transmit to Bob the state $|\phi\rangle$ of another 'ancilla' qubit. She does so by exploiting the correlations in their 'channel state' $|\Psi^-\rangle$ and sending Bob 2 cbits rather than a qubit. The initial ancilla+channel state is $|\phi\rangle \otimes |\Psi^-\rangle$. One can verify that the unitary operator which permutes the state of the ancilla with Bob's qubit can be written as

$$P = 1/2(I \otimes I \otimes I + \sigma_x \otimes I \otimes \sigma_x + \sigma_y \otimes I \otimes \sigma_y + \sigma_z \otimes I \otimes \sigma_z).$$

Therefore,

$$\begin{aligned} |\phi\rangle \otimes |\Psi^-\rangle &= -P(|\Psi^-\rangle \otimes |\phi\rangle) \text{ (using the anti-symmetry of } |\Psi^-\rangle) \\ &= -1/2[|\Psi^-\rangle \otimes |\phi\rangle + (\sigma_x \otimes I)|\Psi^-\rangle \otimes \upsilon_x|\phi\rangle \\ &\quad +(\sigma_y \otimes I)|\Psi^-\rangle \otimes \sigma_y|\phi\rangle + (\sigma_z \otimes I)|\Psi^-\rangle \otimes \sigma_z|\phi\rangle]. \end{aligned}$$

The protocol can now be read directly off this last expression. Alice first measures the Bell operator, collapsing the joint state of the ancilla qubit and hers into one of the four Bell eigenstates, say $(\sigma_y \otimes I)|\Psi^-\rangle$. As a result, Bob's qubit is left in the correlated state $\sigma_y|\phi\rangle$. Alice then communicates to Bob which of the four measurement outcomes she got, and Bob applies the corresponding unitary operation, in this case σ_y, to his qubit. After doing so, he obtains $\sigma_y^2|\phi\rangle = |\phi\rangle$ and, thereby, creates an exact replica of the ancilla qubit's state at his location.

In a sense, teleportation is a form of 'dense coding', because, while a normal classical communication channel would need to carry an infinite (or, at least, arbitrarily large) amount of information to convey to Bob the expansion coefficients of the state $|\phi\rangle$, Alice manages to get by with sending him only 2 cbits. This again raises the question of whether a convincing story can be told here about the 'flow of information' (nonlocally?!, backwards in time?!) between Alice and Bob (see Duwell 2001). But note that there is an important difference from dense coding: if there is information transfer in teleportation at all, it occurs entirely at the quantum level. A successful run of the protocol does not require that Alice know in advance the state $|\phi\rangle$ she is teleporting; nor can it end in Bob's knowing the identity of this state, because (as is well known)

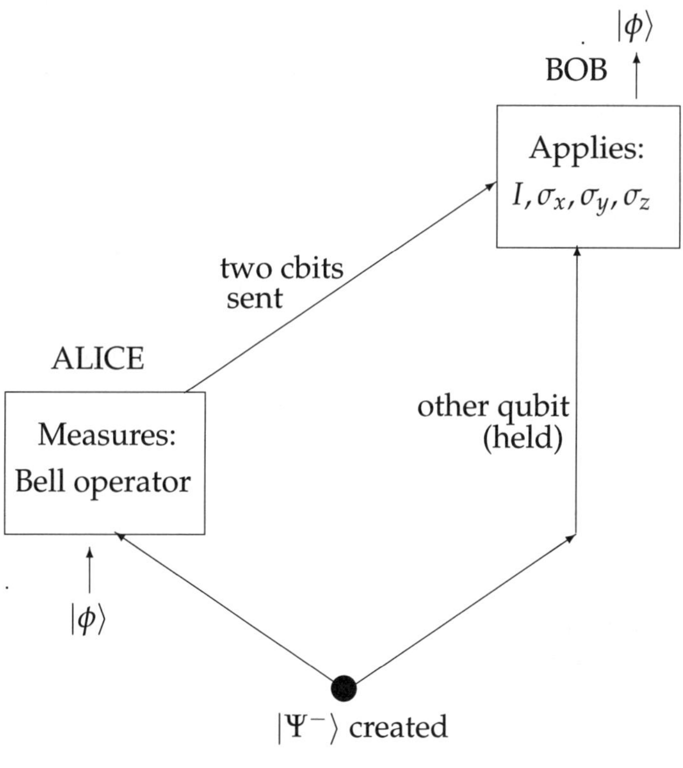

Figure 14.3: Teleportation

there is no way for Bob to determine the quantum state of a single system! Perhaps some new concept of hidden *quantum* information transfer is called for here? (See Cerf & Adami 1997, Deutsch & Hayden 2000)

We can also consider imperfect teleportation, where the state Bob ends up creating at the end of the protocol is not exactly the ancilla's state, but something close. It turns out that imperfect teleportation with a 'fidelity' higher than any classical analogue can achieve is always possible when the singlet channel state Alice and Bob share is replaced by any Bell correlated entangled state (Horodecki *et al.* 1996). However, high fidelity teleportation is *also* possible using the Werner state, which is not Bell correlated (Popescu 1994). One can either take this to be a sign that the Werner state is, after all, nonlocal, or see it as showing that explaining the success of teleportation does not require nonlocality.

This issue can also be attacked more directly by taking a closer look at the correlations of the channel state that are *actually involved* in the teleportation protocol (Żukowski 2000). LHV models will be committed to all 'Bell teleportation inequalities' of the form

$$|\langle f_1 \otimes \sigma_1 \rangle_{|\phi_1\rangle\langle\phi_1|\otimes\rho} + \langle f_1 \otimes \sigma_2 \rangle_{|\phi_1\rangle\langle\phi_1|\otimes\rho} + \langle f_2 \otimes \sigma_1 \rangle_{|\phi_2\rangle\langle\phi_2|\otimes\rho} + \langle f_2 \otimes \sigma_2 \rangle_{|\phi_2\rangle\langle\phi_2|\otimes\rho}| \leq 2,$$

where $f_{1,2}$ are bivalent functions of Alice's Bell operator, $|\phi_{1,2}\rangle$ are alternatives for the teleported state, ρ is the 'quantum channel' state, and $\sigma_{1,2}$ are spin components of Bob's particle. It can be shown that if the teleportation fidelity exceeds a certain threshold, at least one such inequality must be violated, but that no violations occur in the Werner state (Clifton & Pope 2001). While the verdict is still out, this suggests that the ability of a quantum state to facilitate nonclassical teleportation should not be taken as a litmus test for the nonlocality of the state.

4 Entanglement thermodynamics

There is one further trick that can be performed with teleportation that would appear particularly difficult to classically simulate: entanglement swapping. Suppose the ancilla qubit whose state Alice teleports does not actually have a definite state of its own, but is entangled via the singlet state $|\Psi^-\rangle$ with yet another qubit. Let's label that qubit 0, the ancilla qubit 1, and Alice and Bob's channel qubits 2 and 3, respectively. Now follow through the exact same protocol as before, starting instead with the initial state $|\Psi^-(0,1)\rangle \otimes |\Psi^-(2,3)\rangle = -P(|\Psi^-(0,3)\rangle \otimes |\Psi^-(1,2)\rangle)$. Clearly the result, in all cases, will be that qubits $1+2$ will be left in a Bell operator eigenstate, and qubits $0+3$ will now be entangled in the singlet state instead! It appears that there is enough 'substance' to entanglement that it can be 'moved' from one pair of particles to another. Note, however, that this swapping process has not led to a net increase in the entanglement between Alice and Bob; teleportation *destroys* the singlet state entanglement they initially shared in their channel state, as it replaces it with singlet entanglement between 0 and 3.

Since the entanglement of the channel state in teleportation will inevitably get used up (and, similarly, Alice and Bob will use up their shared entanglement in dense coding when she passes her qubit to Bob), it is natural to develop a theory of entanglement as a physical resource for performing further 'informational work'. Entanglement thermodynamics is that theory, and yields constraints on our ability to store, retrieve and manipulate entanglement that are analogous to what classical thermodynamics tells us about heat.

The Fundamental Law of Entanglement Thermodynamics asserts: *Entanglement between systems in two regions cannot be created merely by collective local operations on the systems and classical communication between the regions* (i.e., CLOCC). More precisely, any reasonable measure E of entanglement must satisfy

$$\rho \overset{CLOCC}{\longmapsto} \{\rho_i, p_i\} \implies E(\rho) \geq \sum_i p_i E(\rho_i),$$

where the antecedent of this conditional signifies that Alice and Bob have preprocessed the particles and (possibly) sorted them into subensembles ρ_i with probabilities p_i, and the consequent says that the average entanglement left at the end of the process cannot exceed the initial entanglement that went in. Note that there is a clear analogy here with the Second Law of Classical Thermodynamics: *There is no process the sole effect of which is to extract heat from the colder of two reservoirs and deliver it to the hotter reservoir.* (Plenio and Vedral 1998.) Just as we cannot create refrigeration for free, i.e., without any energy or work input, we cannot produce entanglement for free, i.e., without entangling interactions between the particles.

Motivated by the problems of entanglement storage and retrieval, there are two natural measures of entanglement obeying the fundamental law. Suppose Alice and Bob start by sharing a large collection of pairs of particles in a state $\rho^{\otimes n} = \rho \otimes \ldots \otimes \rho$ (n times). Alice collectively operates on her members of each pair, and Bob on his, with the aim of extracting $k \leq n$ singlet states between them. The *entanglement of distillation* of ρ is the maximum fraction of singlets they can extract from $\rho^{\otimes n}$ in the asymptotic limit as $n \to \infty$:

$$\rho^{\otimes n} \overset{CLOCC}{\longmapsto} |\Psi^-\rangle^{\otimes k}, \;\; D(\rho) = \lim_{n\to\infty} \frac{k_{max}}{n}.$$

(In fact, the CLOCC transformation from $\rho^{\otimes n}$ to $|\Psi^-\rangle^{\otimes k}$ is only required to be perfect in the asymptotic limit.) Similarly, the *entanglement of formation* of ρ is the minimum fraction of singlets Alice and Bob need to invest to create n copies of ρ in the asymptotic limit:

$$|\Psi^-\rangle^{\otimes k'} \overset{CLOCC}{\longmapsto} \rho^{\otimes n}, \;\; F(\rho) = \lim_{n\to\infty} \frac{k'_{min}}{n}.$$

By the fundamental law, it must be the case that $D(\rho) \leq F(\rho)$.

Consider, now, the special case where ρ is a pure entangled state $|\Phi\rangle$. There is a scheme, due originally to Bennett *et al.* (1996), that shows how Alice and Bob can asymptotically distill $|\Phi\rangle^{\otimes n} \overset{CLOCC}{\longmapsto} |\Psi^-\rangle^{\otimes k}$ with an efficiency $S(\rho_\Phi)$, where ρ_Φ is the reduced density operator for either particle in the state $|\Phi\rangle$, and

$$S(\rho_\Phi) = -\mathrm{Tr}(\rho_\Phi \log_2 \rho_\Phi)$$

is the standard von Neumann entropy of ρ_Φ. (Note: $S(\rho_{\Psi^-}) = 1$, as we expect.) This scheme is *reversible*, in the sense that there is a scheme that asymptotically forms $|\Psi^-\rangle^{\otimes k'} \overset{CLOCC}{\longmapsto} |\Phi\rangle^{\otimes n}$ with the same efficiency. (Details of these schemes may be found in Nielsen and Chuang (2000), Section 12.5.2.) By the Fundamental Law, *no schemes can perform better*. The argument for this works in complete analogy with the classical argument for the ideal efficiency of the Carnot cycle (Popescu and Rohrlich 1997). For example, if distillation could be more efficient than the Bennett *et al.* scheme (a similar argument works for formation), then Alice and Bob would be able to start with n copies of $|\Phi\rangle$ and CLOCC transform them into $k > nS(\rho_\Phi)$ singlets. They could then use the reverse of the Bennett *et al.* scheme to CLOCC transform their k singlets back into $k/S(\rho_\Phi) > n$ copies of $|\Phi\rangle$, in violation of the Fundamental Law. It follows that $D(\Phi) = F(\Phi) = S(\rho_\Phi)$.

When ρ is a mixed entangled state, we can define its *bound entanglement* by

$$B(\rho) \equiv F(\rho) - D(\rho) \ (\geq 0).$$

Comparing this with the classical Gibbs-Helmholtz equation,

$$TS = U - A,$$

where U is the internal energy, and A the free energy, we are tempted to think of $F(\rho)$ as the 'internal entanglement' of ρ and $D(\rho)$ as its 'free entanglement'. But what are the analogues of temperature T and entropy S? If we took the 'entanglement entropy' of ρ to be $S(\rho)$, then since the von Neumann entropy of a pure state $|\Psi\rangle$ is zero, it would follow that $F(\Psi) = D(\Psi)$ — something we already know to be true! This suggests that we further define *entanglement temperature*, for mixed states (when $S(\rho) > 0$), by the formula $T(\rho) \equiv \frac{B(\rho)}{S(\rho)}$. Unfortunately, this definition turns out to have counterintuitive consequences (Horodecki *et al.* 1998*a*). So the question remains: Just how deep and convincing can this thermal analogy be made?

As for the values of entanglement of distillation and formation for mixed states, remarkably little is known in general, but what *is* known is fascinating. Though all 2×2 entangled states ρ are distillable, i.e., $D(\rho) > 0$ (Horodecki *et al.* 1997), there are higher-dimensional states, so-called *bound entangled* states, for which $D(\rho) = 0$ (Horodecki *et al.* 1998*b*). However, this does not of itself imply that $B(\rho) > 0$ — and, therefore, that a genuinely *irreversible* entanglement manipulation process exists — because there is no general argument that one cannot get by with investing less and less entanglement in the asymptotic limit for the formation of a mixed entangled state. The first genuinely irreversible entanglement manipulation process was not found until February of this year (Vidal and Cirac 2001), after an initially false start (Horodecki *et al.* 2000*a*). Still more curious is recently found evidence for the possibility of a pair of mixed entangled states $\rho_{1,2}$ such that that $B(\rho_{1,2}) = F(\rho_{1,2}) > 0$ yet $D(\rho_1 \otimes \rho_2) > 0$ (Shor *et al.* 2000). If correct, this would show that bound entangled states can 'catalyze' each other to produce free entanglement!

There is much here for philosophers of science to chew on. Clearly the thermal analogy functions as a useful heuristic for understanding entanglement and harnessing its use. But could there be more to the analogy than that? Since classical thermodynamics is a principle theory, is it appropriate to ask for a constructive underpinning, i.e., micro-theory, for 'entanglodynamics'? Could this motivate a return to the well-trodden path of hidden-variable reconstructions of quantum theory? Or does quantum theory *itself* supply the micro-theory via the theory of local operations? (But if it does, how should we regard the disanalogy that entanglement is still treated as a primitive, unlike its analogue heat, which is reduced to kinetic energy in statistical mechanics?) Finally, and perhaps the biggest philosophical carrot of all, could the thermal analogy be turned into a full-fledged 'interpretation' of quantum theory as a complete theory? As the Horodeckis (2000*b*) have put it:

> It is characteristic that despite the dynamical development of the interdisciplinary domain of quantum information theory, there is no, to our knowledge, impact of the latter on interpretational problems. [...] Does quantum information phenomena provide objective [grounds] for the existence of a 'natural' ontology inherent in the quantum formalism?

Bibliography

Bell, J. S. (1964), 'On the Einstein-Podolsky-Rosen paradox', *Physics* **1**, 195–200.

Bennett, C. H. & Wiesner, S. J. (1992), 'Communication via one- and two-particle operators on Einstein-Podolsky-Rosen states', *Physical Review Letters* **69**, 2881–4.

Bennett, C. H., Bernstein, H., Popescu, S. & Schumacher, B. (1996), 'Concentrating partial entanglement by local operations', *Physical Review A* **53**, 2046–52.

Bennett, C. H., Brassard, G., Crépeau, C., Jozsa, R., Peres, A. & Wootters, W. K. (1993), 'Teleporting an unknown quantum state via dual classical and Einstein-Podolsky-Rosen channels', *Physical Review Letters* **70**, 1895–9.

Bouwmeester, D., Ekert, A. & Zeilinger, A., eds (2000), *The physics of quantum information*, Springer-Verlag, Berlin.

Brassard, G., Cleve, R. & Tapp, A. (1999), 'The cost of exactly simulating quantum entanglement with classical communication', *Physical Review Letters* **83**, 1874–7.

Cerf, N. J. & Adami, C. (1997), 'Negative entropy and information in quantum mechanics', *Physical Review Letters* **79**, 5194–7.

Clifton, R. & Pope, D. (2001), 'On the nonlocality of the quantum channel in the standard teleportation protocol', *Physics Letters A* **292**, 1–11.

Clifton, R., Halvorson, H. & Kent, A. (2000), 'Nonlocal correlations are generic in infinite-dimensional bipartite systems', *Physical Review A* **61**, 042101. Chapter 10 of this volume.

Cushing, J. T. & McMullin, E., eds (1989), *Philosophical Consequences of Bell's Theorem*, Notre Dame University Press, Notre Dame, IN.

Deutsch, D. & Hayden, P. (2000), 'Information flow in entangled quantum systems', *R. Soc. Lond. Proc. Ser. A Math. Phys. Eng. Sci.* **456**, 1759–74.

Duwell, A. (2001), 'Explaining information transfer in quantum teleportation', *Philosophy of Science* **68**, S288–S300.

Einstein, A., Podolsky, B. & Rosen, N. (1935), 'Can quantum-mechanical description of physical reality be considered complete?', *Physical Review* **47**, 777–80.

Fuchs, C. A. & Peres, A. (2000), 'Quantum theory needs no "interpretation"', *Physics Today* **53**, 70–1.

Gisin, N. (2000), Local filtering, in D. Bouwmeester, A. Ekert & A. Zeilinger, eds, 'The Physics of Quantum Information', Springer, NY, pp. 269–71.

Horodecki, M., Horodecki, P. & Horodecki, R. (1997), 'Inseparable two spin-1/2 density matrices can be distilled to a singlet form', *Physical Review Letters* **78**, 574–7.

Horodecki, M., Horodecki, P. & Horodecki, R. (1998*a*), 'Mixed-state entanglement and distillation: Is there a "bound" entanglement in nature?', *Physical Review Letters* **80**, 5239.

Horodecki, M., Horodecki, P. & Horodecki, R. (2000*a*), 'Asymptotic manipulations of entanglement can exhibit genuine irreversibility', *Physical Review Letters* **84**, 4260–63.

Horodecki, P., Horodecki, R. & Horodecki, M. (1998*b*), 'Entanglement and thermodynamical analogies', *Acta Physica Slovaca* **48**, 141–56.

Horodecki, R., Horodecki, M. & Horodecki, P. (1996), 'Teleportation, Bell's inequalities and inseparability', *Physics Letters A* **222**, 21–5.

Horodecki, R., Horodecki, M. & Horodecki, P. (2000*b*), On balance of information in bipartite quantum communication systems: Entanglement-energy analogy, arXiv eprint: quant-ph/0002021.

Lo, H.-K., Popescu, S. & Spiller, T., eds (1998), *Introduction to quantum computation and information*, World Scientific Publishing Co. Inc., River Edge, NJ.

Massar, S., Bacon, D., Cerf, N. & Cleve, R. (2001), 'Classical simulation of quantum entanglement without local hidden variables', *Physical Review A* **63**, 052305.

Maudlin, T. (1994), *Quantum Non-Locality and Relativity*, Blackwell, Oxford.

Mermin, N. D. (1996), Hidden quantum non-locality, in R. Clifton, ed., 'Perspectives on Quantum Reality', Kluwer Academic Publishers, pp. 57–71.

Nielsen, M. A. & Chuang, I. L. (2000), *Quantum computation and quantum information*, Cambridge University Press, Cambridge.

Peres, A. (1996), 'Collective tests for quantum nonlocality', *Physical Review A* **54**, 2685–9.

Plenio, M. & Vedral, V. (1998), 'Teleportation, entanglement and thermodynamics in the quantum world', *Contemporary Physics* **39**, 431–46.

Popescu, S. (1994), 'Bell's inequalities versus teleportation: What is nonlocality?', *Physical Review Letters* **72**, 797–9.

Popescu, S. (1995), 'Bell's inequalities and density matrices: Revealing "hidden" nonlocality', *Physical Review Letters* **74**, 2619–22.

Popescu, S. & Rohrlich, D. (1992), 'Generic quantum nonlocality', *Physics Letters A* **166**, 293–7.

Popescu, S. & Rohrlich, D. (1997), 'Thermodynamics and the measure of entanglement', *Physical Review A* **56**, R3319–21.

Popescu, S. & Rohrlich, D. (1998), The joy of entanglement, in H.-K. Lo, S. Popescu & T. Spiller, eds, 'Introduction to quantum computation and information', World Scientific Publishing Co. Inc., River Edge, NJ.

Preskill, J. (1998), Lecture notes for physics 229: Quantum information and computation,
http://theory.caltech.edu/people/preskill/ph229.

Schrödinger, E. (1935), 'Discussion of probability relations between separated systems', *Proceedings of the Cambridge Philosophical Society* **31**, 555–63.

Schrödinger, E. (1936), 'Probability relations between separated systems', *Proceedings of the Cambridge Philosophical Society* **32**, 446–52.

Shor, P., Smolin, J. & Terhal, B. (2000), Evidence for nonadditivity of bipartite distillable entanglement, arXiv.org eprint: quant-ph/0010054.

Steiner, M. (2000), 'Towards quantifying non-local information transfer: finite-bit non-locality', *Physics Letters A* **270**, 239–44.

Vidal, G. & Cirac, J. I. (2001), 'Irreversibility in asymptotic manipulations of entanglement', *Physical Review Letters* **86**, 5803–6.

Werner, R. F. (1989), 'Quantum states with Einstein-Podolsky-Rosen correlations admitting a hidden-variable model', *Physical Review A* **40**, 4277–81.

Żukowski, M. (2000), 'Bell theorem for nonclassical part of quantum teleportation', *Physical Revew A* **62**, 032101.

Bibliography of the writings of Rob Clifton

Books

1. Editor, *Perspectives on Quantum Reality*, 242 pages, in The University of Western Ontario Series in Philosophy of Science, Kluwer Academic Publishers, 1996.

Research articles

1. 'Generic incomparability of infinite-dimensional entangled states' (with Brian Hepburn and Christian Wüthrich) *Physics Letters A* 303 (2002) 121–4.

2. 'Reconsidering Bohr's reply to EPR' (with H. Halvorson), in T. Placek and J. Butterfield (eds) *Non-locality and Modality*. Kluwer (2002), pp. 3–18. **(Chapter 12)**

3. 'The Subtleties of Entanglement and its Role in Quantum Information Theory', *Philosophy of Science* 69 (2002) S150–S167. **(Chapter 14)**

4. 'No Place for Particles in Relativistic Quantum Theories?' (with Hans Halvorson), *Philosophy of Science* 69 (2002) 1–28. [Reprinted in *The Philosopher's Annual* 25 (2002). Also reprinted in M. Kuhlman, H. Lyre, and A. Wayne (eds), *On-*

tological Aspects of Quantum Field Theory, World Scientific, Singapore, 2002, Chap. 10.] **(Chapter 8)**

5. 'On the Nonlocality of the Quantum Channel in the Standard Teleportation Protocol' (with Damian Pope), *Physics Letters* A 292 (2001) 1–11.

6. 'Insufficient Reason in the "New Cosmological Argument"' (with Kevin Davey), *Religious Studies* 37 (2001) 485–90.

7. 'Are Rindler Quanta Real? Inequivalent Particle Concepts in Quantum Field Theory' (with Hans Halvorson), *The British Journal for Philosophy of Science* 52 (2001) 417–70. **(Chapter 9)**

8. 'Entanglement and Open Systems in Algebraic Quantum Field Theory', (with Hans Halvorson), *Studies in History and Philosophy of Modern Physics* 32 (2001) 1–31. **(Chapter 7)**

9. 'Simulating Quantum Mechanics by Non-Contextual Hidden Variables', (with Adrian Kent), *The Proceedings of the Royal Society of London A* 456 (2000) 2101–14. **(Chapter 13)**

10. 'The Modal Interpretation of Algebraic Quantum Field Theory', *Physics Letters A* 271 (2000) 167–77. **(Chapter 5)**

11. 'Complementarity Between Position and Momentum as a Consequence of Kochen-Specker Arguments', *Physics Letters A* 271 (2000) 1–7. **(Chapter 11)**

12. 'Counting Marbles with "Accessible" Mass Density: A Reply to Bassi and Ghirardi', (with Brad Monton), *The British Journal for Philosophy of Science* 51 (2000) 155–64.

13. 'Generic Bell Correlation Between Arbitrary Local Algebras in Quantum Field Theory', (with Hans Halvorson), *Journal of Mathematical Physics* 41 (2000) 1711–17. **(Chapter 6)**

14. 'Non-local Correlations are Generic in Infinite-dimensional Bipartite Systems', (with Hans Halvorson and Adrian Kent), *Physical Review A* 61 (2000) 042101. **(Chapter 10)**

15. 'Revised Proof for the Uniqueness Theorem for "No Collapse" Interpretations of Quantum Mechanics', (with Jeffrey Bub and Sheldon Goldstein), *Studies in History and Philosophy of Modern Physics* 31 (2000) 95–8. **(Chapter 3)**

16. 'Losing Your Marbles in Wavefunction Collapse Theories', (with Brad Monton), *The British Journal for Philosophy of Science* 50 (1999) 697–717.

17. 'Bipartite Mixed States of Infinite-Dimensional Systems are Generically Nonseparable', (with Hans Halvorson), *Physical Review A* 61 (1999) 012108.

18. 'Changing the Subject: Redei on Causal Dependence in Algebraic Quantum Field Theory', (with Laura Ruetsche), *Philosophy of Science* 66 (Proceedings) (1999) S156–69.

19. 'Beables in Algebraic Quantum Theory', in Jeremy Butterfield and Constantine Pagonis (eds), *From Physics to Philosophy* (Cambridge, Cambridge University Press, 1999) Chap. 2.

20. 'Maximal Beable Subalgebras of Quantum-Mechanical Observables', (with Hans Halvorson), *The International Journal for Theoretical Physics* 38 (1999): 2441–84.

21. 'Valuations on Functionally Closed Sets of Quantum-Mechanical Observables and Von Neumann's No-Hidden-Variables Theorem', (with Jason Zimba), in Dennis Dieks and Pieter Vermaas (eds), *The Modal Interpretation of Quantum Mechanics*, The University of Western Ontario Series in Philosophy of Science 60 (1998): 69–101.

22. 'Lorentz-Invariance in Modal Interpretations', (with Michael Dickson), in Dennis Dieks and Pieter Vermaas (eds), *The Modal Interpretation of Quantum Mechanics*, The University of Western Ontario Series in Philosophy of Science 60 (1998): 9–47. **(Chapter 4)**

23. 'The Bare Theory Has No Clothes', (with Jeffrey Bub and Bradley Monton), in R. Healey and G. Hellman (eds), *Quantum Measurement: Beyond Paradox*, Minnesota Studies in Philosophy of Science XVII (1998): 32–51.

24. 'Superentangled States', (with David Feldman, Hans Halvorson, Michael Redhead, and Alexander Wilce), *Physical Review A* 58 (1998): 135–45.

25. 'On What Being a World Takes Away', *Philosophy of Science* 63 (Proceedings) (1996): S151–8.

26. 'A Uniqueness Theorem for "No-Collapse" Interpretations of Quantum Mechanics', (with Jeffrey Bub), *Studies in History and Philosophy of Modern Physics* 27 (1996): 181–219. **(Chapter 2)**

27. 'The Properties of Modal Interpretations of Quantum Mechanics', *British Journal for Philosophy of Science* 47 (1996): 371–98.

28. 'QuasiBoolean Algebras and Definite Properties in Quantum Mechanics', (with John L. Bell), *The International Journal of Theoretical Physics* 34 (1995): 2409–21.

29. 'Why Modal Interpretations of Quantum Mechanics Must Abandon Classical Reasoning About Physical Properties', *The International Journal of Theoretical Physics* 34 (1995): 1302–12.

30. 'Uniqueness of Prime Factorizations of Linear Operators in Quantum Mechanics', (with Nick Reeder), *Physics Letters A* 204 (1995): 198–204.

31. 'The Definability of Objective Becoming in Minkowski Spacetime', (with Mark Hogarth), *Synthese* 103 (1995): 355–87.

32. 'The Triorthogonal Uniqueness Theorem and Its Irrelevance to the Modal Interpretation of Quantum Mechanics', in K.V. Laurikainen *et al.* (eds), *Proceedings of the Symposium on the Foundations of Modern Physics 1994: 70 Years of Matter Waves* (World Scientific, Singapore, 1995) pp. 45–60.

33. 'Unremarkable Contextualism: Dispositions in the Bohm Theory', (with Constantine Pagonis), *Foundations of Physics* 25 (1995): 281–96.

34. 'Making Sense of the Kochen-Dieks "No-Collapse" Interpretation of Quantum Mechanics Independent of the Measurement Problem', in D. M. Greenberger and A. Zeilinger (eds), *Fundamental Problems of Quantum Theory, Annals of the New York Academy of Sciences* 755 (1995): 570–8.

35. 'Independently Motivating the Kochen-Dieks Modal Interpretation of Quantum Mechanics', *The British Journal for Philosophy of Science* 46 (1995): 33–57. **(Chapter 1)**

36. 'Stapp's Algebraic Proof of Nonlocality', (with Michael Dickson), *Physical Review A* 49 (1994): 4251–6.

37. 'Against Experimental Metaphysics', (with Martin Jones), in P. A. French, T. E. Euling, Jr., and H. K. Wettstein (eds), *Mid-West Studies in Philosophy* Vol. XVIII 'Philosophy of Science' (1993): 295–316.

38. 'Getting Contextual and Nonlocal Elements-of-Reality the Easy Way', *American Journal of Physics* 61 (1993): 443–7.

39. 'Relativity, Quantum Mechanics, and EPR', (with Constantine Pagonis and Itamar Pitowsky), in D. Hull, M. Forbes, and K. Okruhlik (eds), *The Philosophy of Science Association 1992*, Vol. 1 (Philosophy of Science Association, East Lansing, MI) pp. 114–28.

40. 'Hardy's Nonlocality Theorem for *N* Spin-1/2 Particles', (with Constantine Pagonis), *Physics Letters A* 168 (1992): 100–2.

41. 'Locality, Lorentz Invariance, and Linear Algebra: Hardy's Theorem for Two Entangled Spin-*s* Particles', (with Peter Niemann), *Physics Letters A* 166 (1992): 177–94.

42. 'A Second Look at a Recent Algebraic Proof of Nonlocality', (with Michael Redhead and Jeremy Butterfield), *Foundations of Physics Letters* 4 (1991): 395–403.

43. 'The Breakdown of Quantum Nonlocality in the Classical Limit', (with Constantine Pagonis and Michael Redhead), *Physics Letters A* 155 (1991): 441–4.

44. 'Generalization of the Greenberger-Horne-Zeilinger Algebraic Proof of Nonlocality', (with Michael Redhead and Jeremy Butterfield), *Foundations of Physics* 21 (1991): 149–84.

45. 'Noninvasive Measurability, Negative-Result Measurements, and Watched-Pots: Another Look at Leggett's Arguments for the Incompatibility between Macro-Realism and Quantum Mechanics', in P. Lahti and P. Mittelstaedt (eds), *Symposium on the Foundations of Modern Physics 1990: Quantum Theory of Measurement and Related Problems* (World Scientific, Singapore, 1990), pp. 77–88.

46. 'Nonlocal Influences and Possible Worlds', (with Jeremy Butterfield and Michael Redhead), *The British Journal for the Philosophy of Science* 41 (1990): 5–58.

47. 'Toward a Sound Perspective on Modern Physics: Capra's Popularization of Mysticism and Theological Approaches Re-Examined', (with Marilyn Regehr), *Zygon: The Journal of Religion and Science* 25 (1990): 73–104.

48. 'Some Recent Controversy Over the Possibility of Experimentally Determining Isotropy in the Speed of Light', *Philosophy of Science* 56 (1989): 688–696.

49. 'The Rastall Model and Stapp's Proof of Nonlocality', *Foundations of Physics Letters* 2 (1989): 531–47.

50. 'Determinism, Realism, and Stapp's 1985 Proof of Nonlocality', *Foundations of Physics Letters* 2 (1989): 347–69.

51. 'Capra on Eastern Mysticism and Modern Physics: A Critique', (with Marilyn Regehr), *Science and Christian Belief* 1 (1989): 53–74.

52. 'The Compatibility of Correlated CP Violating Systems with Statistical Locality', (with Michael Redhead), *Physics Letters A* 126 (1988): 295–9.

Review and encyclopedia articles

1. 'Cosmology', 'Scientific Determinism', 'Einstein', 'Heisenberg', 'Operationalism', 'Philosophical Problems of Physics', 'Philosophical Problems of Quantum Mechanics', 'Spacetime', 'The Uncertainty Principle', in Ted Honderich (ed), *Oxford Companion to Philosophy* (Oxford, Oxford University Press, 1995).

2. 'A General Survey of the Mathematical Philosophy of Immanual Kant', *Eureka: The Journal of the Cambridge Archimedean Society* 48 (1988): 42–54.

3. 'Bell's Theorem: A Case Study in the Philosophy of Quantum Mechanics', *Phys 13 News* 62 (1987): 6–8.

4. 'Comments on Quantum Reality', *American Journal of Physics* 55 (1987): 1064.

Book reviews

1. Review of A. Peres, *Quantum Mechanics: Concepts and Methods, Foundations of Physics* 25 (1995): 205–9.

2. Review of D. Albert, *Quantum Mechanics and Experience* (with Francisco Flores), *American Scientist September* (1994): 477–8.

3. Review of C. Ray, *Time, Space and Philosophy* (with Mark Hogarth), *Philosophical Books* 34 (1993): 123–5.

4. 'Radical Probabilities', Review of B. C. van Fraassen, Quantum Mechanics: An Empiricist View, *Times Higher Education Supplement* May 1 (1992).

5. Review of R. I. G. Hughes, The Structure and Interpretation of Quantum Mechanics, *Philosophical Books* 32 (1991): 189–91.

6. Critical Notice of J. Leslie, Universes, *The Philosophical Quarterly* 41 (1991): 339–44.

7. 'Elusive Quantum Reality', Review of W. Schommers (ed.), Quantum Theory and Pictures of Reality, *Physics World* June (1990): 45–6.

8. Review of C. Kilmister (ed.), Schrodinger: Centenary Celebration of a Polymath, *Philosophy of Science* 57 (1990): 342–3.

9. Review of J. Barrow and F. Tipler, The Anthropic Cosmological Principle, *Science and Christian Belief* 2 (1990): 41–6.

10. Review of B. d'Espagnat, Reality and the Physicist, *Contemporary Physics* 30 (1989): 396–7.

11. Critical Notice of N. Herbert, Quantum Reality: Beyond the New Physics, *Eidos* 5 (1986): 227–34.

Index

This index gives very few names as entries, namely for pages where some of the person's work is discussed at some length. Citations and other briefer references to names can be traced approximately by using the references at the end of each chapter.

The main results of the papers collected here are listed as subentries to four entries, one for each part, all beginning with 'theorems': namely, the entries 'theorems about modal interpretation', 'theorems about algebraic quantum theory and algebraic quantum field theory', 'theorems about localization and representations', and 'theorems about foundational issues'.

Other standard results are entered by their usual names (sometimes using the name of an author), such as 'no-signalling theorem', 'Reeh-Schlieder theorem' etc.

The index entries of technical terms are mainly to their definition, or first mention, within a paper.